Progress in Molecular and Subcellular Biology

Series Editors: W.E.G. Müller (Managing Editor),
Ph. Jeanteur, Y. Kuchino, A. Macieira-Coelho, R. E. Rhoads

40

Volumes Published in the Series

Progress in Molecular and Subcellular Biology

Volume 26
Signaling Pathways for Translation: Insulin and Nutrients
R.E. Rhoads (Ed.)

Volume 27
Signaling Pathways for Translation: Stress, Calcium, and Rapamycin
R.E. Rhoads (Ed.)

Volume 28
Small Stress Proteins
A.-P. Arrigo and W.E.G. Müller (Eds.)

Volume 29
Protein Degradation in Health and Disease
M. Reboud-Ravaux (Ed.)

Volume 30
Biology of Aging
A. Macieira-Coelho

Volume 31
Regulation of Alternative Splicing
Ph. Jeanteur (Ed.)

Volume 32
Guidance Cues in the Developing Brain
I. Kostovic (Ed.)

Volume 33
Silicon Biomineralization
W.E.G. Müller (Ed.)

Volume 34
Invertebrate Cytokines and the Phylogeny of Immunity
A. Beschin and W.E.G. Müller (Eds.)

Volume 35
RNA Trafficking and Nuclear Structure Dynamics
Ph. Jeanteur (Ed.)

Volume 36
Viruses and Apoptosis
C. Alonso (Ed.)

Volume 38
Epigenetics and Chromatin
Ph. Jeanteur (Ed.)

Volume 40
Developmental Biology of Neoplastic Growth
A. Macieira-Coelho (Ed.)

Subseries: Marine Molecular Biotechnology

Volume 37
Sponges (Porifera)
W.E.G. Müller (Ed.)

Volume 39
Echinodermata
V. Matranga (Ed.)

Alvaro Macieira-Coelho (Ed.)

Developmental Biology of Neoplastic Growth

With 23 Figures, 7 in Color, and 7 Tables

Springer

Professor Dr. Alvaro Macieira-Coelho
INSERM
73 bis, rue du Maréchal Foch
78000 Versailles
France

ISSN 0079-6484
ISBN-10 3-540-25009-3 Springer-Verlag Berlin Heidelberg New York
ISBN-13 978-3-540-25009-8

Library of Congress Control Number: 2005920705

Springer-Verlag is a part of Springer Science+Business Media
springeronline.com

Printed in Germany

Production: Pro Edit GmbH, Heidelberg, Germany
Typesetting: SDS, Leimen, Germany
Cover desing: design & production GmbH, Heidelberg, Germany

Printed on acid free paper 39/3150Re 5 4 3 2 1 0

Preface

Much attention has been directed to changes in cells allowing them to escape growth controls and become an invasive tumor, and to the agents that can cause these changes. Less has been investigated regarding the deviations of normal development that can create an environment favorable for neoplastic growth. However, the idea that tumors result from deviations of normal development has been repeatedly expressed in the past. Waddington (1935) and Needham (1936) proposed that cancers constitute an escape from morphogenetic fields similar to those that guide embryonic development. More recently, Pierce et al. (1978) viewed cancers as a problem of developmental biology and predicted that an understanding of the evolution of the internal milieu through the life span of the organism was necessary for the identification of the mechanisms leading to neoplastic growth. On the other hand, Rubin (1985) envisioned tumors »as part of an inexorable process in which the organism falls behind in its ceaseless effort to maintain order«.

The deviations from the maintenance of order can be expressed through modifications of the microenvironment that become favorable for the transformation of cells through mechanisms at the molecular and cellular levels that are starting to be understood. On the other hand, a cell or cells can become malignant by deviating themselves from their normal developmental pathway and later alter their microenvironment.

Several features described in this volume show that events during embryonic and postnatal developmental stages are important determinants in the initiation and progression of the neoplastic process.

The different histological origin of tumors is suggestive of the role of development: tumors from the earlier stages of the human life span are predominantly mesenchymal, while those manifested later are mainly epithelial. The age incidence of each type of cancer is also supportive of the influence of the developmental stage in the initiation and progression of tumors, and so is the decline in the incidence of cancers during the final stage of the human life. The susceptibility to different carcinogens also depends on the developmental stage.

Some tumors are very obviously influenced by developmental modifications, such as those from the mammary gland, the ovaries or the prostate, where programmed hormonal regulations are critical for normal functions.

The developmental evolution of the thymus can also play a role in tumor growth. On the other hand, there is a higher incidence of cancers in extreme deviations of normal development like those displayed in Werner, Fanconi, Bloom, and other syndromes.

The normal evolution of the adult organism depends inter alia on the interaction between tissues, in particular the integrative and inductive effect of connective tissue on the different organs it infiltrates during embryonic development; indeed, the mesenchyme is a main regulator of the microenvironment in the organism and thus of homeostasis. Deviations from the normal evolution of connective tissue through the human life span can create new regulations that establish field effects favorable for the transformation of epithelial cells. The mechanisms leading to the creation of these field effects are looming; the new findings concerning the regulatory role of the different molecules that constitute the connective tissue and of their degradation products, and the characterization of the functionality of fibroblast populations will further elucidate how deviations of the normal development of connective tissue can influence neoplastic growth. Hence, strong emphasis is given to this subject.

Remnants of embryonic cells that were not eliminated during early postnatal development can be at the origin of tumors in childhood. Later, deregulations of the regeneration of stem cells can be the source of tumor initiation.

Genes involved in embryonic development have been identified as tumor suppressors, when mutated they are involved in neoplastic growth; this constitutes further evidence of an association between regulation of normal development and abnormal growth. Other mechanisms involved in the regulation of normal development, such as telomere-controlled recombination, can misfire and lead to tumor growth.

These are the different aspects of the relationship between deviations from normal development and tumor growth described in this volume. New therapeutic approaches resulting from this concept are suggested. It is seen that an understanding of this relationship is indispensable in order to identify the mechanisms leading to neoplastic growth and win the war against cancers.

Versailles, November 2004 *Alvaro Macieira-Coelho*

References

Needham J (1936) New advances in the chemistry and biology of organized growth. Proc R Soc Lond B Biol Sci 29:1577–1626

Pierce GB, Shikes R, Fink LM (1978) Cancer: a problem of developmental biology. Prentice Hall, Englewood Cliffs

Rubin H (1985) Cancer as a dynamic developmental disorder. Cancer Res 45:2935–2942

Waddington CH (1935) Cancer and the theory of organizers. Nature 135:606–608

Contents

The Hedgehog Signaling Pathway in Cancer

Janni Vestergaard, Mads Bak, Lars Allan Larsen

1 Introduction

The Hedgehog (Hh) signaling pathway plays crucial roles during embryonic development, where it controls cell proliferation, differentiation, and tissue patterning in numerous tissues and organs (Ingham and McMahon 2001; Cohen 2003). Mutations in genes encoding key members of the pathway can result in many human developmental disorders, including defects of the CNS, craniofacial, limb, and skeletal disorders. After embryogenesis, the overall activity of the pathway is diminished. However, recent reports demonstrate the necessity for Hh signaling in proliferation and maintenance of stem cell-like progenitor cells in the adult (Altaba et al. 2004). Interestingly, Hh signaling has also been implicated in the genesis and growth of a variety of human tumors (Pasca and Hebrok 2003), thus creating links between embryonic development, stem cell maintenance, and cancer formation. Several studies have shown reduced growth of tumor cells in response to inhibition of Hh signaling, suggesting a value for inhibiting pathway activity in a clinical setting. Elements of the Hh signaling pathway may thus represent potential drug targets for the treatment of a subset of cancers.

This chapter provides an overview of the Hh signaling pathway and reviews the role of Hh signaling in tumor formation and growth. The potential role of Hh antagonists in cancer treatment is also reviewed and discussed.

2 The Hedgehog Signaling Pathway

Since the discovery of the Hh gene in *Drosophila* in 1980 (Nusslein-Volhard and Wieschaus) and the subsequent identification of the orthologous vertebrate genes in the early 1990s (Echelard et al. 1993), the Hh pathway has been

J. Vestergaard, M. Bak, L.A. Larsen (e-mail: lal@medgen.ku.dk)
Wilhelm Johannsen Centre for Functional Genome Research, Department of Medical Biochemistry and Genetics, University of Copenhagen, Blegdamsvej 3, 2200 Copenhagen, Denmark

Progress in Molecular and Subcellular Biology
Alvaro Macieira-Coelho (Ed.)
Development Biology of Neoplastic Growth

intensively studied, and several novel elements and interactions are discovered each year. Much of our current understanding of the Hh pathway at the molecular level is based on studies in *Drosophila*; many of the facts presented in this section rely largely on work done in this organism. Most aspects of the pathway appear to be conserved from flies to vertebrates. A number of elements of the *Drosophila* Hh pathway have several vertebrate homologues (orthologues); but some of the genes involved in vertebrate Hh signaling have no *Drosophila* counterparts, more layers of regulation and control being added in vertebrate Hh signaling to an already complex network of interacting factors. This section aims to provide the reader an overview of the molecular nature of the Hh signaling pathway and the complex interactions involved. For a more in-depth discussion of the individual elements, their interactions and function in the pathway, there are several excellent reviews (Ingham and McMahon 2001; Cohen 2003; Ogden et al. 2004).

Briefly, the Hh signaling pathway is initiated when extracellular Hh protein binds to and inactivates its receptor Patched (Ptch), a putative membrane-spanning protein (Fig. 1). Binding of Hh to Ptch, relieves Smoothened (Smo) from the inhibitory effect of Ptch. This results in activation of a zinc finger transcription factor Cubitus interruptus (Ci) in *Drosophila* and the Gli proteins in vertebrates (Cohen 2003).

2.1 Synthesis and Processing of Hh Protein

The Hh precursor proteins [*Sonic hedgehog* (*Shh*), *Indian hedgehog* (*Ihh*) and *Desert hedgehog* (*Dhh*) in mammals], undergo autocatalytic cleavage to yield an amino-terminal part (Hh-N) possessing all known signaling activities and a secreted carboxy-terminal part (Hh-C; Lee et al. 1994; Bumcrot et al. 1995). Hh-N is further modified C-terminally by addition of a cholesterol moiety and N-terminally by palmitate (Porter et al. 1996; Pepinsky et al. 1998). Palmitoylation of Hh is mediated by the acyltransferase Skinny hedgehog [Ski, rasp, sightless (sit), central missing (cmn); Amanai and Jiang 2001; Chamoun et al. 2001; Lee and Treisman 2001; Micchelli et al. 2002]. The lipid modification tethers Hh to the cell membrane and is essential for full signaling activity (Pepinsky 1998; Burke et al. 1999; Chamoun et al. 2001; Taylor 2001).

2.2 Movement of Hh Protein

Despite the firm anchoring of the lipid modified Hh in the cell membrane, Hh protein directly induces expression of Hh target genes in *Drosophila* several cell diameters from Hh-producing cells (Basler and Struhl 1994; Tabata and Kornberg 1994; Mullor et al. 1997; Strigini and Cohen 1997) and in ver-

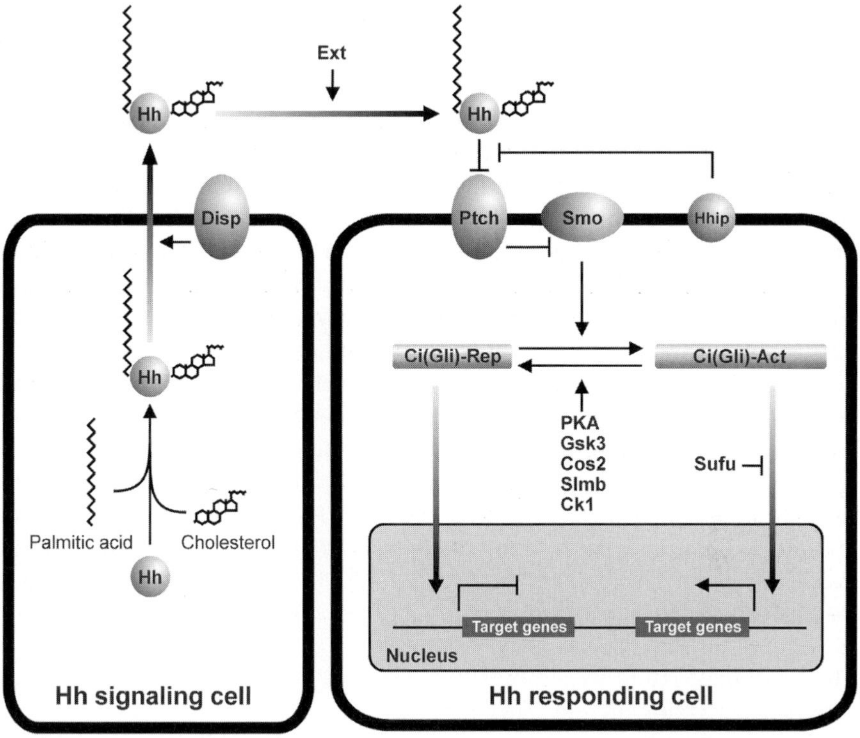

Fig. 1. Secretion of processed and lipid modified Hh from Hh-synthesizing cells is mediated by Disp. In Hh-responding cells, Hh signaling is initiated upon binding of Hh to Ptch. This results in activation of the Ci (Gli) transcription factors and initiation of transcription of Hh target genes (see text for details). *Hh* Hedgehog; *Disp* Dispatched; *Ext* Exostosin; *Ptch* Patched; *Smo* Smoothened; *Hhip* Hedgehog-interacting protein; *Ci* Cubitus interruptus; *Gli* glioma-associated oncogene homologue; *Pka* cyclic AMP-dependent protein kinase A; *Gsk3* glycogen synthase kinase β; Cos2 Costal-2; *Slmb* supernumerary limbs; *Ck1* casein kinase 1; *Sufu* Suppressor of fused. Ci(Gli)-Rep repressor form of Ci/Gli. Ci(Gli)-Act activator form of Ci(Gli.

tebrate embryos, the range of Hh signaling is even greater. In the developing neural tube for example, expression of one of the vertebrate Hh homologues, *Shh*, is restricted to the notochord and floor plate yet generates a dorsal-ventral gradient of Shh protein detectable in the entire ventral neural tube (Incardona et al. 2000; Gritli-Linde et al. 2001). Several mechanisms are involved in the movement of lipid modified Hh-N from the Hh synthesizing cells through the morphogenetic field. *Drosophila dispatched* (*disp*) encodes a putative 12 transmembrane protein, which functions to release membrane-tethered Hh, making it available for long-range signaling (Burke et al. 1999). Two vertebrate *disp* homologues have been identified, *Disp1* (*DispA*) and *Disp2* (*DispB*; Caspary et al. 2002; Ma et al. 2002; Kawakami et al. 2002). *Disp1* is nearly ubiquitously expressed at early stages of development and most aspects of Hh signaling are lost in murine Disp double mutants (Ma et al. 2002). *Disp2*, on the other hand, is not expressed at early developmental stages and its role in vertebrate Hh signaling remains to be established (Ma et al. 2002). Movement of Hh in the extracellular space is regulated by *Drosophila* Tout-velu (Ttv) and its vertebrate homologues exostosins (EXTs; Bellaiche et al. 1998; The et al. 1999). Ttv and EXTs are glycosyltransferases involved in synthesis of heparan sulfate proteoglycans (HSPGs). *Drosophila* cells lacking Ttv are unable to respond to Hh signaling unless they are located directly adjacent to Hh producing cells (Bellaiche et al. 1998). In contrast, mice carrying a hypomorphic allele of *Ext1* display increased Hh signaling range (Koziel et al. 2004). This could indicate different roles of HSPGs in Hh signaling in *Drosophila* and vertebrates, but as the authors note the conflicting results probably reflect the differences in experimental design: a null mutant in *Drosophila* compared to a hypomorphic allele in mice. A dual role for HSPGs in regulating Hh movement has been suggested. Initially, HSPGs bind Hh protein allowing Hh transport from signaling cells, but binding of Hh by HSPGs also sequesters Hh protein thereby limiting its spread (Koziel et al. 2004).

2.3 Reception of Hh Signaling

The binding of Hh to its receptor Ptch (Marigo et al. 1996; Stone et al. 1996) results in activation of Smo. How this is achieved at the molecular level is currently not well understood. Ptch and Smo do not co-localize extensively. While in absence of Hh ligand, Ptch is localized mainly to the plasma membrane, a large proportion of Smo protein resides in endosomal vesicles (Nakano et al. 2004). Upon binding of Hh to Ptch, the Ptch-Hh complex is internalized and targeted for lysosymal degradation (Denef et al. 2000; Torroja et al. 2004). At the same time Smo is stabilized, phosphorylated, moves from internal structures and accumulates at the plasma membrane (Alcedo et al. 2000; Denef et al. 2000; Ingham et al. 2000; Zhu et al. 2003).

A number of other studies have pointed to endocytosis and vesicular trafficking as an important event in Hh signaling. Rab23, encoded by the murine *open-brain (opb)*, has been identified as a negative regulator of the Hh pathway (Eggenschwiler et al. 2001). Rab23 belongs to a family of GTP activated proteins involved in many aspects of subcellular trafficking of vesicles (Somsel and Wandinger-Ness 2000). Furthermore, expression of dominant negative forms of Rab5 – which plays a role in formation of clathrin coated pits, or *shibire (shi)*, that functions in receptor-mediated endocytosis – results in accumulation of both Ptch and Smo at the cell surface (Zhu et al. 2003).

It has been speculated that Ptch could function as a molecule transporter. Ptch shares structural similarity with a bacterial class of molecule transporters and a highly conserved motif in the bacterial family is conserved in Ptch and thus one model suggests that Ptch indirectly inhibits Smo activity by transporting one or more as yet unidentified endogenous small molecules, thereby changing organelle composition or contents (Taipale et al. 2002).

In addition to Ptch, a number of other membrane-associated proteins bind extracellular Hh. Megalin (LRP2, GP330) belongs to the low-density lipoprotein receptor family and has been shown to bind and cause internalization of Shh (McCarthy et al. 2002). Megalin-deficient mice display a phenotype consistent with decreased Hh signaling, but the precise role of megalin in Hh signal transduction is not known.

Hedgehog-interacting protein (Hhip, Hip1, Hip), a putative transmembrane glycoprotein, is expressed in all Hh target tissues and binds all three mammalian Hh homologues with an affinity similar to Ptch (Chuang and McMahon 1999). Transgenic mice ectopically expressing Hhip display a phenotype resembling Ihh mutants (Chuang and McMahon 1999), and Hhip mutants show up-regulation of Hh signaling (Chuang et al. 2003), which suggests that Hhip is most likely involved in attenuating Hh signaling by limiting the amount of ligand available to Ptch.

Growth arrest specific 1 (Gas1) is a glycophosphatidyl-inositol linked protein, which may attenuate Hh signaling by a mechanism similar to that of Hhip (Lee et al. 2001).

2.4 Intracellular Signal Transduction

The ultimate result of Hh signaling is activation of the zinc finger containing transcription factor Cubitus interruptus (Ci) in *Drosophila* and its homologues Gli1, Gli2, and Gli3 in vertebrates (Ingham and McMahon 2001). In absence of Hh ligand, Ci is proteolytically cleaved to yield a transcriptional repressor (Ci-R). When Hh signaling is initiated, the cleavage of Ci (Gli) is inhibited and Ci is activated to generate a full-length transcriptional activator (Ci-A). Cleavage of Ci requires its association with a cytoplasmic protein complex termed the Hedgehog Signaling Complex (HSC). In addition to Ci this complex includes Costal-2 (Cos2), a kinesin-related microtubules-binding protein, and the Fused (Fu) kinase. Phosphorylation of Ci by protein kinase A (PKA) (Wang et al. 1999), glycogen synthase kinase 3 beta (GSK3β) and casein kinase (Ck1; Jia et al. 2002; Price and Kalderon 2002), targets Ci for cleavage, generating the Ci repressor form Ci-R. This process is dependent on supernumerary limbs (slmb), a protein implicated in targeting proteins for degradation through the ubiquitin/proteasome pathway (Jiang and Struhl 1998). Ci-R then enters the nucleus where it represses transcription of Hh target genes. In addition to promoting formation of Ci-R, Cos-2 inhibits Hh target gene expression by retaining Ci in the cytoplasm through binding to microtubules (Robbins et al. 1997; Wang and Holmgren 1999, 2000; G. Wang et al. 2000) or vesicular membranes (Stegman et al. 2004). Additionally, nuclear entry of Ci is restrained by interaction with suppressor of fused [Su(fu)] (Methot and Basler 2000; G. Wang et al. 2000), and by constitutive nuclear export (Chen et al. 1999).

Upon Hh stimulation, Ci cleavage is inhibited and full-length Ci accumulates, but the precise mechanisms promoting the formation of an activator form of Ci is currently not well understood. In addition to Smo, Fu and Cos2 become phosphorylated and the HSC is released from its tether and full-length Ci accumulates in the cytoplasm (Robbins et al. 1997; Nybakken et al. 2002). However, inhibition of Ci cleavage is insufficient to activate the full range of Hh signaling response and additional steps are required for turning Ci into a transcriptional activator (Methot and Basler 1999; G. Wang et al. 2000). Association of HSC with Smo appears to be required for activation of Hh signaling (Jia et al. 2003; Lum et al. 2003; Ogden et al. 2003; Ruel et al. 2003). This association is mediated through binding of Cos2 to the C-terminal cytoplasmic part of Smo. But how this interaction affects Ci is not clear.

In vertebrates, intracellular Hh signal transduction is mediated by the Gli proteins: Gli1, Gli2, and Gli3. The functions of the vertebrate Gli proteins at the molecular level and how each of them are involved in transducing Hh signaling are not completely understood (for recent reviews see for example: Jacob and Briscoe 2003; Cohen 2003). Gli2 and Gli3 share functional domain structure with Ci both containing transcriptional activator and repressor domains (Dai et al. 1999; Sasaki et al. 1999) and evidence suggests that both Gli2 and Gli3 can be cleaved to generate repressor forms (Aza-Blanc et al. 2000; Wang et al. 2000). However, only Gli3 processing has been demonstrated to be blocked in an Hh-dependent manner (Dai et al. 1999; Aza-Blanc et al. 2000). In contrast to Gli2 and Gli3, Gli1 only functions as a transcriptional activator (Hynes et al. 1997; Sasaki et al. 1997; Altaba 1998). Gli1 transcription is induced in response to Hh signaling and it does not appear to mediate the initial Hh signaling response (Bai et al. 2002).

As in *Drosophila*, vertebrate Fu and Su(fu) are involved in activation and nuclear translocation of the Gli proteins (Kogerman et al. 1999; Stone et al. 1999; Murone et al. 2000; Dunaeva et al. 2003).

Several transcriptional targets of the Hh signaling pathway have been identified, including the hedgehog receptor Ptch, and members of the Wnt and Tgfβ family as well as genes involved in cell cycle regulation (see also Sect. 3.4).

3 Hh Signaling in Tumor Formation and Growth

3.1 Defects in Hh Signaling Causes Nevoid Basal-Cell Carcinoma Syndrome

The initial link between Hh signaling and cancer was established by mapping and cloning of *PTCH* as the gene responsible for nevoid basal-cell carcinoma syndrome (NBCCS), also referred to as basal cell nevus syndrome or Gorlin syndrome (Hahn et al. 1996; Johnson et al. 1996).

NBCCS is an autosomal dominant inherited disease characterized by a variety of developmental defects and predisposition to multiple basal cell carcinomas (BCC), medulloblastoma (MB), and several other neoplasms, including ovarian fibroma, meningioma and cardiac fibroma (Shanley et al. 1994; Kimonis et al. 1997; Lo et al. 1999). In most NBCCS patients, a germline mutation in *PTCH* leads to development of BCC during childhood and in some patients more than a thousand BCCs may develop during a lifetime. Areas of the skin, which have been exposed to UV or ionizing radiation, are especially prone to development of BCCs (Kimonis et al. 1997).

3.2 Experiments with Animal Models Link Hh Signaling to Tumorigenesis

Studies based on animal models have provided important experimental evidence for a direct involvement of Hh signal transduction in the formation of various tumor types.

Transgene expression of *SHH*, *GLI1* or *GLI2* induces basal cell carcinoma in mice (Oro et al. 1997; Grachtchouk et al. 2000; Nilsson et al. 2000) and transgene expression of *Gli1* induces epidermal tumors in tadpole skin (Dahmane et al. 1997). Furthermore, transgene expression of *SHH* in the cerebellum of mice can induce MB (Weiner et al. 2002; Rao et al. 2003). These animal studies show that inappropriate activation of Hh signaling is sufficient for development of BCC and MB.

Mouse strains homozygous for targeted deletions of Ptch display severe malformations of the nervous system and heart and die between embryonic days 9 and 10.5 (Goodrich et al. 1997; Hahn et al. 1998). Heterozygous $Ptch^{+/-}$ mice are viable but develop rhabdomyosarcomas and MB dependent on the genetic background (Hahn et al. 1998; Goodrich et al. 1997). Furthermore, exposure of $Ptch^{+/-}$ mice to ultraviolet or ionizing radiation enhances the growth of BCC and MB (Aszterbaum et al. 1999; Pazzaglia et al. 2002). These data are in agreement with the increased incidence of BCC in NBCCS patients in sun-exposed areas of the skin and after radiation therapy treatment for MB (Kimonis et al. 1997) and support a recently proposed model of sequential tumor progression in BCC through the accumulation of genetic alterations (Mancuso et al. 2004).

3.3 Hh Signaling in Sporadic Cancers

The discovery of *PTCH* mutations as the cause of NBCCS suggested that *PTCH* acts as a tumor suppressor in certain cancers and that activation of the Hh signaling pathway could be involved in the sporadic development of the same tumor types as found in NBCCS patients. During recent years several studies have linked mutations in Hh pathway-related genes with BCC, MB and other sporadic tumors. Furthermore, a number of studies have suggested that activation of Hh signaling is pivotal for tumor growth in several cancers including small cell lung cancer, pancreatic cancer, prostate cancer, and cancers in the upper gastrointestinal tract.

3.3.1
A Subset of Sporadic Cancers Have Mutations in Genes Involved in Hh Signaling

Inactivating mutations in *PTCH* and activating mutations in *SMO* are found in 20–30 and 10%, respectively of sporadic BCCs (Gailani et al. 1996; Xie et al. 1998; Evans et al. 2000). In MB, a smaller subset of sporadic tumors have inactivating mutations in *PTCH* (13%) and activating mutations in *SMO* (6%; Vorechovsky et al. 1997a; Xie et al. 1997). *PTCH* mutations have also been reported in a number of other cancers (Table 1). However, more studies are needed in order to clarify the frequency of PTCH mutations in these tumor types.

Mutations in *PTCH2*, the second human ptch homologue, have been associated with BCC and MB, a single mutation in *SHH* has been associated with BCC, MB, and breast carcinomas (Oro et al. 1997) and mutations in *SUFU* have been associated with MB. These reports suggest that mutations in genes encoding other Hh signaling components may play a role in tumor formation, and more research in this area could help elucidate the role of Hh signaling in cancer.

3.3.2
Expression Analysis of Hh Signaling Genes Suggests Diverse Mechanisms of Involvement in Cancer

Expression of central components in Hh signaling has been detected in tumors and cell lines with mutations in *PTCH, SMO* or *SUFU* (Table 1), suggesting a possible role of Hh signaling in tumor growth. This role has been confirmed by growth inhibition of Ptch–/– mouse fibroblasts and human MB by the Hh antagonist *cyclopamine* (see Sect. 4; Taipale et al. 2000; Berman et al. 2002).

Recent studies suggest that Hh signaling may influence tumor growth through at least two different mechanisms: a ligand-independent mechanism caused by mutations in *PTCH, SMO,* and *SUFU,* and a ligand-dependent mechanism in which tumor growth depends on autocrine or paracrine signaling by Hh protein.

Table 1. Molecular changes of Hh related genes in sporadic cancers

Tumor type	Mutations in Hh-related genes Gene (References)	Expression of Hh-related genes Gene (References)
Ameloblastoma		*SHH, PTCH, SMO, GLI1* (Kumamoto et al. 2004)
Basal cell carcinoma (BCC)	*SMO, PTCH, PTCH2,SHH*[a] (Gailani et al. 1996; Oro et al. 1997; Unden et al. 1997; Wolter et al. 1997; Reifenberger et al. 1998; Xie et al. 1998; Smyth et al. 1999; Evans et al. 2000; Couve-Privat et al. 2004)	*PTCH, SMO, HHIP, GLI, GLI2* (Dahmane et al. 1997; Tojo et al. 1999, 2002, 2003; Zaphiropoulos et al. 1999)
Breast carcinomas	*PTCH* (Xie et al. 1997)	
Colon cancer (cell lines)	*PTCH* (Xie et al. 1997)	*SHH, IHH, GLI1 PTCH*[b], *SMO* (Berman et al. 2003; Qualtrough et al. 2004)
Digestive tract tumors (cell lines); originating from esophagus, stomach, and biliary tract	*PTCH* (Maesawa et al. 1998)[c]	*SHH, IHH, PTCH, GLI* (Berman et al. 2003)
Glioma		*SHH, GLI1, GLI2, GLI3, PTCH* (Dahmane et al. 2001)
Invasive transitional cell carcinoma of the bladder	*PTCH* (McGarvey et al. 1998)	
Medulloblastoma/ primitive neuroectodermal tumors	*SUFU, PTCH, PTCH2, SMO* (Pietsch et al. 1997; Raffel et al. 1997; Vorechovsky et al. 1997a; Wolter et al. 1997; Xie et al. 1997; Reifenberger et al. 1998; Smyth et al. 1999; Taylor et al. 2002)	*GLI1, GLI2, GLI3, PTCH* (Reifenberger et al. 1998; Dahmane et al. 2001)
Meningioma	*PTCH* (Xie et al. 1997)	
Odontogenic keratocysts	*PTCH* (Barreto et al. 2000; Ohki et al. 2004)	*SHH, SMO, PTCH, GLI1* (Ohki et al. 2004)
Pancreatic cancer		*SHH, IHH, PTCH, GLI* (Berman et al. 2003; Thayer et al. 2003; Kayed et al. 2004)
Prostate cancer		*SHH, GLI1, PTCH* (Fan et al. 2004; Sanchez et al. 2004)
Small cell lung cancer		*SHH, GLI1* (Watkins et al. 2003)
Squamous cell carcinoma	*PTCH* (Ping et al. 2001)	*GLI1, PTCH, SHH, SMO* (Koike et al. 2002; Nishimaki et al. 2004)
Trichoepitheliomas	*PTCH* (Vorechovsky et al. 1997b)	

[a] SHH mutations was found in 5/33 BCCs in Xeroderma pigmentosum patients
[b] Berman et al. (2003) failed to detect PTCH mRNA expression in 11 colon cancer cell lines
[c] Esophageal squamous cell carcinomas

The ligand-dependent mechanism has been found in small cell lung cancer (SCLC), pancreatic cancer, tumors in the gastrointestinal tract, and prostate cancer (Berman et al. 2003; Thayer et al. 2003; Watkins et al. 2003; Sanchez et al. 2004). Analysis of gene expression in a subset of tumor tissue and cancer cell lines showed that both genes encoding the Hh ligands, *SHH* or *IHH*, and downstream signaling components like *GLI1* were expressed (Table 1).

In agreement with previous results, in vitro and in vivo inhibition assays using cyclopamine demonstrated that tumor growth could be inhibited by this compound, thus indicating a requirement of Hh signaling for tumor growth (Berman et al. 2003; Thayer et al. 2003; Watkins et al. 2003; Sanchez et al. 2004).

Activation of the Hh pathway was apparently independent of mutations in *PTCH* (Watkins et al. 2003) but dependent on Hh ligand, as shown by inhibition of Hh signaling and tumor cell growth by an SHH-specific antibody (Berman et al. 2003; Sanchez et al. 2004). Immunostaining for SHH and GLI1 in SCLC nude mouse xenografts demonstrated SHH-expressing cells adjacent to GLI1-expressing cells, suggesting juxtacrine Hh signaling in SCLC (Watkins et al. 2003).

The Hh ligands *SHH* and *IHH* are expressed in colorectal tumor cell lines (Berman et al. 2003; Qualtrough et al. 2004; van den Brink et al. 2004). However, the role of Hh signaling in colorectal cancer is presently not well understood, due to conflicting results. Berman et al. (2003) reported *GLI1* expression in 3/11 colon-tumor-derived cell lines, but could not detect *PTCH* mRNA expression in any of these cell lines. Furthermore, these cells were negative in an Hh-inducible GLI-luciferase reporter assay, which led to the conclusion that Hh signaling was inactive in the colon cancer cell lines. The results from Berman et al. (2003) are corroborated by experiments of van den Brink et al. (2004), indicating that IHH acts as an antagonist of Wnt signaling during differentiation of colonic epithelial cells. However, these experiments are in conflict with a recent paper demonstrating that *PTCH* and *GLI1* are expressed in five cell lines from colorectal adenoma and adenocarcinoma and that cyclopamine treatment induced apoptosis in these cell lines (Qualtrough et al. 2004). Thus, it is presently unclear whether Hh signaling is involved in formation or growth of a subset of colorectal cancer.

Analysis of Hh signaling in prostate cancer and experiments with a mouse xenograft prostate cancer model showed that *SHH* is expressed in cancer cells in the tumor epithelium, while *GLI1* is expressed in the stroma cells (Fan et al. 2004). These observations suggest a third mechanism of Hh signaling in cancer, in which Hh signaling from cancer cells to the stroma can elicit expression of paracrine tumor growth signals.

3.4 Molecular Mechanisms for Hh-Stimulated Tumor Growth

The observation that Hh signaling is active in some cancers supports the growing appreciation that signaling pathways controlling vertebrate embryonic development may also be important during human carcinogenesis. But how is aberrant Hh signaling able to stimulate tumorigenesis and maintain the cancerogenic phenotype? These are questions to be addressed in this section.

3.4.1 SHH Stimulates Stem-Cell-Like Progenitor Cell Proliferation

Many studies have demonstrated a role for SHH in proliferation and maintenance of stem-cell-like progenitor cells (Kenney and Rowitch 2000; Miyazawa et al. 2000). The molecular mechanisms by which SHH regulates growth of progenitor cells are, however, incompletely understood. In *Drosophila*, Hh is able to stimulate proliferation of ovarian somatic cells by acting specifically on somatic stem cells, which are unable to proliferate in the absence of Hh signaling. In contrast, the number of stem cells roughly doubles upon constitutive activation of the Hh pathway, demonstrating that Hh is a potent stem cell mitogen in *Drosophila* (Zhang and Kalderon 2001). Hh signaling has also been shown to be essential for mammalian embryonic hair follicle development as $Shh^{-/-}$ and $Gli2^{-/-}$ mutant mice exhibit reduced cell proliferation in developing hair follicles (Michno et al. 2003; Mill et al. 2003). In the developing mammalian CNS, *Shh* drives proliferation of many different progenitor cell types including cerebellar granule neuron precursors (Wallace 1999; Wechsler-Reya and Scott 1999), neocortical (Dahmane et al. 2001), retinal (Moshiri and Reh 2004), spinal cord (Rowitch et al. 1999), and midbrain precursors (Dahmane et al. 2001; Britto et al. 2002). Hh signaling is also required for proliferation and maintenance of progenitor cells in the postnatal telencephalon (Machold et al. 2003) and hippocampus (Lai et al. 2003), indicating that Hh signaling may also be important for stem cell maintenance in the adult. A role has also been discovered for SHH in the control of primitive hematopoietic cell proliferation (Bhardwaj et al. 2001) and airway epithelial precursor cells (Watkins et al. 2003).

3.4.2 Do Tumors Arise from Transformation of Stem Cells?

Interestingly, stem cells possess many of the features that constitute the tumor phenotype, including self-renewal, evasion of apoptosis, and essentially unlimited replicative potential, which has raised the question whether the initiation events for tumorigenesis take place in stem cells (Oliver and Wechsler-Reya 2004). Some argue that it would be less complicated to retain these stem cells' characteristics than to reactivate the immortality program in an already differentiated cell (Hanahan and Weinberg 2000; Kopper and Hajdu 2004). Cancer cells also seem to utilize many of the same signaling pathways as stem cells such as Notch, Wnt and Hh (Kopper and Hajdu 2004). The exponential increase in cancer incidence with age and the fact that stem cells persist throughout the life of an individual provide a time frame in which rare mutations can accumulate in a single cell and eventually lead to malignant transformation and tumor formation (Taipale and Beachy 2001). It is therefore tempting to hypothesize that human cancers displaying active Hh signaling could be due to aberrant expansion of stem cell pools where Hh signaling is necessary for proliferation and survival. The presence of stem-cell-like cells within tumors has also been used to suggest that tumors contain cancer stem cells with the capacity to maintain the long-term growth of a tumor in vivo (Dick 2003). In general, tumors are phenotypically heterogenic and tumor cells display various differentiation levels (Kopper and Hajdu 2004). Many brain tumors have been shown to contain cells expressing stem cell markers such as nestin, CD133, and bmi-1 (Ignatova et al. 2002; Hemmati et al. 2003; Singh et al. 2003). It has also been discovered that only a small proportion of the cells within a tumor have tumorigenic capacity (Dick 2003). Evidence for the cancer stem cell theory comes from a recent experiment with breast cancer, where it was found that only a small fraction of tumor cells, which expressed certain markers, were able to form new tumors (Al Hajj et al. 2003). Interestingly, the phenotypic heterogeneity of the new tumor formed resembled the cells of the original tumor, suggesting that these cells, like stem cells, have the ability both to self-renew and to make differentiated progeny that have lost the ability to initiate tumor growth (Al Hajj et al. 2003). Likewise, in leukemia, lung cancer, ovarian cancer, and neuroblastoma, only a small proportion of the tumor cells are clonogenic in culture and in vivo (Reya et al. 2001). If, in fact, cancer stem cells exist, anti-cancer therapies designed to specifically eradicate these cells would be necessary and sufficient to completely stop tumor growth as all other cells in the tumor would have a limited life span (Kopper and Hajdu 2004).

3.4.3 Molecular Targets of Hh Signaling

Important clues to the molecular mechanisms by which Hh signaling may stimulate tumorigenesis come from recent studies identifying target genes for the Hh signaling pathway. Gene expression profiling studies have thus revealed that SHH activates a program of transcription that promotes cell cycle entry and DNA replication (Louro et al. 2002; Oliver et al. 2003). Likewise, stable Gli1 expression leads to altered expression of cell-cycle genes, cell-adhesion genes, signal-transduction genes, and genes regulating apoptosis (Yoon et al. 2002). The Hh pathway inhibitor cyclopamine has been shown to inhibit the growth of a variety of different tumors and tumor-derived cell lines (see Sect. 4). Collectively, these results suggest that Hh signaling activates a program of transcription necessary for formation and growth of a subset of tumors.

In *Drosophila*, Hh has been shown to regulate cell growth and proliferation by promoting transcription of the cell-cycle regulators cyclin D (*cycD*) and cyclin E (*cycE*). Upregulation of *cycE* occurs through direct binding of *Ci* to the *cycE* promoter and mediates the ability of Hh to induce DNA replication while *cycD* upregulation by Hh promotes cellular growth (Duman-Scheel et al. 2002). Several studies with mammalian cells have also demonstrated a role for SHH in the control of key cell-cycle regulators (Kenney and Rowitch 2000; Mill et al. 2003). Gli2 has thus been found to mediate the mitogenic effects of SHH by transcriptional activation of *cyclin D1* and *cyclin D2* in developing hair follicles (Mill et al. 2003). Likewise, Shh promotes sustained cell-cycle progression by positively regulating the cellular levels of *cyclin D1* and *cyclin D2* in cerebellar granule cell precursors. Furthermore, Shh treatment of cerebellar granule cell precursor also resulted in elevation of cellular *cyclin E* levels (Kenney and Rowitch 2000; Oliver et al. 2003), which shows that mammalian cells are comparable to *Drosophila* with respect to the mitogenic effects of Hh at the molecular level. However, in contrast to *Drosophila*, intermediate protein synthesis seems to be required for upregulation of these cyclins in mammalian cells, since cycloheximide treatment prevented the effects of Shh on *cyclin D1*, *cyclin D2*, and *cyclin E* mRNA levels. Therefore, SHH most likely induces synthesis of intermediate proteins, which in turn regulate cyclin expression (Kenney and Rowitch 2000). The transcription factor N-myc has been recently identified as a direct transcriptional target of Shh signaling in cerebellar granule cell precursors. N-myc overexpression in these cells induces expression of *cyclin D1* and promotes proliferation, whereas a dominant-negative N-myc substantially reduces Shh-induced proliferation (Kenney et al. 2003; Oliver et al. 2003).

Further evidence for direct activation of the cell cycle by Hh signaling comes from a study by Barnes and coworkers (2001), who showed that *Xenopus* ptc1 interacted specifically with cyclin B1. Cyclin B1 is part of the M-phase-promoting factor, which translocates to the nucleus at late G2 and is essential for mitotic progression. The interaction between ptc1 and cyclin B1 could be disrupted by ligand-induced activation of the pathway, making the authors suggest the existence of a G2/M checkpoint where ptc1, in the absence of Hh stimulation, sequesters the M-phase-promoting factor, thereby preventing entry into mitosis and cell division (Barnes et al. 2001). The polycomb group gene *Bmi1* has been shown to be required for self-renewal and post-natal persistence of stem cells in the peripheral and central nervous system through repression of the cyclin-dependent kinase inhibitor $p16^{Ink4a}$ (Molofsky et al. 2003). Overexpression of *BMI1* has been found in a substantial fraction of primary human medulloblastomas and, interestingly, addition of Shh or overexpression of *Gli1* in cerebellar granule cell precursors leads to rapid induction of Bmi1, which may constitute yet another way of Hh signaling to affect cell proliferation and tumorigenesis (Leung et al. 2004). Finally, Shh blocks cell-cycle arrest that is mediated by the cyclin-dependent kinase inhibitor $p21^{CIP1/WAF1}$ (Fan and Khavari 1999). Together, these studies provide compelling evidence that increased cell proliferation, a hallmark of tumor formation, can be mediated via direct interaction of the SHH pathway with components of the cell cycle machinery.

Studies indicate that the Hh signaling pathway may also have an effect on cell survival and apoptosis regulation and in this way uphold tumor growth. Reduction of Hh signaling in the chick brain thus leads to a decrease in cell survival, presumably because of increased apoptosis (Ahlgren and Bronner-Fraser 1999). Likewise, removal of Hh responsiveness in murine neural crest cells results in facial truncation attributable to an increase in apoptosis (Jeong et al. 2004). Recently, it has been shown that GLI1 upregulates transcription of the anti-apoptotic proto-oncogene *BCL-2* in keratinocytes (Bigelow et al. 2004). SHH-mediated upregulation of anti-apoptotic factors such as BCL-2 may thus result in an inappropriate viability advantage necessary for cancer development.

Like the Hh signaling pathway, Wnt signaling plays a critical role during development, with effects on proliferation, differentiation, and survival. Many studies demonstrate a role for this pathway in cancer (reviewed in Taipale and Beachy 2001; Sell 2004). Among other neoplastic conditions, Wnt signaling has been implicated in mammary gland neoplasia (Imbert et al. 2001), colorectal cancers, and hepatocellular carcinoma (Karim et al. 2004). Interestingly, several Wnt genes have been shown to be targets of Hh signaling. In the frog embryo, Gli proteins have been found to regulate the expression of many Wnt genes including *Wnt2, 3, 5a, 7B, 7C, 8* and *11*, and *Wnt8* and *11* were found to be required for *Gli2/3* function in posterior mesodermal development (Mullor et al. 2001). Also in BCC displaying increased Hh signaling activity, an increased expression of *WNT2B* and *WNT5A* was found, whereas decreased levels of *WNT4* gene expression were reported (Bonifas et al. 2001). Likewise, *IHH* plays a role in the regulation of colonic Wnt signaling, and expression of *IHH* in colon cancer cells in vitro abrogates endogenous Wnt signaling (van den Brink et al. 2004). Together, these results suggest that crosstalk between Hh and Wnt signaling may play a role in tumor formation and growth.

Forkhead box (FOX) proteins are implicated in a variety of biological processes ranging from embryonic development to adult tissue homeostasis by regulating cell growth, proliferation, differentiation, longevity, and transformation (reviewed in Carlsson and Mahlapuu 2002). Studies show that various FOX genes are targets of Hh signaling in different tissues. During embryogenesis, the FOX genes *HNF-3β* (*FOXA2*), *Mf2* (*FOXD2*), *HFH-8* (*FOXF1*), and *Mfh1* (*FOXC2*) are targets of Hh signaling (Sasaki et al. 1997; Wu et al. 1998; Furumoto et al. 1999; Mahlapuu et al. 2001). A few studies have also investigated the ability of SHH to induce FOX gene expression in skin and the possible involvement of FOX genes in BCC. FOXM1 has thus been shown to be upregulated in BCC and transcription of this FOX gene is inducible by *GLI1* expression in various cell types, demonstrating that *FOXM1* is a likely target gene of the Hh signaling pathway and may play a role in BCC (Teh et al. 2002). Likewise, evidence suggests that also *FOXE1* is a target gene of Hh signaling in epidermis and it has been found to be upregulated in BCC compared to normal epidermis. In addition, the presence of GLI-binding sites in the *FOXE1*-promoter region, together with activation of the *FOXE1* promoter in a reporter assay suggests that *FOXE1* is a direct target of GLI2 (Eichberger et al. 2004).

An interesting gene, newly identified as a transcriptional target of Hh signaling, is the mesenchymal cell marker *Snail*, a zinc finger protein implicated in epithelial-mesenchymal transition in development and during tumor progression. Snail was found to be highly upregulated in *GLI1*-overexpressing cells and its promoter contains four candidate Gli binding sites, making it a potential direct target gene of the Hh signaling pathway (Louro et al. 2002). Snail is highly involved in the acquisition of migratory and invasive properties of tumor cells, partly mediated by transcriptional repression of epithelial markers and upstream regulation of molecules involved in degradation of the basement membrane. Recent evidence suggests that Snail also confers resistance to cell death. These properties makes Snail a marker of a malignant tumor phenotype (reviewed in Nieto 2002; Vega et al. 2004). Hh signaling-mediated upregulation of *Snail* may thus provide cells an increased capability for survival, invasion, and metastasis, all hallmark cancer features.

Insulin-like growth factor 2 (*Igf2*) is upregulated in *Gli1*-overexpressing cells (Louro et al. 2002) and is also consistently overexpressed in rhabdomyosarcoma in *Ptch1*-mutant mice, suggesting that *Igf2* may be a target gene of the Hh signaling pathway. Furthermore, *Igf2* is indispensable for the formation of medulloblastoma and rhabdomyosarcoma in *Ptch1* mutants (Hahn et al. 2000), and may thus play a role in Hh-induced cancer initiation.

4 Small-Molecule Hh Signaling Antagonists – a Therapeutic Opportunity?

Evidence for involvement of Hh signaling in development and growth of several forms of cancer is increasing rapidly (see Sect. 3.3). The apparent requirement of Hh signaling for growth of a subset of tumors points toward a potential therapeutic opportunity based on targeted inhibition of Hh signaling. The discovery of cyclopamine, a specific Hh antagonist, and the potential therapeutic effect of small molecule Hh signaling antagonists like cyclopamine is described in the following.

4.1
Discovery of Small-Molecule Hh Antagonists

The first step toward an anti-cancer drug targeting Hh signaling was taken by the discovery of the plant alkaloid cyclopamine. This teratogenic compound was purified from the lily *Veratrum californicum*, which gives rise to birth of cyclopic lambs when ingested by pregnant sheep (Keeler 1978). It was subsequently shown that cyclopamine causes holoprosencephaly by specific inhibition of Hh signaling through a direct binding to Smo (Cooper et al. 1998; Incardona et al. 1998; Chen et al. 2002a; Fan et al. 2004).

Recently, combinatorial chemistry and high-throughput screening using Shh-responsive cell models have been used to isolate synthetic Hh antagonists with a chemical structure distinctly different from cyclopamine (Chen et al. 2002b; Frank-Kamenetsky et al. 2002; King 2002). Biochemical studies suggest that five of these compounds target Smo directly, possibly with different mechanisms and two of the compounds appear to act on unknown targets downstream of Smo (Chen et al. 2002b; Frank-Kamenetsky et al. 2002).

4.2
Inhibition of Tumor Growth by Hh Antagonists

The use of a synthetic molecule for gene-targeted cancer therapy has many advantages because it is relatively easy to produce and administer and has the potential to cross the blood-brain barrier, which is an attractive property in relation to treatment of MB. Furthermore, cyclopamine administration appears to be without harmful side effects in animal experiments using adult mammals (Keeler 1978).

The potential therapeutic effect of cyclopamine has been investigated in vitro and in vivo. It has been shown that cyclopamine can inhibit the abnormal cell growth of fibroblast cells from *Ptch1*$^{-/-}$ mice (Taipale et al. 2000), several glioblastoma/glioma cell lines (Dahmane et al. 2001), MB cell lines (Berman et al. 2002; Dahmane et al. 2001), squamous cell carcinoma cell lines (Koike et al. 2002), SCLC cell lines (Watkins et al. 2003), pancreatic cancer cell lines (Berman et al. 2003; Thayer et al. 2003; Kayed et al. 2004), cell lines from tumors in the stomach, esophagus, and biliary tract (Berman et al. 2003), colorectal adenoma- and carcinoma-derived cell lines (Qualtrough et al. 2004), and prostate cancer cell lines (Sanchez et al. 2004). Thus, cyclopamine inhibition of tumor growth has been demonstrated in vitro for virtually all cancers in which aberrant Hh signaling has been demonstrated to date. Furthermore, inhibition of tumor growth by cyclopamine has also been demonstrated in vivo by experiments with xenografts in nude mice (Berman et al. 2003; Thayer et al. 2003) and by treatment of tumors in patients with BCCs (Tas and Avci 2004). Experiments with inhibition of tumor cell lines suggest that cyclopamine induces cell-cycle arrest and apoptosis (Watkins et al. 2003; Kayed et al. 2004; Nishimaki et al. 2004; Qualtrough et al. 2004).

Direct testing of inhibition of tumor growth was also performed with the synthetic Hh antagonist CUR61414 using an in vitro BCC model system (Williams et al. 2003). These experiments suggest that the effect of CUR61414 on tumor growth is comparable with cyclopamine.

In conclusion, the initial experiments with Hh antagonists are very promising and hopefully experiments in animal models and clinical trials will follow the same path.

However, several factors may complicate the therapeutic use of Hh antagonists in cancer. Only a subset of tumors seems to express Hh signaling components (see Sect. 3.3.2). It is also likely that cross talk between different developmental pathways (e.g., Hh, Wnt, erythroblastic leukemia viral oncogene homologue [v-erb-b2; ERBB], and IGF pathways) is involved in tumor formation and growth (Gilbertson 2004; Rao et al. 2004). Such signaling redundancy may protect some tumors from the effect of the drug. Furthermore, experiments with activating mutations in *SMO* suggest that such mutations can potentially attenuate the effect of the antagonist (Taipale et al. 2000); activation of the Hh pathway downstream from SMO may also cause resistance to, e.g., cyclopamine. Thus, specific compounds may be without effect in specific patients, or the cancer cells may develop resistance against the compounds. However, these obstacles may be overcome by precise diagnostic classification of tumors, using expression profiling and combinatorial chemotherapy.

5 Future Prospects

An impressive amount of knowledge about the involvement of Hh signaling in tumor formation and growth has accumulated since the discovery of *PTCH* as the disease gene in NBCCS. While the research so far has uncovered many important aspects and laid the foundation for the development of potential anti-cancer drugs, many questions are still unanswered and several surprises can probably be expected.

The Hh signaling pathway is well described and most of the major players have probably been found. Nevertheless, many aspects of Hh signaling are still largely unknown. The molecular function of SMO and the mechanism of signal transduction from SMO to GLI are presently not fully understood. Another unlit area is that of the target genes at the far end of the Hh pathway. Although many examples exist, and recent expression-profiling experiments have revealed novel target genes (Louro et al. 2002; Zhao et al. 2002), systematic experiments are needed in order to illuminate this end of the pathway. Unfortunately, it is very likely that the target genes will differ between tissues and cell populations, which will make this task rather exhausting.

Recent experiments have shown that Hh signaling is involved in a much larger number of cancers than previously expected. It is important that these initial findings are established on a larger sample and that the molecular mechanism of Hh signaling in these cancers is thoroughly investigated. Especially, the role of Hh signaling in colon cancer is presently confusing. All the tumors displaying Hh signaling activity have been found in tissues wherein Hh signaling plays a role during development and in some cases also during mature tissue homeostasis. Therefore, it would be interesting to investigate whether Hh signaling is also involved in tumor growth in other tissues dependent on Hh activity during development, e.g., cancers of the eye, heart, or testis.

It is likely that cross-talk exists between the Hh pathway and other signaling pathways during tumor formation and growth. Mapping and investigations of such cross-talk is important for a more sophisticated understanding of the tumor biology and for design of efficient cancer therapies.

Another important aspect to study further is the possible link between stem cells, Hh signaling, and cancer. This is of interest for several reasons. First of all, it is important to establish if the proliferating capacity of tumors is restricted to a small population of cancer stem cells and if Hh antagonists can eradicate this cell population effectively in vivo. Secondly, if Hh signaling controls stem cell proliferation during normal tissue renewal, then it is important to determine the potential toxic effects of the Hh antagonists, which may be used for future cancer therapy. And last but not least, it is important to establish if there is a cancer risk associated with stem cell therapy of, e.g., Parkinson's disease.

References

Ahlgren SC, Bronner-Fraser M (1999) Inhibition of sonic hedgehog signaling in vivo results in craniofacial neural crest cell death. Curr Biol 9:1304–1314

Al Hajj M, Wicha MS, Benito-Hernandez A, Morrison SJ, Clarke MF (2003) Prospective identification of tumorigenic breast cancer cells. Proc Natl Acad Sci USA 100:3983–3988

Alcedo J, Zou Y, Noll M (2000) Posttranscriptional regulation of smoothened is part of a self-correcting mechanism in the Hedgehog signaling system. Mol Cell 6:457–465

Altaba A (1998) Combinatorial Gli gene function in floor plate and neuronal inductions by Sonic hedgehog. Development 125:2203–2212

Altaba A, Stecca B, Sanchez P (2004) Hedgehog-Gli signaling in brain tumors: stem cells and paradevelopmental programs in cancer. Cancer Lett 20(204):145–157

Amanai K, Jiang J (2001) Distinct roles of Central missing and Dispatched in sending the Hedgehog signal. Development 128:5119–5127

Aszterbaum M, Epstein J, Oro A, Douglas V, LeBoit PE, Scott MP, Epstein EH Jr (1999) Ultraviolet and ionizing radiation enhance the growth of BCCs and trichoblastomas in patched heterozygous knockout mice. Nat Med 5:1285–1291

Aza-Blanc P, Lin HY, Altaba A, Kornberg TB (2000) Expression of the vertebrate Gli proteins in *Drosophila* reveals a distribution of activator and repressor activities. Development 127:4293–4301

Bai CB, Auerbach W, Lee JS, Stephen D, Joyner AL (2002) Gli2, but not Gli1, is required for initial Shh signaling and ectopic activation of the Shh pathway. Development 129:4753–4761

Barnes EA, Kong M, Ollendorff V, Donoghue DJ (2001) Patched1 interacts with cyclin B1 to regulate cell cycle progression. EMBO J 20:2214–2223

Barreto DC, Gomez RS, Bale AE, Boson WL, de Marco L (2000) PTCH gene mutations in odontogenic keratocysts. J Dent Res 79:1418–1422

Basler K, Struhl G (1994) Compartment boundaries and the control of Drosophila limb pattern by hedgehog protein. Nature 368:208–214

Bellaiche Y, The I, Perrimon N (1998) Tout-velu is a Drosophila homologue of the putative tumour suppressor EXT-1 and is needed for Hh diffusion. Nature 394:85–88

Berman DM, Karhadkar SS, Hallahan AR, Pritchard JI, Eberhart CG, Watkins DN, Chen JK, Cooper MK, Taipale J, Olson JM, Beachy PA (2002) Medulloblastoma growth inhibition by hedgehog pathway blockade. Science 297:1559–1561

Berman DM, Karhadkar SS, Maitra A, Montes DO, Gerstenblith MR, Briggs K, Parker AR, Shimada Y, Eshleman JR, Watkins DN, Beachy PA (2003) Widespread requirement for Hedgehog ligand stimulation in growth of digestive tract tumours. Nature 425:846–851

Bhardwaj G, Murdoch B, Wu D, Baker DP, Williams KP, Chadwick K, Ling LE, Karanu FN, Bhatia M (2001) Sonic hedgehog induces the proliferation of primitive human hematopoietic cells via BMP regulation. Nat Immunol 2:172–180

Bigelow RL, Chari NS, Unden AB, Spurgers KB, Lee S, Roop DR, Toftgard R, McDonnell TJ (2004) Transcriptional regulation of bcl-2 mediated by the sonic hedgehog signaling pathway through gli-1. J Biol Chem 279:1197–1205

Bonifas JM, Pennypacker S, Chuang PT, McMahon AP, Williams M, Rosenthal A, De Sauvage FJ, Epstein EH Jr (2001) Activation of expression of hedgehog target genes in basal cell carcinomas. J Invest Dermatol 116:739–742

Britto J, Tannahill D, Keynes R (2002) A critical role for sonic hedgehog signaling in the early expansion of the developing brain. Nat Neurosci 5:103–110

Bumcrot DA, Takada R, McMahon AP (1995) Proteolytic processing yields two secreted forms of sonic hedgehog. Mol Cell Biol 15:2294–2303

Burke R, Nellen D, Bellotto M, Hafen E, Senti KA, Dickson BJ, Basler K (1999) Dispatched, a novel sterol-sensing domain protein dedicated to the release of cholesterol-modified hedgehog from signaling cells. Cell 99:803–815

Carlsson P, Mahlapuu M (2002) Forkhead transcription factors: key players in development and metabolism. Dev Biol 250:1–23

Caspary T, Garcia-Garcia MJ, Huangfu D, Eggenschwiler JT, Wyler MR, Rakeman AS, Alcorn HL, Anderson KV (2002) Mouse dispatched homolog1 is required for long-range, but not juxtacrine, Hh signaling. Curr Biol 12:1628–1632

Chamoun Z, Mann RK, Nellen D, von Kessler DP, Bellotto M, Beachy PA, Basler K (2001) Skinny hedgehog, an acyltransferase required for palmitoylation and activity of the hedgehog signal. Science 293:2080–2084

Chen CH, von Kessler DP, Park W, Wang B, Ma Y, Beachy PA (1999) Nuclear trafficking of Cubitus interruptus in the transcriptional regulation of Hedgehog target gene expression. Cell 98:305–316

Chen JK, Taipale J, Cooper MK, Beachy PA (2002a) Inhibition of Hedgehog signaling by direct binding of cyclopamine to Smoothened. Genes Dev 16:2743–2748

Chen JK, Taipale J, Young KE, Maiti T, Beachy PA (2002b) Small molecule modulation of Smoothened activity. Proc Natl Acad Sci USA 99:14071–14076

Chuang PT, McMahon AP (1999) Vertebrate Hedgehog signalling modulated by induction of a Hedgehog-binding protein. Nature 397:617–621

Chuang PT, Kawcak T, McMahon AP (2003) Feedback control of mammalian Hedgehog signaling by the Hedgehog-binding protein, Hip1, modulates Fgf signaling during branching morphogenesis of the lung. Genes Dev 17:342–347

Cohen MM Jr (2003) The hedgehog signaling network. Am J Med Genet 123A:5–28
Cooper MK, Porter JA, Young KE, Beachy PA (1998) Teratogen-mediated inhibition of target tissue response to Shh signaling. Science 280:1603–1607
Couve-Privat S, Le Bret M, Traiffort E, Queille S, Coulombe J, Bouadjar B, Avril MF, Ruat M, Sarasin A, Daya-Grosjean L (2004) Functional analysis of novel sonic hedgehog gene mutations identified in basal cell carcinomas from xeroderma pigmentosum patients. Cancer Res 64:3559–3565
Dahmane N, Lee J, Robins P, Heller P, Altaba A (1997) Activation of the transcription factor Gli1 and the Sonic hedgehog signalling pathway in skin tumours. Nature 389:876–881
Dahmane N, Sanchez P, Gitton Y, Palma V, Sun T, Beyna M, Weiner H, Altaba A (2001) The Sonic Hedgehog-Gli pathway regulates dorsal brain growth and tumorigenesis. Development 128:5201–5212
Dai P, Akimaru H, Tanaka Y, Maekawa T, Nakafuku M, Ishii S (1999) Sonic Hedgehog-induced activation of the Gli1 promoter is mediated by GLI3. J Biol Chem 19(274):8143–8152
Denef N, Neubuser D, Perez L, Cohen SM (2000) Hedgehog induces opposite changes in turnover and subcellular localization of patched and smoothened. Cell 102:521–531
Dick JE (2003) Breast cancer stem cells revealed. Proc Natl Acad Sci USA 100:3547–3549
Duman-Scheel M, Weng L, Xin S, Du W (2002) Hedgehog regulates cell growth and proliferation by inducing cyclin D and cyclin E. Nature 417:299–304
Dunaeva M, Michelson P, Kogerman P, Toftgard R (2003) Characterization of the physical interaction of Gli proteins with SUFU proteins. J Biol Chem 278:5116–5122
Echelard Y, Epstein DJ, St Jacques B, Shen L, Mohler J, McMahon JA, McMahon AP (1993) Sonic hedgehog, a member of a family of putative signaling molecules, is implicated in the regulation of CNS polarity. Cell 75:1417–1430
Eggenschwiler JT, Espinoza E, Anderson KV (2001) Rab23 is an essential negative regulator of the mouse Sonic hedgehog signalling pathway. Nature 412:194–198
Eichberger T, Regl G, Ikram MS, Neill GW, Philpott MP, Aberger F, Frischauf AM (2004) FOXE1, a new transcriptional target of GLI2 is expressed in human epidermis and basal cell carcinoma. J Invest Dermatol 122:1180–1187
Evans T, Boonchai W, Shanley S, Smyth I, Gillies S, Georgas K, Wainwright B, Chenevix-Trench G, Wicking C (2000) The spectrum of patched mutations in a collection of Australian basal cell carcinomas. Hum Mutat 16:43–48
Fan H, Khavari PA (1999) Sonic hedgehog opposes epithelial cell cycle arrest. J Cell Biol 147:71–76
Fan L, Pepicelli CV, Dibble CC, Catbagan W, Zarycki JL, Laciak R, Gipp J, Shaw A, Lamm ML, Munoz A, Lipinski R, Thrasher JB, Bushman W (2004) Hedgehog signaling promotes prostate xenograft tumor growth. Endocrinology 145:3961–3970
Frank-Kamenetsky M, Zhang XM, Bottega S, Guicherit O, Wichterle H, Dudek H, Bumcrot D, Wang FY, Jones S, Shulok J, Rubin LL, Porter JA (2002) Small-molecule modulators of Hedgehog signaling: identification and characterization of Smoothened agonists and antagonists. J Biol 1:10
Furumoto TA, Miura N, Akasaka T, Mizutani-Koseki Y, Sudo H, Fukuda K, Maekawa M, Yuasa S, Fu Y, Moriya H, Taniguchi M, Imai K, Dahl E, Balling R, Pavlova M, Gossler A, Koseki H (1999) Notochord-dependent expression of MFH1 and PAX1 cooperates to maintain the proliferation of sclerotome cells during the vertebral column development. Dev Biol 210:15–29
Gailani MR, Stahle-Backdahl M, Leffell DJ, Glynn M, Zaphiropoulos PG, Pressman C, Unden AB, Dean M, Brash DE, Bale AE, Toftgard R (1996) The role of the human homologue of Drosophila patched in sporadic basal cell carcinomas. Nat Genet 14:78–81
Gilbertson RJ (2004) Medulloblastoma: signalling a change in treatment. Lancet Oncol 5:209–218

Goodrich LV, Milenkovic L, Higgins KM, Scott MP (1997) Altered neural cell fates and medulloblastoma in mouse patched mutants. Science 277:1109–1113

Grachtchouk M, Mo R, Yu S, Zhang X, Sasaki H, Hui CC, Dlugosz AA (2000) Basal cell carcinomas in mice overexpressing Gli2 in skin. Nat Genet 24:216–217

Gritli-Linde A, Lewis P, McMahon AP, Linde A (2001) The whereabouts of a morphogen: direct evidence for short- and graded long-range activity of hedgehog signaling peptides. Dev Biol 236:364–386

Hahn H, Wicking C, Zaphiropoulous PG, Gailani MR, Shanley S, Chidambaram A, Vorechovsky I, Holmberg E, Unden AB, Gillies S, Negus K, Smyth I, Pressman C, Leffell DJ, Gerrard B, Goldstein AM, Dean M, Toftgard R, Chenevix-Trench G, Wainwright B, Bale AE (1996) Mutations of the human homolog of Drosophila patched in the nevoid basal cell carcinoma syndrome. Cell 85:841–851

Hahn H, Wojnowski L, Zimmer AM, Hall J, Miller G, Zimmer A (1998) Rhabdomyosarcomas and radiation hypersensitivity in a mouse model of Gorlin syndrome. Nat Med 4:619–622

Hahn H, Wojnowski L, Specht K, Kappler R, Calzada-Wack J, Potter D, Zimmer A, Muller U, Samson E, Quintanilla-Martinez L, Zimmer A (2000) Patched target Igf2 is indispensable for the formation of medulloblastoma and rhabdomyosarcoma. J Biol Chem 275:28341–28344

Hanahan D, Weinberg RA (2000) The hallmarks of cancer. Cell 100:57–70

Hemmati HD, Nakano I, Lazareff JA, Masterman-Smith M, Geschwind DH, Bronner-Fraser M, Kornblum HI (2003) Cancerous stem cells can arise from pediatric brain tumors. Proc Natl Acad Sci USA 100:15178–15183

Hynes M, Stone DM, Dowd M, Pitts-Meek S, Goddard A, Gurney A, Rosenthal A (1997) Control of cell pattern in the neural tube by the zinc finger transcription factor and oncogene Gli-1. Neuron 19:15–26

Ignatova TN, Kukekov VG, Laywell ED, Suslov ON, Vrionis FD, Steindler DA (2002) Human cortical glial tumors contain neural stem-like cells expressing astroglial and neuronal markers in vitro. Glia 39:193–206

Imbert A, Eelkema R, Jordan S, Feiner H, Cowin P (2001) Delta N89 beta-catenin induces precocious development, differentiation, and neoplasia in mammary gland. J Cell Biol 153:555–568

Incardona JP, Gaffield W, Kapur RP, Roelink H (1998) The teratogenic Veratrum alkaloid cyclopamine inhibits sonic hedgehog signal transduction. Development 125:3553–3562

Incardona JP, Lee JH, Robertson CP, Enga K, Kapur RP, Roelink H (2000) Receptor-mediated endocytosis of soluble and membrane-tethered Sonic hedgehog by Patched-1. Proc Natl Acad Sci USA 97:12044–12049

Ingham PW, McMahon AP (2001) Hedgehog signaling in animal development: paradigms and principles. Genes Dev 15:3059–3087

Ingham PW, Nystedt S, Nakano Y, Brown W, Stark D, van den HM, Taylor AM (2000) Patched represses the Hedgehog signalling pathway by promoting modification of the Smoothened protein. Curr Biol 19;10:1315–1318

Jacob J, Briscoe J (2003) Gli proteins and the control of spinal-cord patterning. EMBO Rep 4:761–765

Jeong J, Mao J, Tenzen T, Kottmann AH, McMahon AP (2004) Hedgehog signaling in the neural crest cells regulates the patterning and growth of facial primordia. Genes Dev 18:937–951

Jia J, Amanai K, Wang G, Tang J, Wang B, Jiang J (2002) Shaggy/GSK3 antagonizes Hedgehog signalling by regulating Cubitus interruptus. Nature 416:548–552

Jia J, Tong C, Jiang J (2003) Smoothened transduces Hedgehog signal by physically interacting with Costal2/Fused complex through its C-terminal tail. Genes Dev 17:2709–2720

Jiang J, Struhl G (1998) Regulation of the Hedgehog and Wingless signalling pathways by the F-box/WD40-repeat protein Slimb. Nature 391:493–496

Johnson RL, Rothman AL, Xie J, Goodrich LV, Bare JW, Bonifas JM, Quinn AG, Myers RM, Cox DR, Epstein EH, Jr., Scott MP (1996) Human homolog of patched, a candidate gene for the basal cell nevus syndrome. Science 272:1668–1671

Karim R, Tse G, Putti T, Scolyer R, Lee S (2004) The significance of the Wnt pathway in the pathology of human cancers. Pathology 36:120–128

Kawakami T, Kawcak T, Li YJ, Zhang W, Hu Y, Chuang PT (2002) Mouse dispatched mutants fail to distribute hedgehog proteins and are defective in hedgehog signaling. Development 129:5753–5765

Kayed H, Kleeff J, Keleg S, Guo J, Ketterer K, Berberat PO, Giese N, Esposito I, Giese T, Buchler MW, Friess H (2004) Indian hedgehog signaling pathway: expression and regulation in pancreatic cancer. Int J Cancer 110:668–676

Keeler RF (1978) Cyclopamine and related steroidal alkaloid teratogens: their occurrence, structural relationship, and biologic effects. Lipids 13:708–715

Kenney AM, Rowitch DH (2000) Sonic hedgehog promotes G(1) cyclin expression and sustained cell cycle progression in mammalian neuronal precursors. Mol Cell Biol 20:9055–9067

Kenney AM, Cole MD, Rowitch DH (2003) Nmyc upregulation by sonic hedgehog signaling promotes proliferation in developing cerebellar granule neuron precursors. Development 130:15–28

Kimonis VE, Goldstein AM, Pastakia B, Yang ML, Kase R, DiGiovanna JJ, Bale AE, Bale SJ (1997) Clinical manifestations in 105 persons with nevoid basal cell carcinoma syndrome. Am J Med Genet 69:299–308

King RW (2002) Roughing up Smoothened: chemical modulators of Hedgehog signaling. J Biol 1:8

Kogerman P, Grimm T, Kogerman L, Krause D, Unden AB, Sandstedt B, Toftgard R, Zaphiropoulos PG (1999) Mammalian suppressor-of-fused modulates nuclear-cytoplasmic shuttling of Gli-1. Nat Cell Biol 1:312–319

Koike C, Mizutani T, Ito T, Shimizu Y, Yamamichi N, Kameda T, Michimukai E, Kitamura N, Okamoto T, Iba H (2002) Introduction of wild-type patched gene suppresses the oncogenic potential of human squamous cell carcinoma cell lines including A431. Oncogene 21:2670–2678

Kopper L, Hajdu M (2004) Tumor stem cells. Pathol Oncol Res 10:69–73

Koziel L, Kunath M, Kelly OG, Vortkamp A (2004) Ext1-dependent heparan sulfate regulates the range of Ihh signaling during endochondral ossification. Dev Cell 6:801–813

Kumamoto H, Ohki K, Ooya K (2004) Expression of Sonic hedgehog (SHH) signaling molecules in ameloblastomas. J Oral Pathol Med 33:185–190

Lai K, Kaspar BK, Gage FH, Schaffer DV (2003) Sonic hedgehog regulates adult neural progenitor proliferation in vitro and in vivo. Nat Neurosci 6:21–27

Lee CS, Buttitta L, Fan CM (2001) Evidence that the WNT-inducible growth arrest-specific gene 1 encodes an antagonist of sonic hedgehog signaling in the somite. Proc Natl Acad Sci USA 98:11347–11352

Lee JD, Treisman JE (2001) Sightless has homology to transmembrane acyltransferases and is required to generate active Hedgehog protein. Curr Biol 11:1147–1152

Lee JJ, Ekker SC, von Kessler DP, Porter JA, Sun BI, Beachy PA (1994) Autoproteolysis in hedgehog protein biogenesis. Science 266:1528–1537

Leung C, Lingbeek M, Shakhova O, Liu J, Tanger E, Saremaslani P, Van Lohuizen M, Marino S (2004) Bmi1 is essential for cerebellar development and is overexpressed in human medulloblastomas. Nature 428:337–341

Lo ML, Nocini PF, Savoia A, Consolo U, Procaccini M, Zelante L, Pannone G, Bucci P, Dolci M, Bambini F, Solda P, Favia G (1999) Nevoid basal cell carcinoma syndrome. Clinical findings in 37 Italian affected individuals. Clin Genet 55:34–40

Louro ID, Bailey EC, Li X, South LS, McKie-Bell PR, Yoder BK, Huang CC, Johnson MR, Hill AE, Johnson RL, Ruppert JM (2002) Comparative gene expression profile analysis of GLI and c-MYC in an epithelial model of malignant transformation. Cancer Res 62:5867–5873

Lum L, Zhang C, Oh S, Mann RK, von Kessler DP, Taipale J, Weis-Garcia F, Gong R, Wang B, Beachy PA (2003) Hedgehog signal transduction via Smoothened association with a cytoplasmic complex scaffolded by the atypical kinesin, Costal-2. Mol Cell 12:1261–1274

Ma Y, Erkner A, Gong R, Yao S, Taipale J, Basler K, Beachy PA (2002) Hedgehog-mediated patterning of the mammalian embryo requires transporter-like function of dispatched. Cell 111:63–75

Machold R, Hayashi S, Rutlin M, Muzumdar MD, Nery S, Corbin JG, Gritli-Linde A, Dellovade T, Porter JA, Rubin LL, Dudek H, McMahon AP, Fishell G (2003) Sonic hedgehog is required for progenitor cell maintenance in telencephalic stem cell niches. Neuron 39:937–950

Maesawa C, Tamura G, Iwaya T, Ogasawara S, Ishida K, Sato N, Nishizuka S, Suzuki Y, Ikeda K, Aoki K, Saito K, Satodate R (1998) Mutations in the human homologue of the Drosophila patched gene in esophageal squamous cell carcinoma. Genes Chromosomes Cancer 21:276–279

Mahlapuu M, Enerback S, Carlsson P (2001) Haploinsufficiency of the forkhead gene Foxf1, a target for sonic hedgehog signaling, causes lung and foregut malformations. Development 128:2397–2406

Mancuso M, Pazzaglia S, Tanori M, Hahn H, Merola P, Rebessi S, Atkinson MJ, Di MV, Covelli V, Saran A (2004) Basal cell carcinoma and its development: insights from radiation-induced tumors in Ptch1-deficient mice. Cancer Res 64:934–941

Marigo V, Davey RA, Zuo Y, Cunningham JM, Tabin CJ (1996) Biochemical evidence that patched is the Hedgehog receptor. Nature 384:176–179

McCarthy RA, Barth JL, Chintalapudi MR, Knaak C, Argraves WS (2002) Megalin functions as an endocytic sonic hedgehog receptor. J Biol Chem 277:25660–25667

McGarvey TW, Maruta Y, Tomaszewski JE, Linnenbach AJ, Malkowicz SB (1998) PTCH gene mutations in invasive transitional cell carcinoma of the bladder. Oncogene 17:1167–1172

Methot N, Basler K (1999) Hedgehog controls limb development by regulating the activities of distinct transcriptional activator and repressor forms of Cubitus interruptus. Cell 19(96):819–831

Methot N, Basler K (2000) Suppressor of fused opposes hedgehog signal transduction by impeding nuclear accumulation of the activator form of Cubitus interruptus. Development 127:4001–4010

Micchelli CA, The I, Selva E, Mogila V, Perrimon N (2002) Rasp, a putative transmembrane acyltransferase, is required for Hedgehog signaling. Development 129:843–851

Michno K, Boras-Granic K, Mill P, Hui CC, Hamel PA (2003) Shh expression is required for embryonic hair follicle but not mammary gland development. Dev Biol 264:153–165

Mill P, Mo R, Fu H, Grachtchouk M, Kim PC, Dlugosz AA, Hui CC (2003) Sonic hedgehog-dependent activation of Gli2 is essential for embryonic hair follicle development. Genes Dev 17:282–294

Miyazawa K, Himi T, Garcia V, Yamagishi H, Sato S, Ishizaki Y (2000) A role for p27/Kip1 in the control of cerebellar granule cell precursor proliferation. J Neurosci 20:5756–5763

Molofsky AV, Pardal R, Iwashita T, Park IK, Clarke MF, Morrison SJ (2003) Bmi-1 dependence distinguishes neural stem cell self-renewal from progenitor proliferation. Nature 425:962–967

Moshiri A, Reh TA (2004) Persistent progenitors at the retinal margin of ptc+/– mice. J Neurosci 24:229–237

Mullor JL, Calleja M, Capdevila J, Guerrero I (1997) Hedgehog activity, independent of decapentaplegic, participates in wing disc patterning. Development 124:1227–1237

Mullor JL, Dahmane N, Sun T, Altaba A (2001) Wnt signals are targets and mediators of Gli function. Curr Biol 11:769–773

Murone M, Luoh SM, Stone D, Li W, Gurney A, Armanini M, Grey C, Rosenthal A, de Sauvage FJ (2000) Gli regulation by the opposing activities of fused and suppressor of fused. Nat Cell Biol 2:310–312

Nakano Y, Nystedt S, Shivdasani AA, Strutt H, Thomas C, Ingham PW (2004) Functional domains and sub-cellular distribution of the Hedgehog transducing protein Smoothened in Drosophila. Mech Dev 121:507–518

Nieto MA (2002) The snail superfamily of zinc-finger transcription factors. Nat Rev Mol Cell Biol 3:155–166

Nilsson M, Unden AB, Krause D, Malmqwist U, Raza K, Zaphiropoulos PG, Toftgard R (2000) Induction of basal cell carcinomas and trichoepitheliomas in mice overexpressing GLI-1. Proc Natl Acad Sci USA 97:3438–3443

Nishimaki H, Kasai K, Kozaki K, Takeo T, Ikeda H, Saga S, Nitta M, Itoh G (2004) A role of activated Sonic hedgehog signaling for the cellular proliferation of oral squamous cell carcinoma cell line. Biochem Biophys Res Commun 314:313–320

Nusslein-Volhard C, Wieschaus E (1980) Mutations affecting segment number and polarity in *Drosophila*. Nature 287:795–801

Nybakken KE, Turck CW, Robbins DJ, Bishop JM (2002) Hedgehog-stimulated phosphorylation of the kinesin-related protein Costal2 is mediated by the serine/threonine kinase fused. J Biol Chem 277:24638–24647

Ogden SK, Ascano M, Jr., Stegman MA, Suber LM, Hooper JE, Robbins DJ (2003) Identification of a functional interaction between the transmembrane protein Smoothened and the kinesin-related protein Costal2. Curr Biol 13:1998–2003

Ogden SK, Ascano M, Jr., Stegman MA, Robbins DJ (2004) Regulation of Hedgehog signaling: a complex story. Biochem Pharmacol 67:805–814

Ohki K, Kumamoto H, Ichinohasama R, Sato T, Takahashi N, Ooya K (2004) PTC gene mutations and expression of SHH, PTC, SMO, and GLI-1 in odontogenic keratocysts. Int J Oral Maxillofac Surg 33:584–592

Oliver TG, Wechsler-Reya RJ (2004) Getting at the root and stem of brain tumors. Neuron 42:885–888

Oliver TG, Grasfeder LL, Carroll AL, Kaiser C, Gillingham CL, Lin SM, Wickramasinghe R, Scott MP, Wechsler-Reya RJ (2003) Transcriptional profiling of the Sonic hedgehog response: a critical role for N-myc in proliferation of neuronal precursors. Proc Natl Acad Sci USA 100:7331–7336

Oro AE, Higgins KM, Hu Z, Bonifas JM, Epstein EH, Jr., Scott MP (1997) Basal cell carcinomas in mice overexpressing sonic hedgehog. Science 276:817–821

Pasca dM, Hebrok M (2003) Hedgehog signalling in cancer formation and maintenance. Nat Rev Cancer 3:903–911

Pazzaglia S, Mancuso M, Atkinson MJ, Tanori M, Rebessi S, Majo VD, Covelli V, Hahn H, Saran A (2002) High incidence of medulloblastoma following X-ray-irradiation of newborn Ptc1 heterozygous mice. Oncogene 21:7580–7584

Pepinsky RB, Zeng C, Wen D, Rayhorn P, Baker DP, Williams KP, Bixler SA, Ambrose CM, Garber EA, Miatkowski K, Taylor FR, Wang EA, Galdes A (1998) Identification of a palmitic acid-modified form of human Sonic hedgehog. J Biol Chem 273:14037–14045

Pietsch T, Waha A, Koch A, Kraus J, Albrecht S, Tonn J, Sorensen N, Berthold F, Henk B, Schmandt N, Wolf HK, von Deimling A, Wainwright B, Chenevix-Trench G, Wiestler OD, Wicking C (1997) Medulloblastomas of the desmoplastic variant carry mutations of the human homologue of Drosophila patched. Cancer Res 57:2085–2088

Ping XL, Ratner D, Zhang H, Wu XL, Zhang MJ, Chen FF, Silvers DN, Peacocke M, Tsou HC (2001) PTCH mutations in squamous cell carcinoma of the skin. J Invest Dermatol 116:614–616

Porter JA, Young KE, Beachy PA (1996) Cholesterol modification of hedgehog signaling proteins in animal development. Science 274:255–259

Price MA, Kalderon D (2002) Proteolysis of the Hedgehog signaling effector Cubitus interruptus requires phosphorylation by Glycogen Synthase Kinase 3 and Casein Kinase 1. Cell 108:823–835

Qualtrough D, Buda A, Gaffield W, Williams AC, Paraskeva C (2004) Hedgehog signalling in colorectal tumour cells: induction of apoptosis with cyclopamine treatment. Int J Cancer 20(110):831–837

Raffel C, Jenkins RB, Frederick L, Hebrink D, Alderete B, Fults DW, James CD (1997) Sporadic medulloblastomas contain PTCH mutations. Cancer Res 57:842–845

Rao G, Pedone CA, Coffin CM, Holland EC, Fults DW (2003) c-Myc enhances sonic hedgehog-induced medulloblastoma formation from nestin-expressing neural progenitors in mice. Neoplasia 5:198–204

Rao G, Pedone CA, Valle LD, Reiss K, Holland EC, Fults DW (2004) Sonic hedgehog and insulin-like growth factor signaling synergize to induce medulloblastoma formation from nestin-expressing neural progenitors in mice. Oncogene 23:6156–6162

Reifenberger J, Wolter M, Weber RG, Megahed M, Ruzicka T, Lichter P, Reifenberger G (1998) Missense mutations in SMOH in sporadic basal cell carcinomas of the skin and primitive neuroectodermal tumors of the central nervous system. Cancer Res 58:1798–1803

Reya T, Morrison SJ, Clarke MF, Weissman IL (2001) Stem cells, cancer, and cancer stem cells. Nature 414:105–111

Robbins DJ, Nybakken KE, Kobayashi R, Sisson JC, Bishop JM, Therond PP (1997) Hedgehog elicits signal transduction by means of a large complex containing the kinesin-related protein costal2. Cell 90:225–234

Rowitch DH, Jacques B, Lee SM, Flax JD, Snyder EY, McMahon AP (1999) Sonic hedgehog regulates proliferation and inhibits differentiation of CNS precursor cells. J Neurosci 19:8954–8965

Ruel L, Rodriguez R, Gallet A, Lavenant-Staccini L, Therond PP (2003) Stability and association of Smoothened, Costal2 and Fused with Cubitus interruptus are regulated by Hedgehog. Nat Cell Biol 5:907–913

Sanchez P, Hernandez AM, Stecca B, Kahler AJ, DeGueme AM, Barrett A, Beyna M, Datta MW, Datta S, Ruiz IA (2004) Inhibition of prostate cancer proliferation by interference with SONIC HEDGEHOG-GLI1 signaling. Proc Natl Acad Sci USA 101:12561–12566

Sasaki H, Hui C, Nakafuku M, Kondoh H (1997) A binding site for Gli proteins is essential for HNF-3beta floor plate enhancer activity in transgenics and can respond to Shh in vitro. Development 124:1313–1322

Sasaki H, Nishizaki Y, Hui C, Nakafuku M, Kondoh H (1999) Regulation of Gli2 and Gli3 activities by an amino-terminal repression domain: implication of Gli2 and Gli3 as primary mediators of Shh signaling. Development 126:3915–3924

Sell S (2004) Stem cell origin of cancer and differentiation therapy. Crit Rev Oncol Hematol 51:1–28

Shanley S, Ratcliffe J, Hockey A, Haan E, Oley C, Ravine D, Martin N, Wicking C, Chenevix-Trench G (1994) Nevoid basal cell carcinoma syndrome: review of 118 affected individuals. Am J Med Genet 50:282–290

Singh SK, Clarke ID, Terasaki M, Bonn VE, Hawkins C, Squire J, Dirks PB (2003) Identification of a cancer stem cell in human brain tumors. Cancer Res 63:5821–5828

Smyth I, Narang MA, Evans T, Heimann C, Nakamura Y, Chenevix-Trench G, Pietsch T, Wicking C, Wainwright BJ (1999) Isolation and characterization of human patched 2 (PTCH2), a putative tumour suppressor gene in basal cell carcinoma and medulloblastoma on chromosome 1p32. Hum Mol Genet 8:291–297

Somsel RJ, Wandinger-Ness A (2000) Rab GTPases coordinate endocytosis. J Cell Sci 113(2):183–192

Stegman MA, Goetz JA, Ascano M Jr, Ogden SK, Nybakken KE, Robbins DJ (2004) The Kinesin-related protein Costal2 associates with membranes in a Hedgehog-sensitive, Smoothened-independent manner. J Biol Chem 20(279):7064–7071

Stone DM, Hynes M, Armanini M, Swanson TA, Gu Q, Johnson RL, Scott MP, Pennica D, Goddard A, Phillips H, Noll M, Hooper JE, de Sauvage F, Rosenthal A (1996) The tumour-suppressor gene patched encodes a candidate receptor for Sonic hedgehog. Nature 384:129–134

Stone DM, Murone M, Luoh S, Ye W, Armanini MP, Gurney A, Phillips H, Brush J, Goddard A, De Sauvage FJ, Rosenthal A (1999) Characterization of the human suppressor of fused, a negative regulator of the zinc-finger transcription factor Gli. J Cell Sci 112:4437–4448

Strigini M, Cohen SM (1997) A Hedgehog activity gradient contributes to AP axial patterning of the Drosophila wing. Development 124:4697–4705

Tabata T, Kornberg TB (1994) Hedgehog is a signaling protein with a key role in patterning *Drosophila* imaginal discs. Cell 76:89–102

Tabs S, Avci O (2004) Induction of the differentiation and apoptosis of tumor cells in vivo with efficiency and selectivity. Eur J Dermatol 14:96–102

Taipale J, Beachy PA (2001) The Hedgehog and Wnt signalling pathways in cancer. Nature 411:349–354

Taipale J, Chen JK, Cooper MK, Wang B, Mann RK, Milenkovic L, Scott MP, Beachy PA (2000) Effects of oncogenic mutations in Smoothened and Patched can be reversed by cyclopamine. Nature 406:1005–1009

Taipale J, Cooper MK, Maiti T, Beachy PA (2002) Patched acts catalytically to suppress the activity of Smoothened. Nature 418:892–897

Tas S, Avci O (2004) Rapid clearance of psoriatic skin lesions induced by topical cyclopamine. A preliminary proof of concept study. Dermatology 209:126—131

Taylor FR., Wen D, Garber EA, Carmillo AN, BakerDP, Arduini RM, Williams KP, Weinreb PH, Rayhorn P, Hronowski X, Whitty A, Day ES, Boriack-Sjodin A, Shapiro RI, Galdes A, Pepinsky RB (2001) Enhanced potency of human Sonic hedgehog by hydrophobic modification. Biochem 40:4359–4371

Taylor MD, Liu L, Raffel C, Hui CC, Mainprize TG, Zhang X, Agatep R, Chiappa S, Gao L, Lowrance A, Hao A, Goldstein AM, Stavrou T, Scherer SW, Dura WT, Wainwright B, Squire JA, Rutka JT, Hogg D (2002) Mutations in SUFU predispose to medulloblastoma. Nat Genet 31:306–310

Teh MT, Wong ST, Neill GW, Ghali LR, Philpott MP, Quinn AG (2002) FOXM1 is a downstream target of Gli1 in basal cell carcinomas. Cancer Res 62:4773–4780

Thayer SP, di Magliano MP, Heiser PW, Nielsen CM, Roberts DJ, Lauwers GY, Qi YP, Gysin S, Fernandez-del Castillo C, Yajnik V, Antoniu B, McMahon M, Warshaw AL, Hebrok M (2003) Hedgehog is an early and late mediator of pancreatic cancer tumorigenesis. Nature 425:851–856

The I, Bellaiche Y, Perrimon N (1999) Hedgehog movement is regulated through tout veludependent synthesis of a heparan sulfate proteoglycan. Mol Cell 4:633–639

Tojo M, Mori T, Kiyosawa H, Honma Y, Tanno Y, Kanazawa KY, Yokoya S, Kaneko F, Wanaka A (1999) Expression of sonic hedgehog signal transducers, patched and smoothened, in human basal cell carcinoma. Pathol Int 49:687–694

Tojo M, Kiyosawa H, Iwatsuki K, Kaneko F (2002) Expression of a sonic hedgehog signal transducer, hedgehog-interacting protein, by human basal cell carcinoma. Br J Dermatol 146:69–73

Tojo M, Kiyosawa H, Iwatsuki K, Nakamura K, Kaneko F (2003) Expression of the GLI2 oncogene and its isoforms in human basal cell carcinoma. Br J Dermatol 148:892–897

Torroja C, Gorfinkiel N, Guerrero I (2004) Patched controls the Hedgehog gradient by endocytosis in a dynamin-dependent manner, but this internalization does not play a major role in signal transduction. Development 131:2395–2408

Unden AB, Zaphiropoulos PG, Bruce K, Toftgard R, Stahle-Backdahl M (1997) Human patched (PTCH) mRNA is overexpressed consistently in tumor cells of both familial and sporadic basal cell carcinoma. Cancer Res 57:2336–2340

Van den Brink GR, Bleuming SA, Hardwick JC, Schepman BL, Offerhaus GJ, Keller JJ, Nielsen C, Gaffield W, van Deventer SJ, Roberts DJ, Peppelenbosch MP (2004) Indian Hedgehog is an antagonist of Wnt signaling in colonic epithelial cell differentiation. Nat Genet 36:277–282

Vega S, Morales AV, Ocana OH, Valdes F, Fabregat I, Nieto MA (2004) Snail blocks the cell cycle and confers resistance to cell death. Genes Dev 18:1131–1143

Vorechovsky I, Tingby O, Hartman M, Stromberg B, Nister M, Collins VP, Toftgard R (1997a) Somatic mutations in the human homologue of Drosophila patched in primitive neuroectodermal tumours. Oncogene 15:361–366

Vorechovsky I, Unden AB, Sandstedt B, Toftgard R, Stahle-Backdahl M (1997b) Trichoepitheliomas contain somatic mutations in the overexpressed PTCH gene: support for a gatekeeper mechanism in skin tumorigenesis. Cancer Res 57:4677–4681

Wallace VA (1999) Purkinje-cell-derived Sonic hedgehog regulates granule neuron precursor cell proliferation in the developing mouse cerebellum. Curr Biol 9:445–448

Wang B, Fallon JF, Beachy PA (2000) Hedgehog-regulated processing of Gli3 produces an anterior/posterior repressor gradient in the developing vertebrate limb. Cell 100:423–434

Wang G, Wang B, Jiang J (1999) Protein kinase A antagonizes Hedgehog signaling by regulating both the activator and repressor forms of Cubitus interruptus. Genes Dev 13:2828–2837

Wang G, Amanai K, Wang B, Jiang J (2000) Interactions with Costal2 and suppressor of fused regulate nuclear translocation and activity of cubitus interruptus. Genes Dev 14:2893–2905

Wang QT, Holmgren RA (1999) The subcellular localization and activity of *Drosophila* cubitus interruptus are regulated at multiple levels. Development 126:5097–5106

Wang QT, Holmgren RA (2000) Nuclear import of cubitus interruptus is regulated by hedgehog via a mechanism distinct from Ci stabilization and Ci activation. Development 127:3131–3139

Watkins DN, Berman DM, Burkholder SG, Wang B, Beachy PA, Baylin SB (2003) Hedgehog signalling within airway epithelial progenitors and in small-cell lung cancer. Nature 20(422):313–317

Wechsler-Reya RJ, Scott MP (1999) Control of neuronal precursor proliferation in the cerebellum by Sonic Hedgehog. Neuron 22:103–114

Weiner HL, Bakst R, Hurlbert MS, Ruggiero J, Ahn E, Lee WS, Stephen D, Zagzag D, Joyner AL, Turnbull DH (2002) Induction of medulloblastomas in mice by sonic hedgehog, independent of Gli1. Cancer Res 62:6385–6389

Williams JA, Guicherit OM, Zaharian BI, Xu Y, Chai L, Wichterle H, Kon C, Gatchalian C, Porter JA, Rubin LL, Wang FY (2003) Identification of a small molecule inhibitor of the hedgehog signaling pathway: effects on basal cell carcinoma-like lesions. Proc Natl Acad Sci USA 100:4616–4621

Wolter M, Reifenberger J, Sommer C, Ruzicka T, Reifenberger G (1997) Mutations in the human homologue of the Drosophila segment polarity gene patched (PTCH) in sporadic basal cell carcinomas of the skin and primitive neuroectodermal tumors of the central nervous system. Cancer Res 57:2581–2585

Wu SC, Grindley J, Winnier GE, Hargett L, Hogan BL (1998) Mouse Mesenchyme forkhead 2 (Mf2): expression, DNA binding and induction by sonic hedgehog during somitogenesis. Mech Dev 70:3–13

Xie J, Johnson RL, Zhang X, Bare JW, Waldman FM, Cogen PH, Menon AG, Warren RS, Chen LC, Scott MP, Epstein EH Jr (1997) Mutations of the PATCHED gene in several types of sporadic extracutaneous tumors. Cancer Res 57:2369–2372

Xie J, Murone M, Luoh SM, Ryan A, Gu Q, Zhang C, Bonifas JM, Lam CW, Hynes M, Goddard A, Rosenthal A, Epstein EH Jr, de Sauvage FJ (1998) Activating Smoothened mutations in sporadic basal-cell carcinoma. Nature 391:90–92

Yoon JW, Kita Y, Frank DJ, Majewski RR, Konicek BA, Nobrega MA, Jacob H, Walterhouse D, Iannaccone P (2002) Gene expression profiling leads to identification of GLI1-binding elements in target genes and a role for multiple downstream pathways in GLI1-induced cell transformation. J Biol Chem 277:5548–5555

Zaphiropoulos PG, Unden AB, Rahnama F, Hollingsworth RE, Toftgard R (1999) PTCH2, a novel human patched gene, undergoing alternative splicing and up-regulated in basal cell carcinomas. Cancer Res 59:787–792

Zhang Y, Kalderon D (2001) Hedgehog acts as a somatic stem cell factor in the *Drosophila* ovary. Nature 410:599–604

Zhao Q, Kho A, Kenney AM, Yuk DD, Kohane I, Rowitch DH (2002) Identification of genes expressed with temporal-spatial restriction to developing cerebellar neuron precursors by a functional genomic approach. Proc Natl Acad Sci USA 99:5704–5709

Zhu AJ, Zheng L, Suyama K, Scott MP (2003) Altered localization of Drosophila Smoothened protein activates Hedgehog signal transduction. Genes Dev 17:1240–1252

Rho GTPases and Cancer

Pinella Buongiorno, Bharati Bapat

1 Introduction

The Rho (Ras-homologous) family of proteins constitutes a major branch of the Ras superfamily of small GTPases, and is evolutionarily conserved across several phyla. Thus far, 25 members have been identified, and these may be divided into 6 subfamilies based on amino acid sequence identity, structural motifs, and biological function. These include the RhoA-related subfamily (RhoA, RhoB, RhoC), the Rac1-related subfamily (Rac1, Rac1b, Rac2, Rac3, and RhoG), and the Cdc42-related subfamily (Cdc42, G25 K, TC10, TCL, Wrch-1, and Wrch-2; see Table 1). Rho proteins may be distinguished from other Ras superfamily members by the presence of a Rho-type GTPase domain, which shares 30% amino acid sequence identity with the small GTPase domain of Ras proteins. Members of the Rho GTPase family share 40–95% identity within their Rho-type GTPase domains (reviewed in Wennerberg and Der 2004).

Like Ras proteins, most Rho GTPases function as molecular switches that cycle between an inactive GDP-bound state and an active GTP-bound state. The interconversion between the two states is mediated by interaction with regulatory proteins: guanine nucleotide-exchange factors (Rho-GEFs) accelerate the rate of GDP to GTP exchange, and thus promote the activity of Rho proteins, while GTPase-activating proteins (Rho-GAPs) stimulate the intrinsic GTPase activity of Rho proteins, resulting in their inactivation. An additional level of regulation is offered by Rho-GDP dissociation inhibitors (Rho-GDIs), which sequester Rho proteins in the cytoplasm in an inactive state. Active GTP-bound Rho proteins are able to interact with a wide range of downstream effector proteins, through which they mediate diverse cellular functions (see Fig. 1).

An additional determinant of Rho GTPase function is subcellular localization, which may be influenced by several factors, including post-translational isoprenoid modification and the presence of a C-terminal polybasic region

P. Buongiorno, B. Bapat (e-mail: bapat@mshri.on.ca)
Samuel Lunenfeld Research Institute, 9th Floor, Room 992B, Mount Sinai Hospital, 600 University Avenue, Toronto, Ontario, M5G 1X5 Canada

Progress in Molecular and Subcellular Biology
Alvaro Macieira-Coelho (Ed.)
Development Biology of Neoplastic Growth

Table 1. Dysregulation of Rho family members in cancer

Rho family member	Mechanism of dysregulation	Tumor
RhoA-related		
RhoA	Upregulated	Colon[a], breast[a], lung[a], testicular germ cell[b], head and neck squamous cell carcinoma[c]
RhoB	Downregulated	Head and neck squamous cell carcinoma[d], lung[e]
RhoC	Overexpression	Inflammatory breast cancer[f,g], pancreatic ductal adenocarcinoma[h], melanoma[i]
Rac1-related		
Rac1	Upregulated	Breasta
	Hyperactivity	Breast[j], pancreas[k], colon[k]
Rac1b	Alternative splicing	Colon, breast[j,l]
Rac2	Upregulated	Head and neck squamous cell carcinoma[c]
Rac3	Hyperactivity	Breast[m]
RhoG	–	–
Cdc42-related		
Cdc42	Upregulated	Breast[a]
G25K	–	–
TC10	–	–
TCL	–	–
Wrch-2	–	–
Wrch-1	–	–
Rnd		
Rnd1	–	–
Rnd2	–	–
RhoE/Rnd3	–	–
RhoBTB		
RhoBTB-1	–	–
RhoBTB-2	Gene deletion	Breast[n]
	Downregulated	Breast[n]
Miro		
RhoD	–	–
Rif	–	–
TTF/RhoH	Rearrangement	Non-Hodgkin's lymphoma, multiple myeloma[o]
	Point mutation	Diffuse large B-cell lymphomas[p]

[a]Fritz et al. (1999); [b]Kamai et al. (2001); [c]Abraham et al. (2001); [d]Adnane et al. (2002); [e]Mazieres et al. (2004); [f]van Golen et al. (2000); [g]van Golen et al. (1999); [h]Suwa et al. (1998); [i]Clark et al. (2000); [j]Schnelzer et al. (2000); [k]Esufali and Bapat (2004); [l]Jordan et al. (1999); [m]Mira et al. (2000); [n]Hamaguchi et al. (2002); [o]Preudhomme et al. (2000); [p]Pasqualucci et al. (2001)

(PBR). Most Rho proteins may be characterized by a CAAX motif at their C-termini which signals for either farnesyl or geranylgeranyl isoprenoid modification (Adamson et al. 1992). The addition of isoprenoid groups facilitates membrane localization, where the potential for interaction with Rho-GEFs is high. While the C-terminal PBR has similarly been shown to play a role in ensuring proper membrane targeting, it may contain a nuclear localization signal (NLS) that promotes nucleocytoplasmic shuttling of Rho proteins (Lanning et al. 2003). Isoforms of Rac and Rho may be distinguished by the presence or absence of an NLS within their PBRs. Mutational studies have shown that exchanging the NLS-containing PBR of Rac1 with that of RhoA, which lacks an NLS, results in decreased nucleocytoplasmic shuttling of Rac1, and its consequential dysregulation (Lanning et al. 2004). The PBR therefore plays an important role in influencing Rho GTPase activity, allowing different Rho

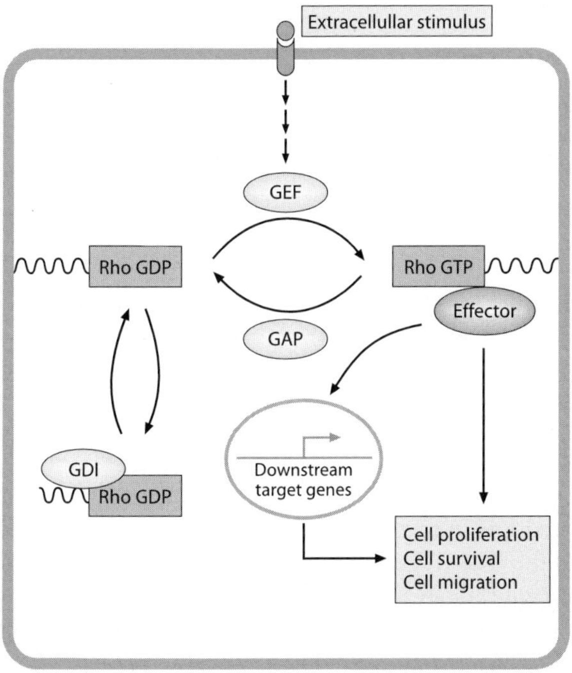

Fig. 1. Regulation of Rho GTPase activity. Rho-GDIs regulate the subcellular localization of Rho GTPases by sequestering GDP-bound Rho proteins in the cytoplasm, or promoting their membrane localization. In response to extracellular stimuli (e.g., growth factor stimulation, integrin engagement), Rho-GEFs activate Rho GTPases by promoting GDP to GTP exchange. GTP-bound Rho proteins interact with a variety of downstream effector proteins, via which they mediate a wide range of cellular activities. Some Rho proteins (e.g., Rac1 GTPase) contain a nuclear localization signal within their polybasic regions that targets them to the nucleus, where they may participate in signaling pathways. Rho protein activity is downregulated by Rho-GAPs that accelerate the intrinsic GTPase activity of Rho proteins, rendering them in a GDP-bound, inactive state

family members to function in distinct subcellular compartments. The proper subcellular targeting of Rho proteins promotes functional interactions with both upstream and downstream binding partners, and thus is crucial in regulating Rho GTPase signaling activity.

Because the different members of the Rho GTPase family are able to interact with a multitude of effectors and are localized in diverse compartments within the cell, Rho proteins participate in an extensive range of cellular processes. While the best characterized of these include regulation of the actin cytoskeleton and cell migraion, they have also been implicated in cell-cycle progression, cell survival, and gene expression. Given that Rho GTPases influence these processes in normal cells, it is likely that aberrant activation of Rho GTPase signaling contributes to their dysregulation in various cancers.

In this chapter, we will discuss the evidence implicating Rho proteins in various stages of cancer development, including tumor initiation, invasion, and metastasis.

2 Rho Proteins and Tumorigenesis

The earliest lines of evidence linking Rho proteins to tumorigenesis derive from focus-formation assays and soft agar experiments, which measure efficiency of transformation by allowing growth-factor- and anchorage-independent growth of cells. Studies indicate that constitutively active Rac1, and to a lesser extent, RhoA, are able to cause malignant transformation of fibroblasts and tumorigenicity in nude mice (Khosravi-Far et al. 1995; Qiu et al. 1995a,b). Moreover, dominant negative Rac1 and RhoA mutants inhibit focus formation by oncogenic Ras in NIH3T3 fibroblasts, but not by an activated, membrane-targeted mutant of Raf kinase (a downstream Ras effector which stimulates mitogen-activated protein kinase, or MAPK, activity). In contrast, constitutively active Rac1 and RhoA mutants were shown to synergize with activated, membrane-targeted Raf in focus-formation assays. Similar studies implicate other members of the Rho GTPase family, namely Cdc42, RhoG, and TC10, in mediating Ras-induced cellular transformation (Qiu et al. 1997; Roux et al. 1997; Murphy et al. 1999). These results indicate that oncogenic Ras activates both the Raf→MAPK pathway and RhoGTPase signaling, independent of one another, and activation of both cascades is necessary for full induction of oncogenic Ras transformation.

3 Elevated Rho GTPase Signaling in Tumorigenesis

Enhanced stimulation of Rho signaling in tumors may be attributed to at least three factors that contribute to elevated Rho activity: (1) genetic aberrations in Rho GTPases, (2) upregulation of Rho protein expression, and (3) accelerated GDP-GTP cycling.

Despite the frequency of activating Ras mutations in human cancers, aberrations in genes that encode Rho proteins are surprisingly rare. Indeed, RhoH represents the only example of a genetically altered Rho family member reported in human cancers to date. In non-Hodgkin's lymphoma, a t(3:4) (q27; p11–13) translocation has been identified which gives rise to a fusion protein consisting of RhoH/TTF GTPase fused to LAZ3/BCL3 (Preudhomme et al. 2000). However, whether the RhoH portion of the fusion protein is responsible for progression of the disease phenotype remains elusive. In addition, point mutations in the 5′ untranslated region (UTR) of the RhoH transcript have been reported in B-cell diffuse large-cell lymphomas (Pasqualucci et al. 2001). It is possible that such mutations affect levels of protein expression, and may potentially contribute to pathogenesis.

While the incidence of activating RhoGTPase mutations in human cancers is low, several members of the RhoGTPase family are often overexpressed in a variety of human tumors (see Table 1). Increased RhoA expression has been reported in tumors of the colon, breast, lung (Fritz et al. 1999), testicular germ cell (Kamai et al. 2001), and in head and neck squamous cell carcinoma (Abraham et al. 2001). Breast tumors also express elevated levels of RhoC protein (van Golen et al. 1999; Kleer et al. 2002) and recent microarray data reveal that RhoC overexpression in inflammatory breast cancer leads to upregulation of multiple genes involved in cell proliferation, invasion, and angiogenesis. The involvement of RhoC downstream genetic targets in these cellular processes supports a role for RhoC in promoting metastasis of breast tumors (Wu et al. 2004).

Protein expression level does not always reflect the active, functional pool of a protein, and thus alterations in expression level do not necessarily result in concomitant signaling changes in vivo. It is therefore important to assess the activation status of a protein in order to correlate expression level and signal transduction. Rho proteins are required to be in a GTP-bound, active state in order to interact with downstream effector proteins, and thus mediate diverse cellular functions. Activated Rac1 levels may be measured using an assay that employs p21-activated-kinase (PAK)-affinity precipitation to »pull down« GTP-bound Rac and Cdc42. Activated Rac1 is observed in colon, breast, and pancreatic cancer cell lines with only minimal levels detected in normal human embryonic kidney cells (Esufali and Bapat 2004), indicating that the pool of active, functional Rac1 in these tumors may be larger than that in normal cells. Other studies have shown increased membrane localization of Rac1 in patients with recurrent breast cancer (Schnelzer et al. 2000). Since active, GTP-bound Rac1 is localized at the cell membrane, these results suggest the presence of hyperactive Rac1 in aggressive mammary tumors. To further emphasize the importance of assessing the activation status of Rho proteins in tumors, hyperactive Rac3 was identified in human metastatic breast tissues, and was found to mediate the persistent induction of PAK activity, resulting in increased proliferation (Mira et al. 2000). Increased activation of Rho proteins within tumors suggests constitutive stimulation of Rho signaling, and hence amplification of the effects normally mediated by Rho GTPases.

Recent studies have identified a Rac1 splice variant, designated Rac1b, which contains a 19 amino acid insertion and is expressed in colorectal and breast tumors (Jordan et al. 1999; Schnelzer et al. 2000). These studies have shown that while the level of Rac1b expression in these tumors is low, the pool of Rac1b protein that is present appears to be predominantly active. The high activation state of Rac1b may be attributed to accelerated GDP/GTP exchange (Fiegen et al. 2004; Schnelzer et al. 2000) and its inability to interact with Rho-GDI (Matos et al. 2003). Normally, Rho-GDIs bind to and sequester GDP-bound Rho proteins in the cytoplasm. The lack of interaction between Rac1b and Rho-GDI results in the predominant localization of Rac1b at the plasma membrane, where it has greater opportunity to interact with membrane-associated activators. Rac1b is thus hyperactive, and mimics the phenotype of a constitutively active mutant. Interestingly, Rac1b is capable of stimulating only a restricted subset of downstream signaling pathways that are mediated by Rac1. For example, Rac1b is not able to induce lamellipodia formation – a hallmark Rac1 function – and is unable to activate PAK or Jun N-terminal kinase (JNK) pathways. However, Rac1b does mediate the stimulation of the transcription factor nuclear factor-kappa B (NF-κB) to the same extent as Rac1 (Jordan et al. 1999). It has been suggested that the 19-amino acid insertion may create a novel effector-binding site in Rac1b, allowing the splice variant to participate in signaling pathways that diverge from those regulated by Rac1. Since Rac1b is preferentially expressed in colorectal and breast tumors, and because its activation state within the cell is high, Rac1b likely participates in signaling pathways that promote the malignant phenotype in these epithelial cells.

4 Upstream Regulation of Rho GTPases

Upstream regulators of Rho GTPases are reportedly subject to genetic alterations in human cancers, and may contribute to dysregulation of Rho signaling. Indeed, several Rho-GEFs, including Dbl, Vav, and Lfc, have been classified as oncoproteins based on in vitro NIH3T3 fibroblast transformation assays (Schmidt and Hall 2002). In vivo, several genetic aberrations of Rho-GEFs have been described. The leukemia-associated Rho guanine nucleotide exchange factor (LARG) gene was isolated as a fusion partner of the mixed lineage leukemia (MLL) gene in acute myeloid leukemia (Kourlas et al. 2000). Full-length LARG was shown to specifically activate RhoA, and its overexpression promotes growth transformation of fibroblasts (Reuther et al. 2001). However, whether the LARG portion of the MLL-LARG fusion protein has any effect on the activation state of RhoA and whether it contributes to the AML phenotype remains to be elucidated.

Another Rho-GEF mutated in human cancers is T-cell invasion and metastasis gene (Tiam), a Rac1-specific activator originally isolated in aggressive T-cell lymphomas (Price and Collard 2001). Several mutant Tiam1 alleles were identified in renal cell carcinomas, one of which harbors an A441G point mutation in the N-terminal pleckstrin homology (PH) domain. This domain is necessary for recruitment of Tiam1 to the plasma membrane, thus the point mutation possibly interferes with membrane localization of the Rac1-specific GEF. While in vitro focus formation assays in fibroblasts have shown that the A441G mutation promotes the transforming capacity of Tiam1 (Engers et al. 2000), the relevance of the mutation in vivo remains to be determined.

To further investigate the contribution of Tiam1 to the formation and progression of tumors in vivo, Tiam1-deficient mice were generated. As expected, these mice displayed reduced Rac1 activity, and were resistant to the development of Ras-induced skin tumors (Malliri et al. 2002). The few tumors that did develop in *Tiam1*-null mice exhibited a reduced growth rate compared to Ras-induced tumors in wild-type mice. Furthermore, embryonic fibroblasts isolated from Tiam1-deficient mice had increased apoptotic activity and impaired proliferation, consistent with in vitro studies that implicate Rac1 in tumor formation. Clearly, these data suggest that Tiam1-mediated activation of Rac1 is necessary for Ras-induced transformation. Interestingly, the few tumors that developed in *Tiam1*-null mice were more aggressive and invasive than those formed in wild-type mice, suggesting that Tiam1 deficiency promotes metastasis of tumor cells. While these data conflict with previous reports supporting the invasion-promoting behavior of Rac1 (Michiels et al. 1995; Quinlan 1999; Braga et al. 2000), other lines of evidence corroborate the anti-invasive properties of Tiam1. Studies performed in Madin-Darby canine kidney (MDCK) cells, for example, show that Tiam1 expression prevents the loss of E-cadherin-mediated adhesion induced by oncogenic Ras, and thereby inhibits the migration of epithelial cells (Hordijk et al. 1997). Subsequent work aimed at clarifying the contradictory data suggests that the effect of Tiam1 on invasiveness and metastasis is dependent on extracellular matrix (ECM) composition and is cell-type-specific (Sander et al. 1998). Alternatively, it is possible that these tumors develop via other mechanisms that are independent of Tiam-1 deficiency. Regardless, it is clear that Tiam1 function influences many aspects of cancer, including apoptosis, growth, and invasion, underlining the relevance of this Rac1-specific GEF in tumor initiation and progression.

Some evidence suggests that Rho-GAPs, which counterbalance Rho-GEF function by inactivating Rho GTPases, may also contribute to neoplastic transformation by acting as potential tumor suppressors. Recently, the Rho-GAP designated DLC2 (deleted in liver cancer 2) has been found to be downregulated in some human hepatocellular carcinomas due to loss of heterozygosity (Ching et al. 2003). Introduction of this RhoA- and Cdc42-specific GAP into mouse fibroblasts was found to suppress Ras signaling and Ras-induced transformation, suggesting that the deletion of DLC2 may promote hepatocarcinogenesis.

Likewise, the gene that encodes p190RhoGAP is frequently deleted in oligodendrogliomas and is thought to act as a putative tumor suppressor. In vitro studies have shown that overexpression of p120RhoGAP decreases Rho activity, and in turn blocks cell proliferation (Wolf et al. 2003). Supporting evidence is provided by animal studies whereby co-introduction of p190RhoGAP and platelet-derived growth factor (PDGF) into mice leads to a decreased incidence of oligodendrogliomas in comparison with mice exposed to PDGF alone. These corroborating observations suggest that deletion of the p120RhoGAP gene may contribute to the glioma phenotype due to loss of its anti-proliferative activity.

5 Downstream Signaling Pathways

5.1 Rho GTPases and Cell-Cycle Progression

The transition of a cell from normal to a transformed state involves dysregulation of cell-cycle progression, which promotes mitogen- and anchorage-independent cell proliferation, ultimately leading to tumor formation. Cyclin-dependent kinases (CDKs) are key components of the cell-cycle machinery that must be activated in order for a cell to enter and progress through the cell cycle. Regulation of CDK activity is dependent on the expression levels of CDK-binding partners, including activating cyclins and deactivating CDK inhibitors (CDKIs). Progression through G1 phase requires upregulation of D-type cyclins (D1, D2, and D3), which associate with their cognate CDK partners, CDK4 or CDK6. E-type cyclins (E1 and E2) form complexes with CDK2, and are also required. Cyclin–CDK complex formation leads to phosphorylation and consequential inhibition of the retinoblastoma protein (pRb), resulting in relief of pRb-mediated repression of E2F transcription factors. E2F activity transactivates genes, for example, cyclinA and thymidine kinase, that are necessary for entry into S phase. Clearly, multiple levels of regulation are in place to prevent the inappropriate progression of cells through G1 and into S phase, the point beyond which the cell becomes committed to proliferation.

Early lines of evidence suggested a role for Rho GTPases in the regulation of G1 to S-phase transition. One particular study has shown that specific inhibition of RhoA, Rac1, and Cdc42 activities in serum-stimulated NIH3T3 fibroblasts completely blocked cell cycle progression into S phase, as well as the phosphorylation of pRb, and induction of cyclinD1 protein (Welsh et al. 2001). Subsequent work corroborated the contribution of Rho GTPases to the regulation of G1 phase progression, revealing that they modulate expression levels of both positive and negative regulators of the cycle, namely cyclinD1 and the CKDIs $p21^{Cip1}$ and $p27^{Kip1}$, respectively.

Upregulation of cyclinD1. A considerable amount of evidence supports a role for Rho GTPases in the upregulation of cyclinD1 expression via both Ras-dependent and Ras-independent pathways. Induction of cyclinD1 transcription by Ras requires sustained activation of the Raf→MAPK pathway in response to growth factors. However, studies have shown that sustained MAPK activity – and hence cyclinD1 transcription – requires simultaneous anchorage-dependent activation of Rho signaling (Roovers et al. 1999, 2003; Welsh et al. 2001; Roovers and Assoian 2003). Specifically, integrin stimulation by the ECM leads to RhoA activation and subsequent induction of Rho-associated coiled-coil-forming kinase (ROCK) activity. These events lead to stress fiber formation, which in turn facilitates integrin clustering, resulting in further activation of Rho signal tansduction and thus sustained MAPK-mediated cyclinD1 transcription. Therefore, growth factor stimulation and integrin-dependent anchorage synergistically induce cyclinD1 transcription and hence cell proliferation via activation of both Ras- and Rho-mediated signaling pathways.

Other studies have shown that Rho GTPases influence the upregulation of cyclinD1 expression independent of Ras signaling as well. In vitro assays demonstrated that overexpression of constitutively active Rac1 or Cdc42 is sufficient to induce cyclinD1 transcription. Moreover, constitutively active Rac1 mutants with inefficient p21-activated kinase 1 (PAK1) binding showed impaired induction of cyclinD1 transcription, suggesting that Rac1-mediated upregulation of cyclinD1 expression involves activation of PAK1 function (Westwick et al. 1997). The PAK1 effector pathway, which may be activated by both Rac1 and Cdc42, triggers stimulation of downstream targets including JNK and p38 MAPK, ultimately resulting in increased activity of Jun and ATF transcription factors (which belong to the AP-1 family of transcription factors). Induction of ATF activity by Rac1 was later found to contribute to activation of cyclinD1 transcription (Joyce et al. 1999), providing a potential link between Rac1-mediated PAK1 stimulation and cyclinD1 expression.

Additional evidence has emerged revealing that induction of cyclinD1 expression by Rac1 also involves activation of the transcription factor NF-κB (Joyce et al. 1999). In most cell types, NF-κB is sequestered in the cytoplasm by an inhibitory protein designated inhibitor of κB (IκB). Upon activation, IκB becomes phosphorylated, and is targeted for degradation by the proteasome. Kinases responsible for the phosphorylation of IκB include IκB kinase α (IKKα) and IκB kinase β (IKKβ). In the absence of IκB, NF-κB translocates to the nucleus where it induces the transcription of target genes. A number of studies have correlated Rac-1 mediated activation of PAK1 with the induction of NF-κB activity. However, these studies disagree as to whether PAK1-mediated activation of NF-κB involves activation of IKKs (Frost et al. 2000; Cammarano and Minden 2001). While it appears as though Rac1 may mediate an increase in cyclinD1 expression via PAK1 activation of NF-κB, further elucidation is necessary to define the link between Rac1 signaling and NF-κB stimulation.

Irrespective of the mechanism, Rho GTPase signaling is clearly involved in driving the upregulation of cyclinD1 expression to induce cell proliferation. It is important to mention that cyclinD1 is also regulated by a number of other signaling pathways, including the Wnt signal transduction cascade, which is frequently dysregulated in cancers. CyclinD1 protein levels are reportedly elevated in approximately 50% of breast carcinomas (Buckley et al. 1993; Bartkova et al. 1994; Gillett et al. 1996), underscoring the relevance of this cell-cycle regulator in tumorigenesis. To further highlight the importance of cyclinD1 dysregulation in cancer development, cyclinD1-null mice were shown to be resistant to formation of breast tumors upon stimulation by oncogenic Ras (Yu et al. 2001). In contrast, transgenic mice that overexpress cyclinD1 in mammary tissue had an increased propensity for developing breast adenocarcinomas (Wang et al. 1994). Taken together, these studies support a causal link between cyclinD1 overexpression and breast tumorigenesis.

Downregulation of CDKIs. An additional level of cell-cycle regulation provided by Rho GTPases involves the downregulation of the CDK inhibitors, $p21^{Cip1}$ (p21) and $p27^{Kip1}$ (p27). When levels of p21 and p27 are high, they bind to and repress the activity of the CDK2-cyclinE complex, halting cell-cycle progression (Coleman et al. 2004). In a variety of cell lines, oncogenic Ras signaling is able to inhibit cell-cycle progression by inducing high levels of p21 (Lloyd et al. 1997; Pumiglia and Decker 1997; Sewing et al. 1997; Woods et al. 1997; Auer et al. 1998; Olson et al. 1998). Studies conducted in Swiss 3T3 fibroblasts revealed that overexpression of a constitutively active RhoA mutant could overcome cell-cycle arrest induced by oncogenic Ras, through suppression of p21 transcription (Olson et al. 1998). While in Ras-transformed fibroblasts and colon cancer cell lines downregulation of p21 transcription by RhoA does not seem to involve ROCK activity (Sahai et al. 1999, 2001; Roovers and Assoian 2003), similar studies performed in other cell types indicate otherwise (Lai et al. 2002; Zuckerbraun et al. 2003). These observations suggest that the mechanism by which RhoA mediates its effects on p21 transcription can be quite diverse and is cell-type-dependent. Irrespective of the mechanism, RhoA-induced progression through G_1 via inhibition of p21 is dependent upon integrin-mediated adhesion in normal cells. Under normal conditions, then, RhoA conveys information about the cellular environment to the cell-cycle machinery, promoting anchorage-dependent cell proliferation in response to Ras stimulation. Of course, aberrant induction of RhoA signaling in transformed cells (for example, in colon, breast, and lung cancer;cf. Table 1) eliminates the need for integrin-dependent activation of RhoA. Thus, in tumors, RhoA facilitates oncogenic Ras-induced cell cycle progression in an anchorage-independent fashion.

RhoA was found to have similar effects on p27 levels within the cell. While RhoA inactivation leads to an increase in p27 levels (Hirai et al. 1997; Weber et al. 1997; Laufs et al. 1999; Rivard et al. 1999), constitutively active RhoA suppresses p27 expression (Hu et al. 1998; Laufs et al. 1999; Rivard et al. 1999). Indeed, evidence suggests that RhoA activation is both necessary and sufficient for inhibition of p27 in response to growth-factor stimulation in a variety of cell types (Hirai et al. 1997). RhoA induced downregulation of p27 appears to be associated with increased CDK2-cyclinE activity and progression through G1 phase (Hirai et al. 1997), although the mechanism by which this occurs remains unclear. Studies performed in breast cancer cells suggest that RhoA may regulate p27 protein expression at the level of mRNA translation. RhoA was reported to suppress p27 mRNA translation by means of a Rho-responsive element in the 3'UTR of p27 mRNA, contributing to Ras-induced transformation (Vidal et al. 2002). Therefore, by affecting levels of cyclinD1, p21, and p27, Rho GTPases play an important role in the regulation of cell-cycle progression and hence cell proliferation. The contribution of Rho proteins to these events underscores their relevance in early stages of cancer development.

5.2 Rho GTPases and Cell Survival

An essential feature of tumor cells is the enhanced ability to resist pro-apoptotic signals and evade cell death. Several lines of evidence support a role for Rho GTPases in regulating cellular response to both pro- and anti-apoptotic triggers. For example, studies have shown that active Rac1 is able to prevent cell death induced by both serum deprivation (Ruggieri et al. 2001) and loss of integrin-mediated adherence to the ECM (Coniglio et al. 2001). Constitutively active Rac and Rho mutants are also able to block Ras-induced apoptosis, while dominant negative mutants seem to potentiate the pro-apoptotic effects of Ras. Moreover, a Rac mutant with impaired ability to activate NF-κB was unable to suppress Ras-induced cell death. These data suggest that Rac is able to induce anti-apoptotic effects in response to oncogenic Ras, likely via NF-κB activation (Joneson and Bar-Sagi 1999). A possible effector function of Rac that may be involved in inducing NF-κB-mediated cell survival is activation of NADPH oxidase. Rac interacts with NADPH oxidase in a variety of cell types resulting in the production of superoxide (O2−), and consequential activation of NF-κB. Studies using Rac effector mutants with inefficient NADPH binding, and hence impaired superoxide generation, suggest that Rac-mediated inhibition of apoptosis is dependent on its ability to produce superoxide. Furthermore, experiments performed in bladder carcinoma cell lines reveal that Rac activation results in increased superoxide anion concentration, and a concomitant increase in tumor cell resistance to Ras-induced apoptosis (Pervaiz et al. 2001). Therefore, these observations imply that Rho proteins may influence cell survival via superoxide-dependent stimulation of NF-κB.

Among the downstream targets of NF-κB transcriptional activity are genes that encode proteins with anti-apoptotic functions, including cyclooxygenase-2 (COX-2). Overexpression of COX-2, which occurs in approximately 80% of human colorectal cancers, confers resistance to apoptosis via a number of mechanisms, including inhibition of the mitochondrial apoptotic pathway, and suppression of the membrane death receptor apoptotic cascade (reviewed in Sinicrope and Gill 2004). Recent evidence implicates a role for Rho GTPases in the transcriptional regulation of COX-2 expression. RhoA, Rac1, and Cdc42 were shown to increase COX-2 protein levels through stimulation of NF-κB activity, in a cell type- and tumor-specific manner (Benitah et al. 2003). Furthermore, the ability of RhoA to modulate COX-2 expression appears to involve activation of the NF-κB pathway by ROCK. Whether Rho GTPase-mediated induction of COX-2 expression correlates with enhanced resistance to apoptosis requires further investigation. Taken together, this evidence supports a role for Rho protein-mediated stimulation of NF-κB in cell survival. Additionally, other studies propose that the anti-apoptotic effects of Rho GTPases are dependent on PAK1 activity. This proposal relies on the observation that activated PAK1 phosphorylates Bad, a component of the cell death machinery, thereby suppressing its pro-apoptotic function (Schurmann et al. 2000; Tang et al. 2000).

While Rho GTPases play a role in protecting the cell from apoptosis in response to cellular stress, they may also act conversely, conferring increased sensitivity to pro-apoptotic signals. Previous reports have established the involvement of RhoB in promoting cell death. Evidence derives from analysis of Ras-transformed RhoB-null mice, which show increased resistance to apoptosis induced by cytotoxic stress. The resistant phenotype could be reversed by ectopic expression of RhoB, indicating that RhoB mediates the pro-apoptotic response of neoplastically transformed cells to DNA damage (Liu et al. 2001a). Moreover, the ability of RhoB to promote apoptosis seems to inhibit Ras-induced transformation and malignant progression (Jiang et al. 2004). Consistent with these observations, RhoB-deficient mice have an increased propensity for developing Ras-induced skin tumors, suggesting a putative tumor-suppressor role for RhoB (Liu et al. 2001b). Recent reports of decreased RhoB expression in human lung cancer progression further corroborate the importance of RhoB suppressive activity. The loss of RhoB expression in lung tumor cells may be attributed to Ras activation of the phosphatidylinositol-3-kinase/Akt pathway (Mazieres et al. 2004). As a whole, the evidence presented here suggests that cellular activities mediated by Rho GTPases in response to both anti- and pro-apoptotic signals may contribute to the ability of cells to evade cell death.

5.3 Rho GTPases and Invasion/Metastasis

Latter stages of cancer development involve the acquisition of migratory and invasive properties, which promote the metastasis of tumors. Given that these processes involve reorganization of the actin cytoskeleton, a well-established function of Rho GTPases, it is likely that Rho proteins contribute to the metastatic phenotype. Indeed, substantial evidence supports a role for Rho GTPases in key metastatic events, including alterations in intercellular adhesion, cell motility, and ECM remodeling.

Rho GTPases regulate the formation and maintenance of specialized adhesive junctions, such as adherens junctions and tight junctions, that are essential for the establishment of epithelial cell polarity. Activated Rac1 and Cdc42 mediate intercellular adhesion by promoting the association between adherens junctions (which consist of E-cadherin/catenin complexes) and the actin cytoskeleton, via sequestration of the inhibitory IQGAP1. In contrast, RhoA antagonizes the effects of Rac1 and Cdc42, leading to destabilization of cell–cell adhesions (reviewed in Wheelock and Johnson 2003). Therefore, the relative levels of Rac1/Cdc42 and RhoA activity play an important role in determining epithelial cell morphology. In this regard, studies using MDCK cells have shown that sustained activation of oncogenic Ras signaling downregulates Rac1 activity, leading to a concomitant increase in RhoA activation. Consequential RhoA-induced stress fiber formation promotes epithelial-to-mesenchymal transition (EMT), which is characterized by the loss of cell–cell contacts – and hence disruption of epithelial cell polarity – and increased motility. Consistent with these observations, reconstitution of Rac1 activity in Ras-transformed mesenchymal cells results in decreased RhoA activation, and restoration of the epithelial phenotype (Zondag et al. 2000). Furthermore, transforming growth factor-β1 (TGF-β) is thought to stimulate EMT through a RhoA-dependent mechanism. A study has shown that inhibition of RhoA or ROCK suppresses TGF-β-induced EMT, suggesting that TGF-β-mediated activation of RhoA/ROCK signaling stimulates stress fiber formation and a mesenchymal phenotype (Bhowmick et al. 2001).

Conflicting evidence has emerged surrounding the effects of Rac1 on intercellular adhesion. Similar studies conducted in both Ras-transformed primary epithelial cells and human keratinocytes revealed that constitutively active Rac1 expression disrupts E-cadherin-mediated adhesion, and that dominant negative Rac1 promotes cell–cell adhesion by stabilizing E-cadherin/catenin complexes (Quinlan 1999; Braga et al. 2000). These apparent inconsistencies suggest that cell type and cell context may be important in determining whether Rac1 positively or negatively regulates intercellular adhesion. Indeed, in vitro studies have shown that Rac1 signaling may either promote or inhibit cell–cell adhesion – and thus cell migration – depending on the composition of the ECM. For example, Rac1 activation in Ras-transformed MDCK cells stabilizes E-cadherin-mediated adhesion and inhibits cell migration on a fibronectin or laminin substrate. In contrast, activated Rac1 promotes the migration of the same cells when the substrate is changed to collagen (Sander et al. 1998). These observations suggest that integrin-mediated signaling plays an important role in eliciting either pro-adhesion or pro-migration effects in response to Rac1 activation.

The increasingly motile phenotype acquired by tumor cells upon disruption of adhesive junctions and cell polarity may be further enhanced by Rho GTPase-mediated regulation of ezrin/radixin/moesin (ERM) proteins. ERM proteins link the actin cytoskeleton to the plasma membrane via CD44, a receptor that binds components of the ECM, and thereby promote cell motility. Studies have demonstrated that ROCK phosphorylates all three ERM proteins downstream of RhoA signaling, resulting in increased ERM-cytoskeleton interactions (Matsui et al. 1998). Furthermore, Rac1 and Cdc42 activation induces PAK2-mediated phosphorylation of the ERM-related protein, merlin (Shaw et al. 2001). Merlin is the product of the neurofibromatosis type II (NF2) tumor-suppressor gene, the deletion of which gives rise to highly metastatic tumors (McClatchey et al. 1998). Although the tumor-suppressive function of merlin is unclear, its active form is thought to antagonize ezrin. Rac1 and Cdc42-induced phosphorylation of merlin results in its inhibition (Shaw et al. 2001; Xiao et al. 2002), and possibly promotes ERM-cytoskeleton interactions. Since ERM proteins localize to cortical actin structures near the plasma membrane that are involved in cell migration, such as membrane ruffles and lamellipodia (discussed below), Rho GTPase-mediated promotion of ERM–cytoskeleton associations may contribute to the increased migration of cells. The relevance of ERM involvement in cell motility is emphasized by reports of ezrin and CD44 overexpression in highly metastatic pancreatic (Akisawa et al. 1999) and colon tumors (Clarke et al. 2000), respectively.

The invasiveness of tumor cells likely depends on the specific functions of Rho GTPases that act in concert to coordinate cellular responses necessary for the migratory behavior of cells. Rho proteins induce morphological changes that promote motility via rearrangement of the actin cytoskeleton: Rac1 is responsible for the extension of lamellipodia, or actin-rich membrane protrusions, at the leading edge of migrating cells; RhoA activation promotes forward movement by causing the body of the migratory cell to contract and the lagging end to retract; and Cdc42 is involved in determining the direction of movement through the formation of filopodia that act as sensors of the extracellular environment (reviewed in Ridley 2001). In order to traverse tissue barriers, migratory tumor cells must be able to penetrate the basement membrane. This requires degradation of the ECM, mainly through the upregulation of matrix metalloproteinases (MMPs) and downregulation of tissue inhibitors of metalloproteinases (TIMPs). Some studies have shown that Rho GTPases contribute to ECM degradation, and hence tissue invasion, by regulating the activities of MMPs. In vitro invasion assays revealed that constitutively active Rac1 induces the activation of MMP-2 in fibrosarcomas and enhances invasion in an MMP-2-dependent manner (Zhuge and Xu 2001). In contrast, expression of active Rac1 in human renal cell carcinoma cell lines was found to upregulate TIMP-1 and TIMP-2 activities and prevent invasion in vitro (Engers et al. 2001), suggesting that the effects of Rac1 on ECM degradation are cell type-specific. Taken together, these data suggest that Rho GTPases enhance the metastatic potential of tumor cells by eliciting various cellular responses that promote the invasive phenotype.

6 Cross Talk Between Rac1 and Wnt Signaling Pathways

Recent evidence has emerged implicating Rac1 as a novel mediator of the canonical Wnt signaling pathway. The Wnt pathway coordinates various cellular events during embryonic development via activation of Wnt target genes, which are mainly involved in cell proliferation and cell fate determination (Giles et al. 2003). Canonical Wnt pathway players interact to regulate the activity of β-catenin, the key effector of the pathway. In the absence of Wnt stimulation, a complex comprised of adenomatous polyposis coli (APC), axin, and glycogen synthase kinase-3β (GSK-3β) targets β-catenin for proteasomal degradation, thereby maintaining low levels of cytosolic β-catenin and inhibiting activation of Wnt target genes. Induction of the Wnt signaling cascade in normal cells results from engagement of the frizzled receptor with its cognate Wnt ligand, which signals to inhibit the degradation of β-catenin. As a consequence, stabilized β-catenin accumulates in the cytosol and subsequently translocates into the nucleus, where it complexes with T-cell factor (TCF)/lymphoid-enhancing factor (LEF) family of transcription factors to co-activate the expression of Wnt target genes, including cyclinD1 and c-myc. Aberrant activation of Wnt signaling may arise due to mutations in key players of the pathway, most commonly inactivating APC mutations and activating mutations of β-catenin. Consequently, inappropriate transcription of Wnt target genes ensues, leading to uncontrolled cell proliferation and cell transformation. Dysregulation of Wnt signaling represents one of the earliest events in progression towards many types of malignancies, including colorectal and mammary cancers (Giles et al. 2003).

Evidence that underscores the complexity of Wnt signaling continues to emerge, challenging the conception of a linear model for Wnt pathway activation. Indeed, the Wnt pathway has evolved into a network, encompassing both canonical (β-catenin-mediated) and non-canonical (β-catenin-independent) components, including the Wnt/Ca^{2+} and planar-cell-polarity (PCP) pathways. Several studies have shown that Rho GTPases regulate the PCP branch, and are involved in the control of cell motility during vertebrate gastrulation (Habas et al. 2003). Furthermore, the identification of novel functional interactions for key canonical Wnt players implicates other pathways, such as TGF-β (Labbe et al. 2000), retinoic acid receptor (RAR; Easwaran et al. 1999),

MAPK (Ishitani et al. 1999), and estrogen receptor α (ER α; Kouzmenko et al. 2004) in aberrant canonical Wnt signaling.

Recent studies have revealed a novel cross-talk between Rac1 signal transduction and the canonical Wnt pathway. Active Rac1 was shown to enhance the signaling activity of β-catenin in colon cancer cell lines with inherent dysregulation of the canonical Wnt pathway, as determined by the TCF/LEF-responsive transcription assay (Esufali and Bapat 2004). In contrast, expression of a dominant negative Rac1 mutant in colon cancer cells was found to markedly inhibit Wnt signaling. In human embryonic kidney cells, in which the Wnt pathway is intact, constitutively active Rac1 expression causes up to 40-fold induction of Wnt activity in the presence of stabilized β-catenin. This induction could be completely blocked by the expression of a dominant negative TCF-4 mutant, suggesting that β-catenin and TCF-4 complex formation is required for Rac1-mediated transcriptional activity. To further corroborate these observations, Rac1 was found to regulate the transcription of a Wnt target gene, cyclinD1, in a β-catenin/TCF-4-dependent manner. Some insights into the mechanism via which active Rac1 mediates its stimulatory effects include the observations that active, GTP-bound Rac1 physically interacts with β-catenin and TCF-4, and overexpression of active Rac1 results in increased levels of cytosolic and nuclear β-catenin. Interestingly, mutation of the Rac1 PBR, which prevents its nuclear localization, results in decreased nuclear accumulation of β-catenin, and abolishes the ability of Rac1 to stimulate the transcriptional co-activator function of β-catenin. Taken together, these data suggest a model whereby dysregulation of β-catenin may potentiate cross-talk with active Rac1, resulting in increased expression of cytosolic β-catenin, possibly by re-distribution of the membrane pool or further stabilization of β-catenin protein. Furthermore, active Rac1 augments nuclear accumulation of β-catenin, possibly by physically associating with β-catenin and/or TCF-4. It is proposed that these events culminate in the amplification of β-catenin signaling activity, resulting in enhanced transcriptional activation of a specific subset of Wnt target genes important in cancer progression (see Fig. 2). These novel insights suggest an additional mechanism by which Rac1 realizes its oncogenic potential and influences various aspects of malignancy.

7 Rho GTPases as Therapeutic Targets

Considering the multiple lines of evidence implicating Rho family members in various cancer-promoting processes, it seems plausible to target the signaling activities of Rho proteins for therapeutic intervention. In this regard, efforts have been made in the design of inhibitors that interfere with the function of Rho proteins or their effectors. The ROCK inhibitor Y-27632, which blocks the kinase activity of ROCK, has been found to suppress the invasive behavior of tumor cells in vivo. In animal models, Y-27632 treatment prevents

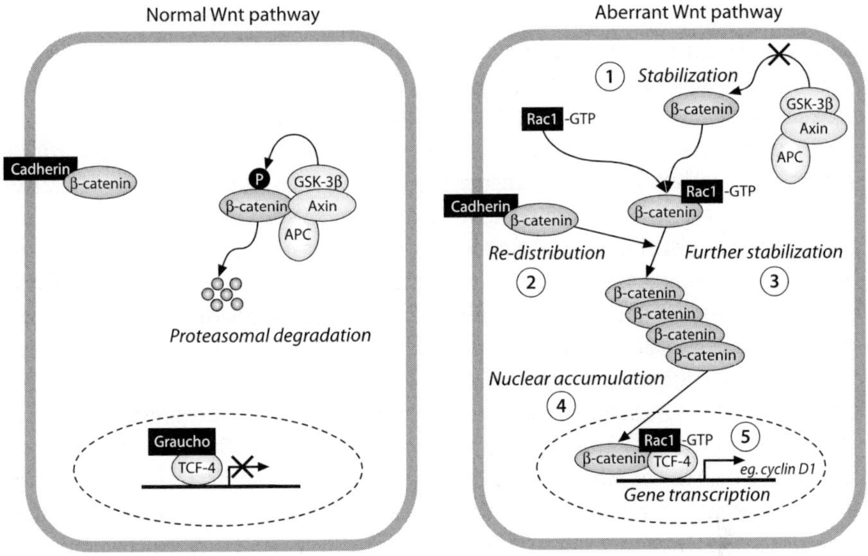

Fig. 2. Model for Rac1-mediated amplification of dysregulated canonical Wnt signaling. *Left* Normally, cytosolic and nuclear pools of β-catenin are maintained at low levels by being targeted for proteasomal degradation by phosphorylation by GSK-3β, which exists in a complex with APC and axin. The remaining β-catenin is bound to E-cadherin at sites of cell–cell contact, supporting cellular adhesion. In the absence of nuclear β-catenin, TCF-mediated transcription is repressed by association with the Groucho family of transcription factors. *Right* Mutations in APC, Axin, or β-catenin hinder proteasomal degradation of β-catenin, which results in its accumulation in the cytoplasm and nuclear translocation (*1*). In the nucleus β-catenin complexes with TCF-4 and activates transcription of Wnt target genes. Dysregulation of β-catenin may potentiate cross talk with active Rac1 GTPase, resulting in increased expression of cytosolic β-catenin, possibly by re-distribution of the membrane pool (*2*) or further stabilization of β-catenin protein (*3*). Rac1-GTP augments nuclear accumulation of β-catenin (*4*) and physical association of Rac1 with β-catenin and/or TCF-4 may facilitate this process. This culminates in the amplification of β-catenin signaling activity, resulting in enhanced transcriptional activation of perhaps a specific subset of Wnt target genes important in cancer progression (*5*). (Reprinted with permission rom oncogene, Esufali and Bapat, copyright 2004, Macmillan Publishers Ltd.)

the dissemination of highly metastatic hepatoma cells (Itoh et al. 1999) and reduces the spreading of human prostate cancer cells (Somlyo et al. 2000). While these observations highlight the potential of targeting ROCK activity as an anti-cancer strategy, further work is necessary to determine its clinical relevance.

A recent study has reported the design of a Rac1-specific inhibitor, NSC23766, that blocks the interaction between Rac1 and Tiam1, and hence prevents Tiam1-mediated activation of Rac1. In vitro evidence suggests that the inhibitor is able to repress the proliferation, anchorage-independent growth, and invasive phenotype of human prostate cancer cells (Gao et al.

2004). However, these results are preliminary, and additional studies are required to assess the effectiveness of NSC23766 as Rac1 inhibitor in vivo.

Evidence suggests that farnesyl transferase inhibitors (FTIs), which were originally developed to target RAS oncoproteins, influence the function of Rho proteins as well. Most members of the Ras superfamily undergo isoprenoid modification by the addition of either a farnesyl or a geranylgeranyl group at their C-terminal ends. Such isoprenoid modifications are necessary for proper membrane localization and function of both Ras and Rho GTPases. Because Ras oncoproteins are subject to farnesylation, FTIs were designed to inhibit the isoprenoid modification and hence the activity of oncogenic Ras. Interestingly, several studies revealed that FTIs induce a gain of function in RhoB, which could account for the growth inhibition and cell death observed in FTI-treated Ras-transformed cells (reviewed in Prendergast 2001). The ability of FTIs to induce RhoB activity relies on the fact that cells express two different isoforms of RhoB: one isoform is geranylgeranylated (-GG), whereas the other is farnesylated (-F). While the treatment of Ras-transformed cells with FTIs blocks the formation of RhoB-F, the levels of RhoB-GG increase due to the geranylgeranylation of any newly synthesized RhoB. The shift in the pattern of RhoB isoprenoid modification is associated with elevated expression and mislocalization of RhoB-GG, which possibly affects its activity. The induction of RhoB-GG levels upon FTI treatment was found to inhibit cell-cycle progression and promote apoptosis in various human carcinoma cell lines. Moreover, RhoB expression was found to be crucial for FTI-mediated induction of cell death in Ras-transformed mouse embryonic fibroblasts. Clinical trials are currently underway to evaluate the usefulness of FTIs as potential anti-cancer drugs.

Given the effectiveness of FTIs in targeting Ras and RhoB activities, geranylgeranyltransferase inhibitors (GGTIs) have similarly been designed to inhibit the functions of geranylgeranylated Rho GTPases. GGTIs have been shown to induce cell-cycle arrest of tumor cells in vitro (Pollack et al. 1999) and to suppress tumor growth in mouse models (Sun et al. 1998). Efforts aimed at determining the downstream targets of GGTIs have implicated Rac1 and Rac3 as potential mediators of their inhibitory effects (Joyce and Cox 2003). While both in vitro and in vivo studies support the relevance of interfering with the activities of Rho proteins and their effectors as a potential anti-cancer strategy, further work is required to validate the transfer of such strategies into the clinical setting.

8 Future Directions

The interplay between distinct signaling pathways within the Rho GTPase circuit, and between Rho proteins and external signal transduction cascades underscores the complexity of the Rho GTPase network. Different Rho family members act in concert and also cooperate with other signaling pathways to coordinate various cellular events, including cell cycle progression, cell survival, intercellular adhesion, and ECM remodeling. Multiple lines of evidence have linked aberrant activation of Rho signaling with dysregulation of these cancer-promoting events, suggesting the involvement of Rho proteins in various stages of tumor progression. Rho proteins are particularly relevant in that they appear to influence a wide range of different cancers. However, most of the evidence derives from in vitro analyses, and may not always accurately reflect the physiological relevance of aberrant Rho protein activity. It will therefore be important to corroborate the effects of Rho GTPases in tumorigenesis using in vivo animal models and human tissue samples. Efforts should also be aimed at elucidating the specific effector pathways that are dysregulated by aberrant Rho GTPase signaling, such that we may gain further insight into the mechanism by which they influence tumorigenesis. Determining the specific downstream effectors that mediate the cellular effects of Rho proteins might allow for the development of anti-cancer therapies that target their specific activities. Particular attention should also be paid to Rho proteins that are expressed preferentially in tumors, for example, Rac1b, and upstream regulators that target specific Rho GTPases in the design of anti-cancer strategies.

Acknowledgements. We would like to thank Susmita Esufali for critical review of the manuscript. B.B. is the recipient of Grant No. MOP_64225 (Canadian Institute for Health Research). P.B. is the recipient of Ontario Graduate Scholarship, and the Ontario Student Opportunity Trust Funds.

References

Abraham MT, Kuriakose MA, Sacks PG, Yee H, Chiriboga L, Bearer EL, Delacure MD (2001) Motility-related proteins as markers for head and neck squamous cell cancer. Laryngoscope 111:1285–1289

Adamson P, Marshall CJ, Hall A, Tilbrook PA (1992) Post-translational modifications of p21rho proteins. J Biol Chem 267 20033–20038

Adnane J, Muro-Cacho C, Mathews L, Sebti SM, Munoz-Antonia T (2002) Suppression of rho B expression in invasive carcinoma from head and neck cancer patients. Clin Cancer Res 8:2225–2232

Akisawa N, Nishimori I, Iwamura T, Onishi S, Hollingsworth MA (1999) High levels of ezrin expressed by human pancreatic adenocarcinoma cell lines with high metastatic potential. Biochem Biophys Res Commun 258:395–400

Auer KL, Park JS, Seth P, Coffey RJ, Darlington G, Abo A, McMahon M, Depinho RA, Fisher PB, Dent P (1998) Prolonged activation of the mitogen-activated protein kinase pathway promotes DNA synthesis in primary hepatocytes from p21Cip-1/WAF1-null mice, but not in hepatocytes from p16INK4a-null mice. Biochem J 336(Pt 3):551–560

Bartkova J, Lukas J, Muller H, Lutzhoft D, Strauss M, Bartek J(1994) Cyclin D1 protein expression and function in human breast cancer. Int J Cancer 57:353–361

Benitah SA, Valeron PF, Lacal JC (2003) ROCK and nuclear factor-kappaB-dependent activation of cyclooxygenase-2 by Rho GTPases: effects on tumor growth and therapeutic consequences. Mol Biol Cell 14:3041–3054

Bhowmick NA, Ghiassi M, Bakin A, Aakre M, Lundquist CA, Engel ME, Arteaga CL, and Moses HL (2001) Transforming growth factor-beta1 mediates epithelial to mesenchymal transdifferentiation through a RhoA-dependent mechanism. Mol Biol Cell 12:27–36

Braga VM, Betson M, Li X, Lamarche-Vane N (2000) Activation of the small GTPase Rac is sufficient to disrupt cadherin-dependent cell-cell adhesion in normal human keratinocytes. Mol Biol Cell 11:3703–3721

Buckley MF, Sweeney KJ, Hamilton JA, Sini RL, Manning DL, Nicholson RI, deFazio A, Watts CK, Musgrove EA, Sutherland RL (1993) Expression and amplification of cyclin genes in human breast cancer. Oncogene 8:2127–2133

Cammarano MS, Minden A (2001) Dbl and the Rho GTPases activate NF kappa B by I kappa B kinase (IKK)-dependent and IKK-independent pathways. J Biol Chem 276:25876–25882

Ching YP, Wong CM, Chan SF, Leung TH, Ng DC, Jin DY, Ng IO (2003) Deleted in liver cancer (DLC) 2 encodes a RhoGAP protein with growth suppressor function and is underexpressed in hepatocellular carcinoma. J Biol Chem 278:10824–10830

Clark EA, Golub TR, Lander ES, Hynes RO (2000) Genomic analysis of metastasis reveals an essential role for RhoC. Nature 406:532–535

Clarke G, Ryan E, O'Keane JC, Crowe J, Mathuna PM (2000) Mortality association of enhanced CD44v6 expression is not mediated through occult lymphatic spread in stage II colorectal cancer. J Gastroenterol Hepatol 15:1028–1031

Coleman ML, Marshall CJ, Olson MF (2004) RAS and RHO GTPases in G1-phase cell-cycle regulation. Nat Rev Mol Cell Biol 5:355–366

Coniglio SJ, Jou TS, Symons M (2001) Rac1 protects epithelial cells against anoikis. J Biol Chem 276:28113–28120

Easwaran V, Pishvaian M, Salimuddin, Byers S (1999) Cross-regulation of beta-catenin-LEF/TCF and retinoid signaling pathways. Curr Biol 9:1415–1418

Engers R, Zwaka TP, Gohr L, Weber A, Gerharz CD, Gabbert HE (2000) Tiam1 mutations in human renal-cell carcinomas. Int J Cancer 88:369–376

Engers R, Springer E, Michiels F, Collard JG, Gabbert HE (2001) Rac affects invasion of human renal cell carcinomas by up-regulating tissue inhibitor of metalloproteinases (TIMP)-1 and TIMP-2 expression. J Biol Chem 276:41889–41897

Esufali S, Bapat B (2004) Cross-talk between Rac1 GTPase and dysregulated Wnt signaling pathway leads to cellular redistribution of beta-catenin and TCF/LEF-mediated transcriptional activation. Oncogene 23:8260–8271

Fiegen D, Haeusler LC, Blumenstein L, Herbrand U, Dvorsky R, Vetter IR, Ahmadian MR (2004) Alternative splicing of Rac1 generates Rac1b, a self-activating GTPase. J Biol Chem 279:4743–4749

Fritz G, Just I, Kaina B(1999) Rho GTPases are over-expressed in human tumors. Int J Cancer 81:682–687

Frost JA, Swantek JL, Stippec S, Yin MJ, Gaynor R, Cobb MH (2000) Stimulation of NFkappa B activity by multiple signaling pathways requires PAK1. J Biol Chem 275:19693–19699

Gao Y, Dickerson JB, Guo F, Zheng J, Zheng Y (2004) Rational design and characterization of a Rac GTPase-specific small molecule inhibitor. Proc Natl Acad Sci USA 101:7618–7623

Giles RH, van Es JH, Clevers H (2003) Caught up in a Wnt storm: Wnt signaling in cancer. Biochim Biophys Acta 1653:1–24

Gillett C, Smith P, Gregory W, Richards M, Millis R, Peters G, Barnes D (1996) Cyclin D1 and prognosis in human breast cancer. Int J Cancer 69:92–99

Habas R, Dawid IB, He X (2003) Coactivation of Rac and Rho by Wnt/Frizzled signaling is required for vertebrate gastrulation. Genes Dev 17:295–309

Hamaguchi M, Meth JL, von Klitzing C, Wei W, Esposito D, Rodgers L, Walsh T, Welcsh P, King MC, Wigler MH (2002) DBC2, a candidate for a tumor suppressor gene involved in breast cancer. Proc Natl Acad Sci USA 99:13647–13652

Hirai A, Nakamura S, Noguchi Y, Yasuda T, Kitagawa M, Tatsuno I, Oeda T, Tahara K, Terano T, Narumiya S et al (1997) Geranylgeranylated rho small GTPase(s) are essential for the degradation of p27Kip1 and facilitate the progression from G1 to S phase in growth-stimulated rat FRTL-5 cells. J Biol Chem 272:13–16

Hordijk PL, ten Klooster JP, van der Kammen RA, Michiels F, Oomen LC, Collard JG (1997) Inhibition of invasion of epithelial cells by Tiam1-Rac signaling. Science 278:1464–1466

Hu ZY, Madamanchi NR, Rao GN (1998) cAMP inhibits linoleic acid-induced growth by antagonizing p27(kip1) depletion, but not interfering with the extracellular signal-regulated kinase or AP-1 activities. Biochim Biophys Acta 1405:139–146

Ishitani T, Ninomiya-Tsuji J, Nagai S, Nishita M, Meneghini M, Barker N, Waterman M, Bowerman B, Clevers H, Shibuya H, Matsumoto K (1999) The TAK1-NLK-MAPK-related pathway antagonizes signalling between beta-catenin and transcription factor TCF. Nature 399:798–802

Itoh K, Yoshioka K, Akedo H, Uehata M, Ishizaki T, Narumiya S (1999) An essential part for Rho-associated kinase in the transcellular invasion of tumor cells. Nat Med 5:221–225

Jiang K, Sun J, Cheng J, Djeu JY, Wei S, Sebti S (2004) Akt mediates Ras downregulation of RhoB, a suppressor of transformation, invasion, and metastasis. Mol Cell Biol 24:5565–5576

Joneson T, Bar-Sagi D (1999) Suppression of Ras-induced apoptosis by the Rac GTPase. Mol Cell Biol 19:5892–5901

Jordan P, Brazao R, Boavida MG, Gespach C, Chastre E (1999) Cloning of a novel human Rac1b splice variant with increased expression in colorectal tumors. Oncogene 18:6835–6839

Joyce D, Bouzahzah B, Fu M, Albanese C, D'Amico M, Steer J, Klein JU, Lee RJ, Segall JE, Westwick JK et al (1999) Integration of Rac-dependent regulation of cyclin D1 transcription through a nuclear factor-kappaB-dependent pathway. J Biol Chem 274:25245–25249

Joyce PL, Cox AD (2003) Rac1 and Rac3 are targets for geranylgeranyltransferase I inhibitor-mediated inhibition of signaling, transformation, and membrane ruffling. Cancer Res 63:7959–7967

Kamai T, Arai K, Tsujii T, Honda M, Yoshida K (2001) Overexpression of RhoA mRNA is associated with advanced stage in testicular germ cell tumour. BJU Int 87:227–231

Khosravi-Far R, Solski PA, Clark GJ, Kinch MS, Der CJ (1995) Activation of Rac1, RhoA, and mitogen-activated protein kinases is required for Ras transformation. Mol Cell Biol 15:6443–6453

Kleer CG, van Golen KL, Zhang Y, Wu ZF, Rubin MA, Merajver SD (2002) Characterization of RhoC expression in benign and malignant breast disease: a potential new marker for small breast carcinomas with metastatic ability. Am J Pathol 160:579–584

Kourlas PJ, Strout MP, Becknell B, Veronese ML, Croce CM, Theil KS, Krahe R, Ruutu T, Knuutila S, Bloomfield CD, Caligiuri MA (2000) Identification of a gene at 11q23 encoding a guanine nucleotide exchange factor: evidence for its fusion with MLL in acute myeloid leukemia. Proc Natl Acad Sci USA 97:2145–2150

Kouzmenko AP, Takeyama K, Ito S, Furutani T, Sawatsubashi S, Maki A, Suzuki E, Kawasaki Y, Akiyama T, Tabata T, Kato S (2004) Wnt/beta-catenin and estrogen signaling converge in vivo. J Biol Chem 279:40255–40258

Labbe E, Letamendia A, Attisano L (2000) Association of Smads with lymphoid enhancer binding factor 1/T cell-specific factor mediates cooperative signaling by the transforming growth factor-beta and wnt pathways. Proc Natl Acad Sci USA 97:8358–8363

Lai JM, Wu S, Huang DY, Chang ZF (2002) Cytosolic retention of phosphorylated extracellular signal-regulated kinase and a Rho-associated kinase-mediated signal impair expression of p21(Cip1/Waf1) in phorbol 12-myristate-13- acetate-induced apoptotic cells. Mol Cell Biol 22:7581–7592

Lanning CC, Ruiz-Velasco R, Williams CL (2003) Novel mechanism of the co-regulation of nuclear transport of SmgGDS and Rac1. J Biol Chem 278:12495–12506

Lanning CC, Daddona JL, Ruiz-Velasco R, Shafer SH, Williams CL (2004) The Rac1 C-terminal polybasic region regulates the nuclear localization and protein degradation of Rac1. J Biol Chem 279:44197–44210

Laufs U, Marra D, Node K, Liao JK (1999) 3-Hydroxy-3-methylglutaryl-CoA reductase inhibitors attenuate vascular smooth muscle proliferation by preventing rho GTPase-induced down-regulation of p27(Kip1) J Biol Chem 274:21926–21931

Liu A, Cerniglia GJ, Bernhard EJ, Prendergast GC (2001a) RhoB is required to mediate apoptosis in neoplastically transformed cells after DNA damage. Proc Natl Acad Sci USA 98:6192–6197

Liu AX, Rane N, Liu JP, Prendergast GC (2001b) RhoB is dispensable for mouse development, but it modifies susceptibility to tumor formation as well as cell adhesion and growth factor signaling in transformed cells. Mol Cell Biol 21:6906–6912

Lloyd AC, Obermuller F, Staddon S, Barth CF, McMahon M, Land H (1997) Cooperating oncogenes converge to regulate cyclin/cdk complexes. Genes Dev 11:663–677

Malliri A, van der Kammen RA, Clark K, van der Valk M, Michiels F, Collard JG (2002) Mice deficient in the Rac activator Tiam1 are resistant to Ras-induced skin tumours. Nature 417:867–871

Matos P, Collard JG, Jordan P (2003) Tumor-related alternatively spliced Rac1b is not regulated by Rho-GDP dissociation inhibitors and exhibits selective downstream signaling. J Biol Chem 278:50442–50448

Matsui T, Maeda M, Doi Y, Yonemura S, Amano M, Kaibuchi K, Tsukita S (1998) Rho-kinase phosphorylates COOH-terminal threonines of ezrin/radixin/moesin (ERM) proteins and regulates their head-to-tail association. J Cell Biol 140:647–657

Mazieres J, Antonia T, Daste G, Muro-Cacho C, Berchery D, Tillement V, Pradines A, Sebti S, Favre G (2004) Loss of RhoB expression in human lung cancer progression. Clin Cancer Res 10:2742–2750

McClatchey AI, Saotome I, Mercer K, Crowley D, Gusella JF, Bronson RT, Jacks T (1998) Mice heterozygous for a mutation at the Nf2 tumor suppressor locus develop a range of highly metastatic tumors. Genes Dev 12:1121–1133

Michiels F, Habets GG, Stam JC, van der Kammen RA, Collard JG (1995) A role for Rac in Tiam1-induced membrane ruffling and invasion. Nature 375:338–340

Mira JP, Benard V, Groffen J, Sanders LC, Knaus UG (2000) Endogenous, hyperactive Rac3 controls proliferation of breast cancer cells by a p21-activated kinase-dependent pathway. Proc Natl Acad Sci USA 97:185–189

Murphy GA, Solski PA, Jillian SA, Perez de la Ossa P, D'Eustachio P, Der CJ, Rush MG (1999) Cellular functions of TC10, a Rho family GTPase: regulation of morphology, signal transduction and cell growth. Oncogene 18:3831–3845

Olson MF, Paterson HF, Marshall CJ (1998) Signals from Ras and Rho GTPases interact to regulate expression of p21Waf1/Cip1. Nature 394:295–299

Pasqualucci L, Neumeister P, Goossens T, Nanjangud G, Chaganti RS, Kuppers R, Dalla-Favera R (2001) Hypermutation of multiple proto-oncogenes in B-cell diffuse large-cell lymphomas. Nature 412:341–346

Pervaiz S, Cao J, Chao OS, Chin YY, Clement MV (2001) Activation of the RacGTPase inhibits apoptosis in human tumor cells. Oncogene 20:6263–6268

Pollack IF, Bredel M, Erff M, Hamilton AD, Sebti SM (1999) Inhibition of Ras and related guanosine triphosphate-dependent proteins as a therapeutic strategy for blocking malignant glioma growth. II. Preclinical studies in a nude mouse model. Neurosurgery 45:1208–1214; discussion 1214–1205

Prendergast GC (2001) Actin' up: RhoB in cancer and apoptosis. Nat Rev Cancer 1:162–168

Preudhomme C, Roumier C, Hildebrand MP, Dallery-Prudhomme E, Lantoine D, Lai JL, Daudignon A, Adenis C, Bauters F, Fenaux P, et al (2000) Nonrandom 4p13 rearrangements of the RhoH/TTF gene, encoding a GTP-binding protein, in non-Hodgkin's lymphoma and multiple myeloma. Oncogene 19:2023–2032

Price LS, Collard JG (2001) Regulation of the cytoskeleton by Rho-family GTPases: implications for tumour cell invasion. Semin Cancer Biol 11:167–173

Pumiglia KM, Decker SJ (1997) Cell cycle arrest mediated by the MEK/mitogen-activated protein kinase pathway. Proc Natl Acad Sci USA 94:448–452

Qiu RG, Chen J, Kirn D, McCormick F, Symons M (1995a) An essential role for Rac in Ras transformation. Nature 374:457–459

Qiu RG, Chen J, McCormick F, Symons M (1995b) A role for Rho in Ras transformation. Proc Natl Acad Sci USA 92:11781–11785

Qiu RG, Abo A, McCormick F, Symons M (1997) Cdc42 regulates anchorage-independent growth and is necessary for Ras transformation. Mol Cell Biol 17:3449–3458

Quinlan MP (1999) Rac regulates the stability of the adherens junction and its components, thus affecting epithelial cell differentiation and transformation. Oncogene 18:6434–6442

Reuther GW, Lambert QT, Booden MA, Wennerberg K, Becknell B, Marcucci G, Sondek J, Caligiuri MA, Der CJ (2001) Leukemia-associated Rho guanine nucleotide exchange factor, a Dbl family protein found mutated in leukemia, causes transformation by activation of RhoA. J Biol Chem 276:27145–27151

Ridley AJ (2001) Rho GTPases and cell migration. J Cell Sci 114:2713–2722

Rivard N, Boucher MJ, Asselin C, L'Allemain G (1999) MAP kinase cascade is required for p27 downregulation and S phase entry in fibroblasts and epithelial cells. Am J Physiol 277:C652–C664

Roovers K, Assoian RK (2003) Effects of rho kinase and actin stress fibers on sustained extracellular signal-regulated kinase activity and activation of G(1) phase cyclin-dependent kinases. Mol Cell Biol 23:4283–4294

Roovers K, Davey G, Zhu X, Bottazzi ME, Assoian RK (1999) Alpha5beta1 integrin controls cyclin D1 expression by sustaining mitogen-activated protein kinase activity in growth factor-treated cells. Mol Biol Cell 10:3197–3204

Roovers K, Klein EA, Castagnino P, Assoian RK (2003) Nuclear translocation of LIM kinase mediates Rho-Rho kinase regulation of cyclin D1 expression. Dev Cell 5:273–284

Roux P, Gauthier-Rouviere C, Doucet-Brutin S, Fort P (1997) The small GTPases Cdc42Hs, Rac1 and RhoG delineate Raf-independent pathways that cooperate to transform NIH3T3 cells. Curr Biol 7:629–637

Ruggieri R, Chuang YY, Symons M (2001) The small GTPase Rac suppresses apoptosis caused by serum deprivation in fibroblasts. Mol Med 7:293–300

Sahai E, Ishizaki T, Narumiya S, Treisman R (1999) Transformation mediated by RhoA requires activity of ROCK kinases. Curr Biol 9:136–145

Sahai E, Olson MF, Marshall CJ (2001) Cross-talk between Ras and Rho signalling pathways in transformation favours proliferation and increased motility. EMBO J 20:755–766

Sander EE, van Delft S, ten Klooster JP, Reid T, van der Kammen RA, Michiels F, Collard JG (1998) Matrix-dependent Tiam1/Rac signaling in epithelial cells promotes either cell-cell adhesion or cell migration and is regulated by phosphatidylinositol 3-kinase. J Cell Biol 143:1385–1398

Schmidt A, Hall A(2002) Guanine nucleotide exchange factors for Rho GTPases: turning on the switch. Genes Dev 16:1587–1609

Schnelzer A, Prechtel D, Knaus U, Dehne K, Gerhard M, Graeff H, Harbeck N, Schmitt M, Lengyel E (2000) Rac1 in human breast cancer: overexpression, mutation analysis, and characterization of a new isoform, Rac1b. Oncogene 19:3013–3020

Schurmann A, Mooney AF, Sanders LC, Sells MA, Wang HG, Reed JC, Bokoch GM (2000) p21-activated kinase 1 phosphorylates the death agonist bad and protects cells from apoptosis. Mol Cell Biol 20:453–461

Sewing A, Wiseman B, Lloyd AC, Land H (1997) High-intensity Raf signal causes cell cycle arrest mediated by p21Cip1. Mol Cell Biol 17:5588–5597

Shaw RJ, Paez JG, Curto M, Yaktine A, Pruitt WM, Saotome I, O'Bryan JP, Gupta V, Ratner N, Der CJ et al (2001) The Nf2 tumor suppressor, merlin, functions in Rac-dependent signaling. Dev Cell 1:63–72

Sinicrope FA, Gill S (2004) Role of cyclooxygenase-2 in colorectal cancer. Cancer Metastasis Rev 23:63–75

Somlyo AV, Bradshaw D, Ramos S, Murphy C, Myers CE, Somlyo AP (2000) Rho-kinase inhibitor retards migration and in vivo dissemination of human prostate cancer cells. Biochem Biophys Res Commun 269:652–659

Sun J, Qian Y, Hamilton AD, Sebti SM (1998) Both farnesyltransferase and geranylgeranyltransferase I inhibitors are required for inhibition of oncogenic K-Ras prenylation but each alone is sufficient to suppress human tumor growth in nude mouse xenografts. Oncogene 16:1467–1473

Suwa H, Ohshio G, Imamura T, Watanabe G, Arii S, Imamura M, Narumiya S, Hiai H, Fukumoto M (1998) Overexpression of the rhoC gene correlates with progression of ductal adenocarcinoma of the pancreas. Br J Cancer 77:147–152

Tang Y, Zhou H, Chen A, Pittman RN, Field J (2000) The Akt proto-oncogene links Ras to Pak and cell survival signals. J Biol Chem 275:9106–9109

Van Golen KL, Davies S, Wu ZF, Wang Y, Bucana CD, Root H, Chandrasekharappa S, Strawderman M, Ethier SP, Merajver SD (1999) A novel putative low-affinity insulin-like growth factor-binding protein, LIBC (lost in inflammatory breast cancer), and RhoC GTPase correlate with the inflammatory breast cancer phenotype. Clin Cancer Res 5:2511–2519

Van Golen KL, Wu ZF, Qiao XT, Bao LW, Merajver SD (2000) RhoC GTPase, a novel transforming oncogene for human mammary epithelial cells that partially recapitulates the inflammatory breast cancer phenotype. Cancer Res 60:5832–5838

Vidal A, Millard SS, Miller JP, Koff A (2002) Rho activity can alter the translation of p27 mRNA and is important for RasV12-induced transformation in a manner dependent on p27 status. J Biol Chem 277:16433–16440

Wang TC, Cardiff RD, Zukerberg L, Lees E, Arnold A, Schmidt EV (1994) Mammary hyperplasia and carcinoma in MMTV-cyclin D1 transgenic mice. Nature 369:669–671

Weber JD, Hu W, Jefcoat SC Jr, Raben DM, Baldassare JJ (1997) Ras-stimulated extracellular signal-related kinase 1 and RhoA activities coordinate platelet-derived growth factor-induced G1 progression through the independent regulation of cyclin D1 and p27. J Biol Chem 272:32966–32971

Welsh CF, Roovers K, Villanueva J, Liu Y, Schwartz MA, Assoian RK (2001) Timing of cyclin D1 expression within G1 phase is controlled by Rho. Nat Cell Biol 3:950–957

Wennerberg K, Der CJ (2004) Rho-family GTPases: it's not only Rac and Rho (and I like it). J Cell Sci 117:1301–1312

Westwick JK, Lambert QT, Clark GJ, Symons M, van Aelst L, Pestell RG, Der CJ (1997) Rac regulation of transformation, gene expression, and actin organization by multiple, PAK-independent pathways. Mol Cell Biol 17:1324–1335

Wheelock MJ, Johnson KR (2003) Cadherin-mediated cellular signaling. Curr Opin Cell Biol 15:509–514

Wolf RM, Draghi N, Liang X, Dai C, Uhrbom L, Eklof C, Westermark B, Holland EC, Resh MD (2003) p190RhoGAP can act to inhibit PDGF-induced gliomas in mice: a putative tumor suppressor encoded on human chromosome 19q13.3. Genes Dev 17:476–487

Woods D, Parry D, Cherwinski H, Bosch E, Lees E, McMahon M (1997) Raf-induced proliferation or cell cycle arrest is determined by the level of Raf activity with arrest mediated by p21Cip1. Mol Cell Biol 17:5598–5611

Wu M, Wu ZF, Kumar-Sinha C, Chinnaiyan A, Merajver SD (2004) RhoC induces differential expression of genes involved in invasion and metastasis in MCF10A breast cells. Breast Cancer Res Treat 84:3–12

Xiao GH, Beeser A, Chernoff J, Testa JR (2002) p21-activated kinase links Rac/Cdc42 signaling to merlin. J Biol Chem 277:883–886

Yu Q, Geng Y, Sicinski P (2001) Specific protection against breast cancers by cyclin D1 ablation. Nature 411:1017–1021

Zhuge Y, Xu J (2001) Rac1 mediates type I collagen-dependent MMP-2 activation. role in cell invasion across collagen barrier. J Biol Chem 276:16248–16256

Zondag GC, Evers EE, ten Klooster JP, Janssen L, van der Kammen RA, Collard JG (2000) Oncogenic Ras downregulates Rac activity, which leads to increased Rho activity and epithelial-mesenchymal transition. J Cell Biol 149:775–782

Zuckerbraun BS, Shapiro RA, Billiar TR, Tzeng E (2003) RhoA influences the nuclear localization of extracellular signal-regulated kinases to modulate p21Waf/Cip1 expression. Circulation 108:876–881

Developmental Biology of Fibroblasts and Neoplastic Disease

D. Brouty-Boyé

1 Introduction

Multiple genomic anomalies such as mutations, suppressions, and gene amplifications are believed to be essential intracellular events that transform a certain tissue independently of its surrounding cellular and biochemical microenvironment. However, tumors are not simply a cluster of transformed cells but, like their normal counterparts, are embedded in a complex supportive connective tissue. Earlier reports eloquently analyzed by van den Hooff in 1988, indicated that connective tissue could be involved in tumor development. These studies postulated a continuous network of strong interdependencies between transformed cells and the supportive components – cells and extracellular matrix – during tumor pathogenesis (van den Hooff 1986). One of the connective tissue cell types predicted to play a key role in this network is the *fibroblast*. During the past two decades, significant progress has been made in regard to fibroblast biology. There is now ample evidence to consider them active partners and not just accessory cells in tumorigenesis, and also in many non-neoplastic diseases. This new perception finds ample support as more studies emphasize the multiple and adaptable potentialities of adult tissue fibroblasts and how their interventions influence cellular processes.

2 What Are Fibroblasts?

Fibroblasts – confusingly called mesenchymal cells due to their embryonic tissue origin – are the connective tissue cells that elaborate the physical support of other tissues: epithelia/parenchyma, myeloid tissues, capillaries, nerves, and muscle cells. Fibroblasts also form capsules and membranes

D. Brouty-Boyé
INSERM U602, Institut André Lwoff, Hôpital Paul Brousse, 12 avenue Paul Vaillant-Couturier, 94807 Villejuif cedex, France

Progress in Molecular and Subcellular Biology
Alvaro Macieira-Coelho (Ed.)
Development Biology of Neoplastic Growth

that surround organs, and the ligaments and tendons that bind bones to each other or to muscles. They are also known for their capacity to give rise to the special bone and cartilage cells that support body organs as well as the fat-storing cells (or fat tissues) that not only cushion body organs but also insulate them and provide reserve energy. Therefore, while from a strict basic supportive role, fibroblast is a designation that belies diversity, it is seen that much structural and functional diversity indeed depends on this cell.

2.1 Matrix Diversity in Situ

For support, fibroblasts produce substantial quantities of extracellular matrix (ECM). *Collagen* is the most abundant ECM protein that fibroblasts synthesize under distinct forms according to the type of tissue they support, surround, and hold in place. Type I collagen is the most widely distributed; it is the bone matrix and scar tissue. It has great tensile strength and is the main component of ligaments and tendons. Type II is the collagen of articular cartilage. Type III is the reticular connective tissue that holds blood cells in place in hematopoietic tissues. It is also the granulation tissue when tissue heals by repair. Abundance of collagen deposition – *fibrosis* – is a hallmark of chronically activated tissues developing in response to previous tissue damage or disease and in most epithelial tumors, where an additional type V is frequently present in high amounts.

Fibronectin is another major ECM component that fibroblasts produce. It is distributed in all connective tissue and is specialized in the mediation of a wide variety of cellular interactions with the matrix. Indeed, the macromolecule contains a certain number of functional domains each with a high affinity for a variety of substrates – collagen/gelatin, heparan sulfate, and fibrin – and specific cell adhesion surface receptors, namely, integrins (Pankov and Yamada 2002). Alternative RNA splicing of the primary transcript from the single fibronectin gene results in the generation of as many 20 distinct isoforms. The investigation of isoform tissue specificity shows that the alternatively spliced fibronectin isoform (FNi) containing the extra-domain A (ED-A) is present in most adult healthy tissues, whereas tumoral tissues preferentially produce the ED-B FNi that is naturally present in fetal tissues. It is not, however, confined to tumoral tissues but also seen in the neighboring connective tissues close and distant to tumors, and in a variety of non-tumoral pathologies all associated with chronic tissue remodeling (Kaczmarek et al. 1994; Berndt et al. 1995). Although fibroblasts appear as one of the preferential sources of ED-B FNi, it is also seen in the invasive front of tumors and associated with the revascularization process occurring under tumoral and physio-pathological circumstances (Koshmel et al. 1999; Kriegsmann et al. 2004). ED-B FNi appears, thus, generally associated with the migratory activity of diverse cell types in diverse tissue-remodeling processes.

It is now well understood that ECM turnover relies, in part, on a family of zinc-dependent proteinases, the matrix metalloproteinases (MMPs) that fibroblasts unselectively produce in variable amounts in all types of tissues. MMPs were initially recognized for their capacity of collectively degrading all kinds of ECMs during tissue remodeling and also processing a number of bioactive macromolecules. The proteolytic ECM degradation results in the liberation of peptides, *matrikines*, that constitute new signals for the surrounding cells (Pasco et al. 2004). Recently, Schor's group (Schor and Schor 2001; Schor et al. 2003) identified for the first time a matrikine generated genetically from fibronectin and able to regulate cell activity by cryptic, cytokine-like bioactivities not expressed by the intact molecule. It was termed migration-stimulating factor (MSF) for promoting the migration of fetal fibroblasts into native collagen-type I gel substrate. It is produced not only by fetal fibroblasts but also by fibroblasts from cancer patients. MSF is shown to be abundantly present in tumors and it localizes in tumor cell clusters, tumor-associated microvascular endothelial cells and fibroblasts. This distribution pattern is actually similar to that for ED-B FNi and consistent with the notion that ED-B FNi expression is associated with migratory cell functions during embryonic and fetal development and induced in tissue changes associated with wound healing and solid tumors.

2.2 Diversity of Fibroblast Phenotypes in Situ

Quiescent fibroblasts are predominant during adult life. They are histologically characterized by a condensed nuclear chromatin and scant pale staining cytoplasm. When activated, fibroblasts are identifiable by their spindled shape or wide cytoplasmic processes, ample basophilic cytoplasm, and prominent nucleoli. Apart from morphological and histological features, fibroblasts lack specific markers, which certainly contributes to the long held notion of them as a uniform class of cells.

2.2.1 Cytoskeleton Composition

Characterization of cytoskeletal *intermediate filament* (IF) protein is actually the most common way to distinguish the diverse cell types present in a tissue. The basis is that these proteins are cell type-specific and the expression in tumor cells closely parallels that of the normal precursor cell. Keratin-containing IFs are present in epithelial cells and composed of at least 20 distinct proteins with tissue-specific expression patterns. Desmin filaments are present in striated and smooth muscle cells. Glial acid fibrillar protein (GAFP) is present in astrocytes and other glial cells, whereas neurofilament protein (NFP) is distributed in neurons. The IF protein vimentin is contrastingly more widely distributed and present not only in fibroblasts but also in all other mesenchyme-derived cells – vascular endothelial cells and blood leucocytes – and in neuroectoderm-derived cells as well.

Despite their biochemical diversity, IFs play similar roles in providing intracellular elastic support and maintaining shape in cells. They are prominent at sites where cells are subjected to mechanical restraint such as in epithelial and cardiac muscle cells, along cellular extensions in axons and dendrites and throughout smooth muscle cells globally, to transmit and dispatch tensile stresses throughout the cells.

Microfilaments are another class of cytoskeletal elements. They are made up of actin and provide the strength for cell to move and generate contractile forces, commonly in association with myosin-containing filaments. There are six distinct forms of actin proteins: two present in microfilaments of every kind of cells and four restricted to cells from smooth, striated, and cardiac muscle.

Antibodies to the distinct actins combined with IF proteins have significantly improved fibroblast typing and help identify muscle actin-expressing fibroblasts, the *myofibroblasts* (MFs). In 1971 Gabbiani first described these specialized fibroblasts during wound healing in granulation tissues on the basis of well-developed microfilament bundles whose ultrastructural organization resembled that of muscle cells (Gabbiani et al. 1971). By using monoclonal antibodies to various IF proteins together with muscle cell-specific actin isoforms Gabbiani's group subsequently identified four distinct types of MFs: the »V-type« that contains vimentin (V) and ubiquitous actin microfilament bundles arising in early formed granulation tissue and actually proposed as potential precursor of the other types; the »VA-type« with additional alpha-smooth muscle actin (ASMA); the »VD-type« with muscle desmin (D); and the »VAD-type« with a mixture of vimentin, ASMA, and desmin. The last three types of MFs are shown to abundantly accumulate during the phase of wound contraction (Skalli et al. 1989). Many ASMA MFs also acquire additional muscular proteins, i.e., smooth muscle (SM) myosin heavy chains, SM22, tropomyosin, and caldesmon. ASMA is actually the most common protein used for identifying muscle-like fibroblasts.

These in situ observations support the original Gabbiani's concept that fibroblasts bearing muscular proteins are the source of the force producing wound contraction and eventually connective tissue retraction during diverse fibrotic processes, including tumoral development. Indeed, the VASMA-type MFs – commonly designated MFs – are shown to irreversibly develop in variable amounts in chronic fibrosis from diverse organs and in the neighboring tissues of most epithelial tumors, notably those from breast (Schürch et al. 1981; Sappino et al. 1990).

Apart from pathological tissues, muscular fibroblasts are also naturally present in tissue locations – lung alveolar and testis septa, kidney capsule, intestinal pericrypts, and uterine mucosa – where a certain degree of tension is needed for the tissue to function. The muscular protein content of fibroblasts appears well adjusted to the tension forces they have to exert in the tissue: more muscle proteins are present where more strength and contractile force are required. Very recently, Gabbiani's group has well clarified the functional role of ASMA in microfilaments (Tomasek et al. 2002; Hinz and Gabbiani 2003). They elegantly demonstrated in a series of studies that mechanical stress is a prerequisite condition for microfilament bundles to express ASMA. Acquisition of ASMA but not other muscle actins permits fibroblasts to increase their contractility. ASMA neo-synthesis is triggered by the mechanical tension caused by the newly formed collagen–fibronectin-rich ECM that is transmitted through specific membrane junctions where ECM and microfilaments interconnect. Reciprocal modulations of ASMA content in microfilaments and ECM products provide thus a means for a nonmuscle cell to regulate contractile activity without changing shape. Upon stress release, ASMA neo-synthesis stops and fibroblasts relax.

2.2.2
Functionalities of Cytoskeletal Proteins

An important notion emerging from the studies on cytoskeleton is that cytoskeletal proteins are not merely cell type-specific markers but functional proteins. They regulate the organization of filaments as well as the dynamic interactions between filaments during cellular events associated with cell reshaping and movement. For instance, the IF protein nestin, originally described as a predominant neuronal stem/progenitor cell marker (Messam et al. 2000), is actually more widely distributed than first suspected. Its expression normally absent during adult life is activated in response to injury in diverse tissues and under regenerating conditions in association with neovascuralization (Klein et al. 2003). Nestin actually regulates IF structure during the mitotic process (Chou et al. 2003). ASMA protein, which is absent in motile fibroblasts, serves as a brake for slowing down cell movement (Ronnov-Jessen and Peterson 1996).

There is another contractile cell type – the myoepithelial (ME) cell – present in mammary, sweat, and salivary secretory glands. ME cells lie between the secretory luminal epithelial cells and the investing basement membrane (BM) that separates fibroblasts from epithelial cells. They make up the BM by producing its specific components, collagen type IV and the cell adhesion-related macromolecule laminin. They also contribute to maintaining the polarity of luminal cells and expelling the secretory products. ME cells carry both an epithelial (polarity, tight cell junction, keratins K14 and K17 in IFs) and fibroblastic (V and ASMA proteins) phenotype. They are absent in most invasive breast tumors whereas they are present in non-invasive tumors consistent with the notion that they mechanically help restrain growth of luminal epithelial cells. The extensive work of Peterson's group (Peterson et al. 2001) on human mammary epithelium lends suspicion to a possible ME cell affiliation of MFs, at least for some of them, in breast tumors.

3 Evidence of Fibroblast Diversity in a Two-Dimensional (2-D) Culture Model

Studying fibroblasts in cultures ex vivo largely substantiates the notion that fibroblasts from different locations present distinct traits and functionalities. Cell culturing in 2-D is a standard procedure for most research purposes and particularly well suited for studying »activated« fibroblasts. Indeed, tissue culture plastic substrate closely replicates the *mechanical stress* that activates fibroblasts in situ. Rigid substrates triggers the development of bundles of actin microfilaments – *stress fibers* – (Fig. 1) that provides the force for fibroblasts to adhere, move, proliferate, fill the surface, and return to a resting state. Varying the substrate's rigidity and adhesive properties are known to have profound impact on fundamental cellular processes, i.e., cell shape, growth, differentiation, and tumorigenicity (Brouty-Boyé et al. 1980).

In the search for distinct, specific characteristics associated with fibroblasts, we started a strategy as far back as 15 years ago by investigating the phenotypic and functional changes in human fibroblast isolates from various anatomic locations and pathological status in culture ex vivo (Table 1). The study was based on screening a broad spectrum of functional proteins related to cytoskeleton, matrix, adhesion, communication, growth, inflammation, and cancer.

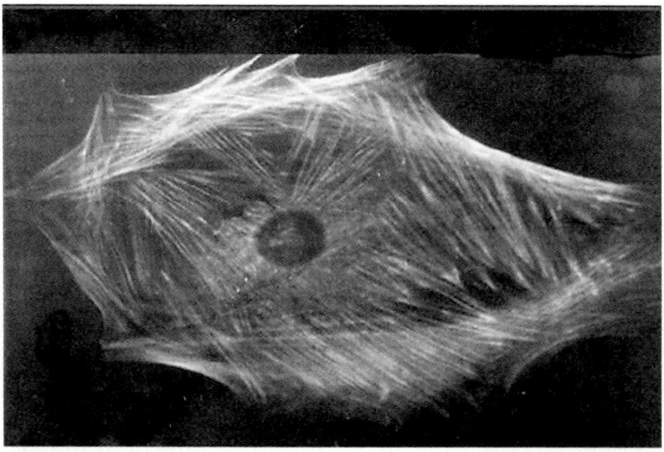

Fig. 1. Development of stress fibers in human fibroblasts adherent to rigid culture plastic

Table 1. Tissue origin and pathological status of fibroblast isolates

Tissue type	Clinical status	Histopathology
Breast	Normal	Normal
Spleen	Normal	Normal
Femoral bone marrow	Non-pathological	Normal
Knee synovium	Non-pathological	Normal
Distant to breast tumor	Non-tumoral	Normal
Distant-to-liver metastasis[a]	Non-tumoral	Normal
Adjacent to breast tumor	Non-tumoral	Normal
Irradiated breast skin	Fibrosis	Fibrosis
Breast tumor[b]	Benign	Normal
Breast carcinoma[c]	Malignant	Tumoral
Spleen	Myeloid metaplasia	Tumoral

[a] Mostly colon metastasis
[b] Fibroadenoma, fibrocystic lesions, and sclerosing adenosis
[c] Ductal carcinoma of grade II and III and lobular carcinoma (detailed in Brouty-Boyé and Magnien 1994)

3.1 Differences Associated with Cytoskeletal and ECM Components

Fibroblast isolates from most normal healthy organs predominantly show an ordinary V-type phenotype upon activation on rigid substrate. They form typical stress fibers and produce an EDA-FNi-rich ECM. There are occasionally a few ASMA cells that reflect the natural propensity of fibroblasts for acquiring the ASMA phenotype (Fig. 2). Moreover, increased numbers of ASMA

Fig. 2. Presence of a few ASMA MFs in culture of normal breast-derived fibroblasts

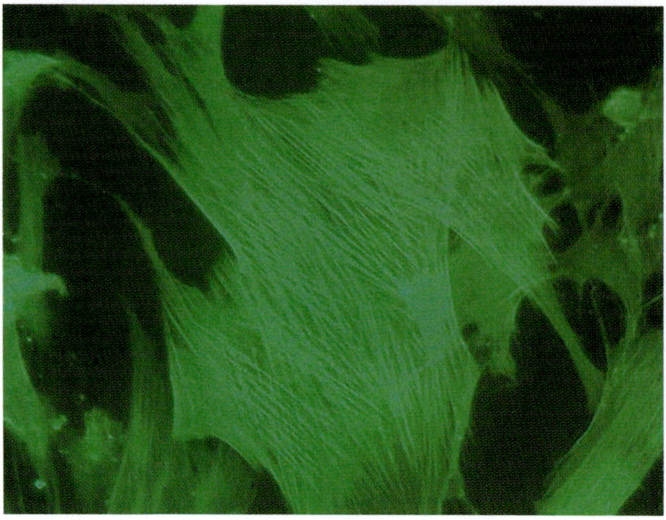

Fig. 3. Homogeneous staining of fibroblasts isolated from human spleen with monoclonal antibody to ASMA

cells generally develop in further cell subcultures on plastic even after cloning (Desmoulière et al. 1992).

There are, however, exceptions – spleen and liver isolates – where nearly all fibroblasts demonstrate the ASMA phenotype straightaway (Fig. 3). Rare desmin-expressing cells are occasionally seen, especially in spleen isolates, but they usually rapidly vanish with further cell subculturing. When the profile of SM proteins is analyzed, splenic ASMA MFs appear as one of the strongest muscle cell type as compared with others from different types of tissues (Charbord, pers. comm.).

Analysis of cultured fibroblast isolates from pathological (non-tumoral and tumoral) tissues confirms the occurrence of new ASMA MFs seen in situ. They are, however, generated in variable proportions from all kind of fibrotic tissues, except in tumor isolates where they consistently appear in much higher numbers. This is documented in fibroblast isolates from epithelial tumor and myelometaplasic tissue as well (Brouty-Boyé et al. 1991a,b, Brouty-Boyé and Magnien 1994, Brouty-Boyé et al. 1998). In all cases, ASMA MFs predominantly produce ED-B FNi-rich ECM (Fig. 4).

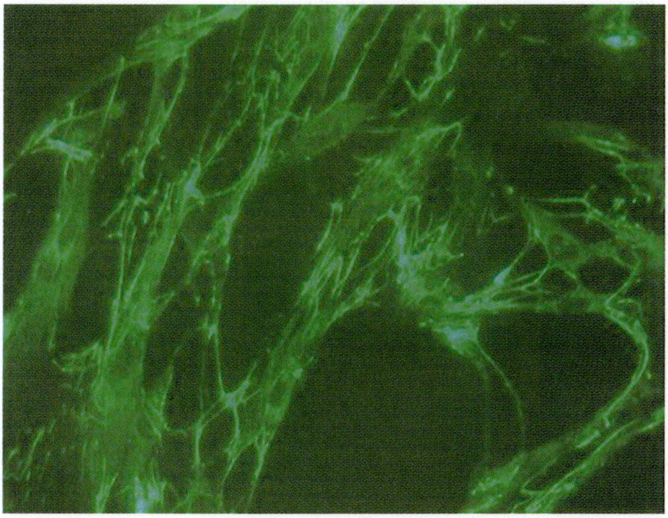

Fig. 4. Production of extracellular EDB-fibronectin in fibroblasts isolated from tissue distant to breast carcinoma

3.2
Expression of Other Non-Mesenchymal Cytoskeletal Proteins

Fibroblasts unexpectedly and differentially express other cell lineage cytoskeletal IF proteins. For instance, keratin-containing IFs are present in cultured MFs from spleen and liver (Fig. 5a), but are consistently absent in fibroblasts from most other locations. In contrast, nearly all types of fibroblasts contain the neural IF proteins GFPA and nestin (Brouty-Boyé, unpubl.; Fig. 5b,c).

It is only recently that cytokeratin K8-expressing MFs have been identified in the human spleen (Steiniger and Barth 2000; Steiniger et al. 2001). They are seen to localize in the extensive area between the white pulp and the red pulp in replacement of missing sinus lining. As numerous lymphocytes are seen accumulating, MFs may serve attract lymphocytes and guide them from the blood into the white pulp. In contrast, MFs are absent from resting liver and only seen apposed to newly formed hepatocytes during liver regeneration (Friedman 2000). Similarly to splenic MFs, migratory cell guidance is perhaps one cellular function in which the new MFs emerging during liver regeneration are involved. Indeed, they are shown to direct the migration of proliferating hepatocytes along a linear track in an experimental reconstituted coculture model (Brouty-Boyé, unpubl.; Fig. 6).

Fig. 5. Homogeneous staining of liver-derived fibroblasts with monoclonal anti-body to **a** pan-cytoK, **b** GFAP, and **c** nestin

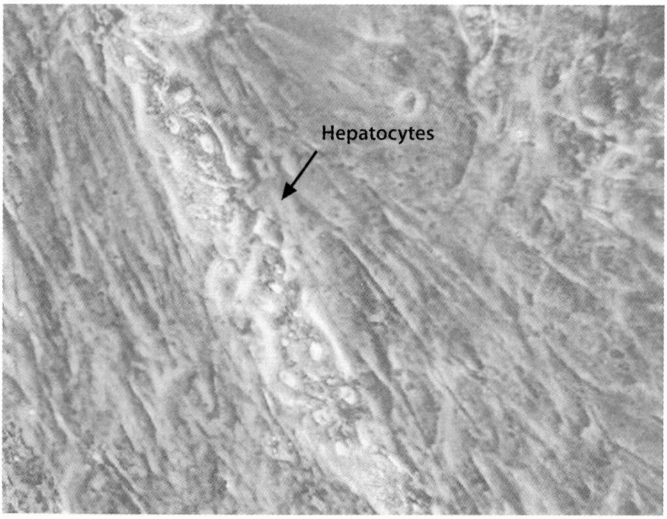

Fig. 6. Three-week old culture of human hepatocytes with liver-derived fibroblasts. *Arrow* shows lining of hepatocyte progeny expanding onto the fibroblastic monolayer

3.3 Behavioral Dissimilarities Among MFs

The new MFs generated from different pathological tissues do not behave the same way in culture. For instance, the ASMA/ED-B MFs generated from different kinds of non-tumoral fibrotic tissue persist for several cell generations, as actually do natural ASMA/ED-A MFs. In contrast, MFs from tumoral tissues do not persist as long, and are even consistently associated with poor cell survival ex vivo when derived from the tumor itself, especially from breast tumor. The environmental dependency of pathological fibroblasts to maintain ASMA/ED-B FNi expression seems to rely on a complex array of cytokines. For instance, TGFβ, FGF-2, and GM-CSF are able to prolong the ASMA/ED-B phenotype in culture, whereas IL-2 has an opposite effect (Brouty-Boyé and Magnien 1994; Liu and Brouty-Boyé 1994). The cytokine modulations do not necessarily require stimulatory growth effects and certainly have less efficiency in tumoral fibroblasts.

Dissimilarities are also evident in the way the diverse types of MFs interact with tumor cells. Indeed, fibroblasts from non-tumoral and tumoral breast tissues affect differently the transplant of breast tumor cells while, in turn, tumor cells influence fibroblast growth differently, depending on tissue source (Brouty-Boyé and Raux 1993; Adam et al. 1994). These studies are consistent with the notion of a strong interdependency between epithelial cells and fibroblasts during tumor development.

3.4 Diversity Associated with Cell Surface and Secreted Molecules

The conventional belief that fibroblasts are rather passive and innocuous cells led many to focus their studies on paracrine signals elaborated by other cell types during pathological processes, disregarding the fibroblasts' own contribution. Cultured fibroblasts do actually express and produce a broad spectrum of surface and secreted molecules that underline the various ways by which fibroblasts are able to communicate and directly interact with diverse cell types, including hematopoietic, immune/inflammatory, vascular, and epithelial cells (Fig. 7) The diversity of these molecules indicates that fibroblasts interfere in several basic processes including cell-type-specific proliferation/differentiation, inflammation, angiogenesis, tissue remodeling, occurring under physio-pathological circumstances including tumor development. In addition, the differential expression of several groups of molecules seen in fibroblasts from different anatomic sites reflects an expression pattern adjusted to the specific functions of the organ to which the fibroblasts belong. Indeed, in experimental reconstituted coculture models, fibroblasts isolated from bone marrow are shown to selectively promote the proliferation/differentiation of bone marrow-derived myeloid progenitors into myeloid mono-granulocytic lineage cells, whereas fibroblasts from the spleen selectively drive the differentiation of circulating blood progenitors into lymphoid natural-killer

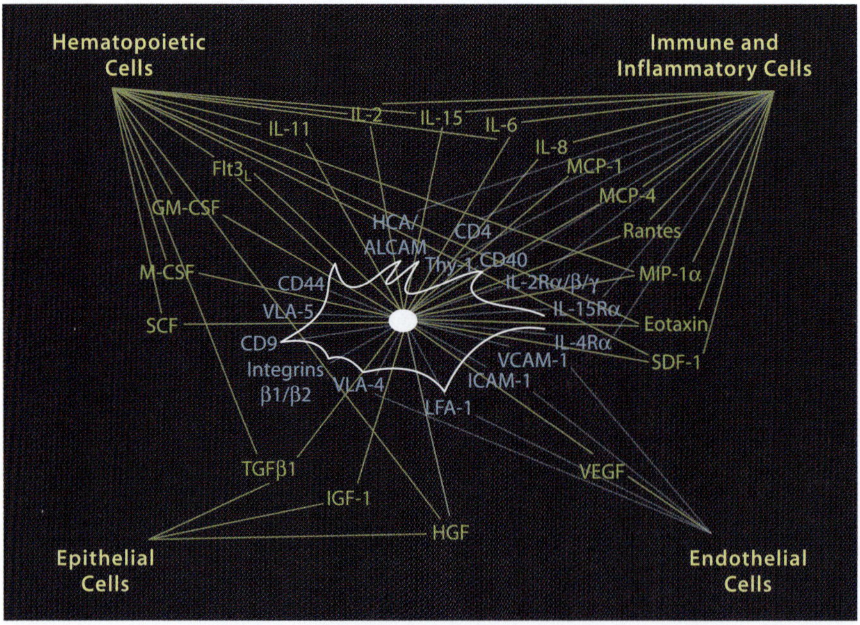

Fig. 7. Network of surface and secreted molecules produced in human tissue-derived fibroblasts

cells (Brouty-Boyé et al. 2001; Briard et al. 2002). Accordingly, liver-derived fibroblasts do not promote hematopoietic cell proliferation/differentiation, but contribute to the proliferation of isolated hepatocytes while directing their spatial organization (Brouty-Boyé, unpubl.).

There are qualitative and quantitative differences in diverse groups of molecules in pathological tissue-derived fibroblasts compared with normal ones. For instance, the chemokine repertoire of fibroblasts is selectively modulated by the pathologic state of the tissue (Brouty-Boyé et al. 2002). As chemokines determine the type of immune/inflammatory cells that are attracted to the reactive site, the nature and the magnitude of leukocyte recruitment may be severely affected during the immune/inflammatory response to tissue damage. However, these differences may not all be associated with the pathological status of the tissue source but rather under the influence of non-pathological parameters, i.e., cell structure, metabolism, growth rate, and isolation and culture conditions.

4 Fibroblasts from Non-Tumoral and Tumoral Breast Tissue

4.1 Quantitative Variation of Gene Expression

The consistency of MF phenotype and behavioral changes that fibroblasts present during pathological situations let us screen in the early 1990s a small number of transcripts known to be generally related to growth control, inflammation, and cancer in fibroblasts from different sources of breast tissues (Spanakis and Brouty-Boyé 1995). The intention was not to discover underlying biological mechanisms or create an inventory of gene expression but to estimate the amounts of changes that fibroblasts undergo during breast pathogenesis and the kind of genes involved. The quantitative multivariate analysis of the variance (MANOVA) of gene expression we used in this pilot study was based on the principle that the expression of a gene may depend simultaneously on several interrelated biological factors. Therefore, the level of a given transcript is considered as a complex function of multiple continuous and discrete biological variables, i.e., structure, growth, antigenicity, pathology – and of the combination of these variables.

Multiple differences in gene expression among normal, non-tumoral pathological, and tumoral fibroblasts are the most striking result of this study. The differences in the transcript level of groups of genes are actually found to correlate with biological variations; for instance, the rate at which cells grow (evaluated by the doubling time of the cell populations) or the ASMA MF content (defined by the percentage of ASMA-positive cells and ED-B expression level). Once these theoretically reversible and not necessarily pathological variations are removed, it appears that fibroblasts from normal and path-

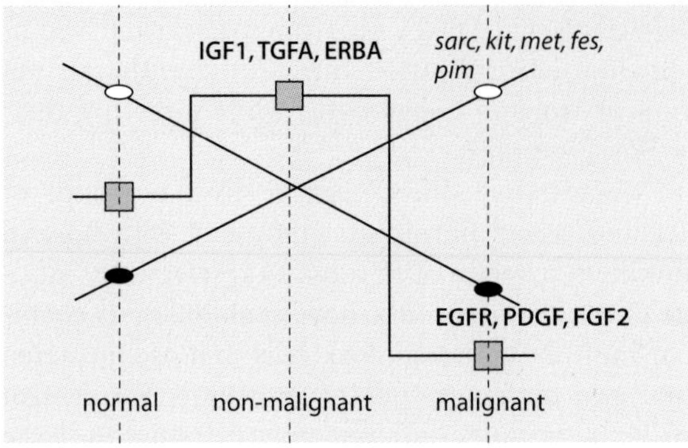

Fig. 8. Variations of gene expression in breast fibroblasts during tumor development. Depicted are the transcripts consistently over- or underexpressed in malignant tumor-derived fibroblasts consequently to progressive (*circles*) and discrete (*square*) variations in gene expression

ological breast tissues present, in any case, multiple irreversible (genetic) differences in gene expression. The differences may actually arise from continuous and discrete changes in gene expression throughout tumor development (Fig. 8). This implies that tumoral or normal fibroblasts/MFs are not fixed entities but extremely heterogeneous classes of multi-gene phenotypes. Cells, cell populations, and tissues may be visualized as the result of gene-product pools the content of which change quantitatively and qualitatively at all times, instead of by single (or more) irreversible genetic change(s) that spurs tumor development.

4.2 Discrimination of Fibroblast Subtypes by Multivariate Analysis of Gene Expression

Discriminant analysis of quantitative variations of gene expression provides a useful method for determining the phenotypic relation among fibroblasts from different classes of tissue (Spanakis and Brouty-Boyé 1997). Merging dendograms of phenotypes reveal that fibroblasts from radiation fibrosis and non-tumoral fibrotic breast tissues present the greatest phenotypic similarity (Fig. 9). The next close relatives of fibroblasts from fibrotic tissues are those of neighboring tumor tissues or benign lesions. Moreover, skin breast fibroblasts appear closer to the mammary gland family than do malignant tumor fibroblasts. This study underlines the extraordinary phenotypic dissimilarity between malignant tumoral fibroblasts and the other pathological subtypes. The fibroblast heterogeneity may arise from peculiar traits that cells acquired to perform specific functions and/or from distinct precursor cell types.

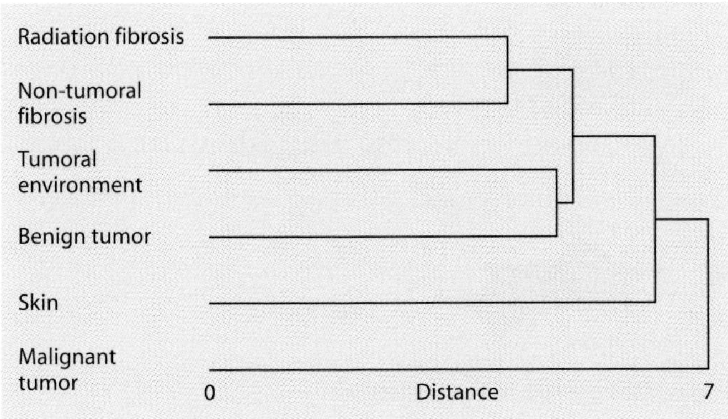

Fig. 9. Dendogram merging the phenotypically similar groups of fibroblasts according to the intercluster distance

Our analyses were performed, at the time, on a limited number of transcripts selected because of their known involvement in growth and differentiation control, inflammatory reaction, and in breast biology and/or cancer. The recent powerful technologies of cDNA micro-arrays and bioinformatics applied by Brown's group at Stanford on numerous and randomly selected genes definitely substantiate the phenotypic heterogeneity of fibroblasts from different body locations. The topographic diversity of fibroblasts is perhaps based on the same Hox gene code that determines the development of body parts during embryogenesis (Chang et al. 2002). Moreover, the transcriptome of fibroblasts physiologically activated by serum confirms their primacy as central communication operators, by implication in a repertoire of basic developmental morphogenetic processes, namely, cell direction, cell shape, cell movement, cell growth, cell death, and ECM synthesis (Iyer et al. 1999). Consistent with our previous pilot studies, it appears that reciprocal activations of a well-coordinated series of genes in fibroblasts and epithelial cells are occurring all through tumor development (Chang et al. 2004).

5 What Really Are Fibroblasts?

5.1 Current Opinions

The newly formed fibroblasts of major connective repair, pathological or not, are assumed to arise in large part from local ordinary fibroblasts. There is now accumulating evidence for other potential cellular sources (Fig. 10).

Indeed, fibroblasts may be recruited from the contractile microvascular *pericytes* while repairing connective tissue and smooth muscles of damaged

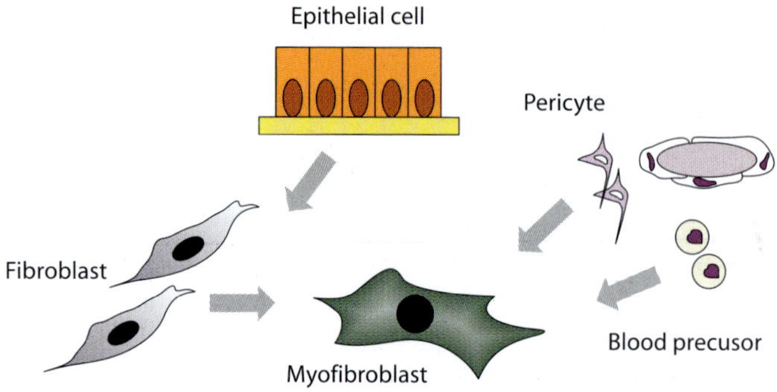

Fig. 10. Potential sources of MFs present in injured tissues

small blood vessels. In liver, for instance, the pericytes – *stellate cells* – present in sinusoids and veins, are able to express the MF phenotype once in culture. They are actually proposed as one major cellular source of ECM overproduction in fibrotic liver diseases (Friedman 2000).

Another alternative source may be progenitors circulating in the blood that are recruited at sites of the tissue damage. The exact identity of these circulating precursors, however, remains unclear. They may be a CD14 (+) monocyte subset designated at first *monocytoid cell* by Labat et al. in 1991 and later on *fibrocyte* by Bucala et al. in 1994. These blood-borne cells are able to generate ex vivo morphologically fibroblast-like cells. The expanded progeny carries a mixed macrophage- and fibroblast-like phenotype based on the expression of a certain number of blood cell CD antigens, vimentin protein in IFs, and on the production of collagen-fibronectin ECM. The fibrocytes are actually acting as multipotent stem cells, able to generate spontaneously or under appropriate inductive conditions ex vivo diverse phenotypes evoking mesenchymal lineage cells, i.e., chondroblasts, bone-resorbing osteoclasts, endothelial cells, macrophages, T blood cells, and ASMA MFs – and even cells from distinct lineage pathways, including epithelial cells, hepatocytes, and neuronal cells (Labat 2001; Metz 2003; Zhao et al. 2003). Recently, fibrocytes are reported to be present in tumors from the pancreas, breast, and cervix uterus. These findings were based, however, on the expression of a single marker, the hematopoietic progenitor cell antigen CD34 (Barth et al. 2002a–c). Recruited fibrocytes are also suspected to be a source of fibrogenic cells in chronic inflammatory diseases (Pilling et al. 2003).

The link between circulating fibrocytes and the recently identified mesenchymal stem cells (MSCs) present in bone marrow is unclear. The phenotype of circulating and even bone marrow MSCs, themselves, is actually poorly defined (Javazon et al. 2004). It is generally accepted that MSCs are devoid of hematopoietic and endothelial markers (CD11b, CD31, and CD45), express numerous surface adhesion molecules (CD44, CD49e/VLA-5, and CD62/E-selectin), and are positive for MHC-I. Variable expressions of other surface antigens, i.e., CD90/Thy-1, CD117/c-kit, SH2/CD105 or endoglin, SH3, SH4/CD73, and STRO-1, are also present. However, the MSC surface antigen repertoire is similarly present in all fibroblasts from various tissue sources and, in some cases, associated with mesenchymal differentiating potentialities (Fiorito et al. 2004). The most reliable biological criterion used to define MSCs is actually the capacity to give rise in culture to the three types of mesenchymal tissues: bone, cartilage, and adipose tissue. The persistence of antigenic and biological characteristics of MSCs in fibroblasts from body locations other than bone marrow strongly suggests a close relation with perhaps local MSCs.

Lately, many reviews highlight the possibility that local epithelia/parenchyma may be involved in the formation of new fibroblasts. Epithelial-mesenchymal transition (EMT) is actually an essential developmental mechanism in establishing the body plan of many multicellular organisms. The basis of the process is the disassembling of epithelial units and cellular reshaping. This implicates the loss of the epithelial phenotype, i.e., polarity, intercellular junctions, and matrix adherence, in association with new cytoskeletal organization, to globally facilitate cell movement and migration.

It is well acknowledged that EMT is part of the formation of the branching morphogenesis of lung, kidney, and mammary gland epithelium. It is not, therefore, a paradox to observe evidence of EMT in fibrosis developing in such organs during adult life. In kidney, it seems clear that not all fibroblasts generate MFs and that at least part of them originate from epithelial cells (Strutz and Müller 2000; Kalluri and Neilson 2003). EMT implication is also well recognized in the invasive and metastatic properties of tumoral cells (Thiery 2002). Consequently, tumoral EMT cells may be confusedly depicted by the misleading term »stromal MFs,« especially in the characterization of breast tumor microenviroment (Allinen et al. 2004). The existence of EMT may also help to explain the somatic mutations frequently detected in the tissues surrounding invasive breast tumors (Kurose et al. 2002).

5.2 Revisiting Fibroblast Diversity

The possibility that fibroblasts originate from precursors other than local ordinary fibroblasts is important to take into account in regard to the phenotypic and functional diversity of fibroblasts under particular circumstances and types of tissues. In spleen, for instance, MFs may have a closer relation to cells from the monocyte-macrophage lineage than previously thought. Their privileged interactions with surrounding blood cells and the specific constitutive expression of hematopoietic factor, i.e., the NK cell inducer IL-15, may well be imprints of a blood-cell precursor origin. In addition, the presence of epithelial keratins may reflect a yet undefined function of the proteins analogous to the ASMA content in microfilaments. In contrast, EMT may better help to explain the keratin content of liver-derived MFs that is revealed under activated conditions, i.e., cell culturing or liver regeneration in situ.

The presence of neural GAFP and nestin proteins in nearly all types of tissue-derived fibroblasts, bone marrow MSCs, and their circulating monocyte-like precursors as well (Labat et al. 2000; Woodbury et al. 2000) may suggest a common embryonic origin, the neural crest. There is now experimental evidence suggesting that adult stem cells are able to give rise to mature cells of a tissue different from the tissue in which they reside. For instance, it has been shown that neural stem cells are able to generate cells of blood, muscle, liver, lung, stomach, and intestine in culture (Vescovi et al. 2001). The reverse conversion is also well documented.

6 Attempts to Study Fibroblast-Epithelial Cell Interaction in a Novel Dimension

Although research on fibroblasts interacting with 2-D substratum has provided many insights into their functional diversity, this approach does not take in account a fundamental regulatory element, the 3-D architecture in which cells live in situ, where they are completely surrounded by matrix (Grinnell 2003). The functional interactions existing between fibroblasts, ECM and the other tissue cell types may not be equivalent and may be regulated differently in 3-D as opposed to 2-D.

In an early study (Brouty-Boyé et al. 1994), we used a rudimentary 3-D coculture model to investigate the direct influence of fibroblasts on tumor cell growth and differentiation in a more tumoral-tissue-like context (Beaupain 1999). The reconstituted tissue (Fig. 11a) was initially composed of tumor cells mixed with tumor-derived fibroblasts from the same anatomic site (breast). After a few days or weeks in culture, the tissue contained highly atypical layers of tumor cells with no evidence of organized epithelial structures (Fig. 11b). In contrast, when the tissue is formed with the same tumor cells but mixed with fibroblasts from normal breast tissue, there are numerous areas of organized structures that are bordered by typical polarized and mucus-secreting epithelial cells (Fig. 11c). One striking result is that fibroblast-mediated changes in epithelial structure take place at times fibroblasts are no longer present in the tissue in association with tissue growth. The specific mechanisms underlying epithelial cell growth and the restoration of normal epithelial morphology and secretory function were at that time only speculative, i.e., involving ECM composition, intercellular adhesion, and protease and/or growth factor secretion. Nevertheless, the data strengthened the perception of tumors as conditional phenotypes influenced or controlled by the close cellular environment of the tumor cells.

The work of Bissell together with Peterson's group (Bissell et al. 2003) expertly improved our understanding of how the contact of epithelial cells with surrounding matrix is important in disclosing the correct cell architecture that ultimately regulates epithelial functions. It was long known that ep-

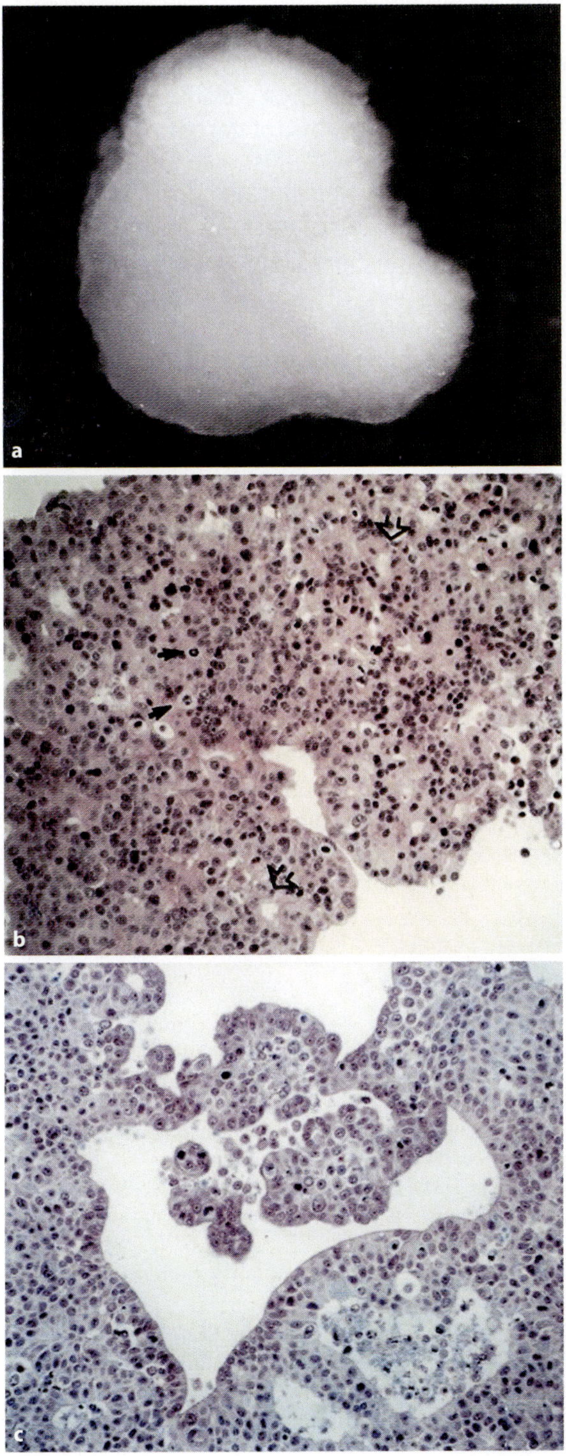

Fig. 11. Human breast tissue reconstituted in **a** 3-D culture system. Histological HES-stained section of a 10-day-old tissue reconstituted with tumor cells and fibroblasts from **b** tumoral and **c** normal breast tissue. Note in b mitoses (*solid arrows*) and pseudo-glandular-like formations (*open arrows*) compared with the full development of papillary and tubular structures bordered by polarized epithelial cells in c

ithelial cells do not survive well and function properly in standard 2-D culture, even in contact with matrix components. However, when implanted in a malleable BM component-rich matrix, epithelial cells are able to reconstitute functionally organized 3-D structures characteristic of the corresponding tissue they originate from (Myers et al. 2003). As well illustrated in Bissell's work, tumoral breast cells are unable to develop epithelial structures even before they become fully transformed. It is, however, possible to circumvent the architectural tumoral cell defect by using functional inhibitors of adhesion (β1 integrin) and growth factor (EGF) receptors, which individually restore the normal glandular structure and function (milk production).

It is striking that the substitution of fibroblasts of tumor type with normal fibroblasts permits the reversion of tumor cells toward normal mammary structures. The use of tissue fibroblasts therapeutically to repair defective microenvironment and to restore normal tissue function appears a promising approach. The current therapeutic strategies to cure diseased tissues focus on the use of marrow-derived mesenchymal or totipotent hematopoietic stem cells that have the potentialities to generate all kind of cell types. However, in contrast to marrow stem cells, tissue fibroblasts present properties and functions specific to the tissue they surround and hold in place. Such types of tissue adaptation may reveal fibroblasts as having more therapeutic effectiveness than marrow stem cells, and should thus be studied in more detail.

As pertinently emphasized by Jacks and Weinberg (2002), the study of cancer (and other tissue diseases) in two dimensions seems »quaint, if not archaic.« Three-dimensional cultures should help clarify the critical developmental interactions of cells with their neighbors and their microenvironment and how interventions influence cellular processes.

References

Adam L, Crépin M, Lelong J-C, Spanakis E, Israël L (1994) Selective interactions between mammary epithelial cells and fibroblasts in co-culture. Int J Cancer 59:262–268

Allinen M, Beroukhim R, Cai L, Brennan C, Lahti-Domenici J, Huang H, Portet D, Hu M, Chin L, Richardson A, Schnitt S, Sellers WR, Polyak K (2004) Molecular characterization of the tumor microenvironment in breast cancer. Cancer Cell 6:17–32

Barth PJ, Ramaswany A, Moll R (2002a) CD34+ fibrocytes in normal cervical stroma, cervical intraepithelial neoplasia III, and invasive squamous cell carcinoma of the cervix uteri. Virchows Arch 441:564–568

Barth PJ, Ebrahimsade S, Hellinger A, Moll R (2002b) CD34+ fibrocytes in neoplastic pancreatic lesions. Virchows Arch 440:128–133

Barth PJ, Ebrahimsade S, Ramaswamy A, Moll R (2002c) CD34+ fibrocytes in invasive ductal carcinoma in situ, and benign breast lesions. Virchows Arch 440:298–303

Beaupain R (1999) A method for three-dimensional coculture of cancer cells combined to any other type of cells maintained organotypically. Methods Cell Sci 21:25–30

Berndt A, Koshmel H, Mandel U, Gabler U, Luo X, Celeda D, Zardi L, Katenkamp D (1995) TGFβ and bFGF synthesis and localization in Dupuytren's disease (nodular palmar fibromatosis) relative to cellular activity, myofibroblast phenotype and oncofetal variants of fibronection. Histochem J 27:1014–1020

Bissell MJ, Rizki A, Mian IS (2003) Tissue architecture: the ultimate regulator of breast epithelial function. Curr Opin Cell Biol 15:753–762

Briard D, Brouty-Boyé D, Azzarone B, Jasmin C (2002) Fibroblasts from human spleen regulate NK cell differentiation from blood CD34 (+) progenitors via cell surface IL-15. J Immunol 168:4326–4332

Brouty-Boyé D, Magnien V (1994) Myofibroblast and concurrent ED-B fibronectin phenotype in human stromal cells cultured from non-malignant and malignant breast tissue. Eur J Cancer 30A:66–73

Brouty-Boyé D, Raux H (1993) Differential influence of stromal fibroblasts from different breast tissue on human breast tumour cell growth in nude mice. Anticancer Res 13:1087–1090

Brouty-Boyé D, Tucker RW, Folkman J (1980) Transformed and neoplastic phenotype: reversibility during culture by cell density and cell shape. Int J Cancer 26:501–507

Brouty-Boyé D, Raux H, Chraïbi-Hajji F, Magnien V, Sallé V (1991a) Sorting of subclasses of epithelial and fibroblastic cells in primary cultures of benign and malignant breast tissues by use of fluorescent probes monospecific for the cytoskeleton and the extracellular matrix. J Cell Pharmacol 3:103–110

Brouty-Boyé D, Raux H, Azzarone B, Tamboise A, Tamboise E, Béranger S, Magnien V, Pihan I, Zardi L, Israël L (1991b) Fetal myofibroblast-like cells isolated from post-radiation fibrosis in human breast cancer. Int J Cancer 47:697–702

Brouty-Boyé D, Mainguené C, Magnien V, Israël L, Beaupain R (1994) Fibroblast-mediated differentiation in human breast carcinoma cells (MCF-7) grown as nodules in vitro. J Int Cancer 56: 731–735

Brouty-Boyé D, Doucet C, Clay D, Le Bousse-Kerdilès M-C, Lampidis TJ, Azzarone B (1998) Phenotypic diversity in human fibroblasts from myelometaplasic and non-myelometaplasic hematopoietic tissues. Int J Cancer 76:767–773

Brouty-Boyé D, Briard D, Azzarone B, Le Bousse-Kerdilès M-C, Clay D, Pottin-Clémenceau C, Jasmin C (2001) Effects of human fibroblasts from myelometaplasic and non-myelometaplasic hematopoietic tissues on CD34+ stem cells. Int J Cancer 92:484–488

Brouty-Boyé D, Pottin-Clémenceau C, Doucet C, Jasmin C, Azzarone B (2002) Chemokines and CD40 expression in human fibroblasts. Eur J Immunol 30:914—919

Bucala R, Spiegel LA, Chesney J, Hogan M, Cerami A (1994) Circulating fibrocytes define a new leukocyte subpopulation that mediates tissue repair. Mol Med 1:71–81

Chang HY, Chi J-T, Dudoit S, Bondre C, van de Rijn M, Botstein D, Brown PO (2002) Diversity, topographic differentiation, and positional memory in human fibroblasts. Proc Natl Acad Sci USA 99:12877–12882

Chang HY, Sneddon JB, Alizadeh AA, Sood R, West RB, Montgomery K, Chi JT, van de Rijn M, Botstein D, Brown PO (2004) Gene expression signature of fibroblast serum response predicts human progression: similarities between tumors and wound. PLS Biol 2:1–9

Chou Y-H, Khuon S, Herrmann H, Goldman RD (2003) Nestin promotes the phosphorylation-dependent disassembly of vimentin intermediate filaments during mitosis. Mol Biol Cell 14:1468–1478

Desmoulière A, Rubbia-Brandt L, Abdiu A, Walz T, Macieira-Coelho A, Gabbiani G (1992) Alpha-smooth muscle actin is expressed in a subpopulation of cultured and cloned fibroblasts and is modulated by gamma-interferon. Exp Cell Res 20:64–73

Fiorito S, Magrini L, Adrey J, Maihlé D, Brouty-Boyé D (2005) Inflammatory status and cartilage regenerative potential of synovial fibroblasts from patients with osteoarthritis and chondropathy. Rheumatol (44:164–171)

Friedman SL (2000) Molecular regulation of hepatic fibrosis, an integrated cellular response to tissue injury. J Biol Sci 275:2247–2250

Gabbiani G, Ryan GB, Majno G (1971) Presence of modified fibroblasts in granulation tissue and their possible role in wound contraction. Experientia 27:549–550

Grinnell F (2003) Fibroblast biology in three-dimensional collagen matrices. Trends Cell Biol 13:264–269

Hinz B, Gabbiani G (2003) Mechanisms of force generation and transmission by myofibroblasts. Curr Opin Biotech 14:538–546

Iyer VR, Eisen MB, Ross DT, Schuler G, Moore T, Lee JCF, Trent JM, Staut LM, Hudson J Jr, Boguski MS, Lashkari D, Shalon D, Botstein D, Brown PO (1999) The transcriptional program in the response of human fibroblasts to serum. Science 283:83–87

Jacks T, Weinberg RA (2002) Taking the study of cancer cells to a new dimension. Cell 111:923–925

Javazon EH, Beggs KJ, Flake AW (2004) Mesenchymal stem cells: paradoxes of passaging. Exp Hematol 32:414–425

Kaczmarek J, Castellani P, Nicolo G, Spina B, Allemani G, Zardi L (1994) Distribution of oncofetal fibronectin isoforms in normal, hyperplasic and neoplastic human breast tissues. Int J Cancer 59:11–16

Kalluri R, Neilson EG (2003) Epithelial-mesenchymal transition and its implications for fibrosis. J Clin Invest 112:1776–1784

Klein T, Ling Z, Heimberg H, Madsen OD, Heller RS, Serup P (2003) Nestin is expressed in vascular endothelial cells in the adult human pancreas. J Hist Cytochem 51:697–706

Koshmel H, Berndt A, Strassburger S, Borsi L, Rouselle P, Mandel U, Hyckel P, Zardi L, Katenkamp D (1999) Distribution of laminin and fibronectin isoforms in oral mucosa and oral squamous cell carcinoma. Br J Cancer 81:1071–1079

Kriegsmann J, Berndt A, Hansen T, Borsi L, Zardi L, Brauer R, Petrow PK, Otto M, Kirkpatrick CJ, Gay S, Kosehl H (2004) Expression of fibronectin variants and oncofetal glycosylated fibronectin in the synovial membranes of patients with arthritis and osteoarthritis Rheumatol Int 1:25–33

Kurose K, Gilley K, Matsumoto S, Watson PH, Zhou X-P, Eng C (2002) Frequent somatic mutations in *PTEN* and TP53 are mutually exclusive in the stroma of breast carcinomas. Nature Gen 32:355–357

Labat ML (2001) Stem cells and the promise of eternal youth: embryonic versus adult stem cells. Biomed Pharmacother 55:179—185

Labat ML, Milhaud G, Pouchelet M, Boireau P (2000) On the track of a human circulating mesenchymal stem cell of neural crest origin. Biomed Pharmacother 54:146–162

Labat ML, Bringuier AF, Arys-Philippart C, Arys A, Wellens F. (1994) Monocytic origin of fibrosis. In vitro transformation of HLA-DR monocytes into neo-fibroblasts: inhibitory effect of all-trans retinoic acid on this process. Biomed Pharmacother 48:103–11

Liu L, Brouty-Boyé D (1994) Action of various growth factors and cytokines on α-smooth muscle actin expression by fibroblasts from postradiation breast fibrosis. Proc Am Assoc Cancer Res 35:648

Messam CA, Hou J, Major EO (2000) Coexpression of nestin in neural and glial cells in the developing human CNS defined by a human-specific anti-nestin antibody Exp Neurol 161:585–596

Metz CN (2003) Fibrocytes: a unique cell population implicated in wound healing CMLS. Cell Mol Life Sci 60:1342–1350

Myers CA, Liu H, Lee E, Bissell MJ (2003) Three-dimensional cultures of human breast epithelial cells to achieve *in vivo*-like architecture and function. In: Celis J (ed) Cell biology: a laboratory handbook, 3rd edn. Elsevier Science, London (in press)

Pankov R, Yamada KM (2002) Fibronectin at a glance. J Cell Sci 115:3861–3866

Pasco S, Ramont L, Maquart F-X, Monboisse J-C (2004) Control of melanoma progression by various matrikines from basement membrane macromolecules. Oncol Hematol 49:221–233

Peterson OW, Nielson HL, Gudjonson T, Villadsen R, Ronnov-Jessen L, Bissell MJ (2001) The plasticity of human breast carcinoma cells is more than epithelial to mesenchymal conversion. Breast Cancer Res 3:213–214

Pilling D, Buckley CD, Salmon M, Gomer RH (2003) Inhibition of fibrocyte differentiation by serum amyloid P. J Immunol 171:5537–5546

Ronnov-Jessen L, Petersen OW (1996) A function for filamentous α-smooth muscle actin: retardation of motility in fibroblasts. J Cell Biol 134:67–80

Sappino AP, Schürch W, Gabbiani G (1990) Biology of disease. Differentiation repertoire of fibroblastic cells: expression of cytoskeletal proteins as marker of phenotypic modulations. Lab Invest 63:144–161

Schor SL, Schor AM (2001) Tumour-stroma interactions. Phenotypic and genetic alterations in mammary stroma: implications for tumour progression. Breast Cancer Res 3:373–379

Schor SL, Ellis IA, Sarah JJ, Baillie R, Seneviratne K, Clausen J, Motegi K, Vojtesek, B, Kankova K, Furrie E, Sales MJ, Schor AM, Kay RA (2003) Migration-stimulating factor: a genetically truncated onco-fetal fibronectin isoform expressed by carcinoma and tumor-associated stromal cells. Cancer Res 63:8827–8836

Schürch W, Seemayer TA, Lagacé R (1981) Stromal myofibroblasts in primary invasive and metastatic carcinomas. A combined immunological, light and electron microscopic study. Virchows Arch A Pathol Anat Histol 391:125–139

Skalli O, Schürch W, Seemayer T, Lagacé R, Montandon D, Pittet B, Gabbiani G (1989) Myofibroblasts from diverse pathologic settings are heterogeneous in their content of actin isoforms and intermediate filament proteins. Lab Invest 60:275–285

Spanakis E, Brouty-Boyé D (1995) Quantitative variation of proto-oncogene and cytokine gene expression in isolated breast fibroblasts. Int J Cancer 61:698–705

Spanakis E, Brouty-Boyé D (1997) Discrimination of fibroblast subtypes by multivariate analysis of gene expression. Int J Cancer 71:402–409

Steiniger B, Barth P (2000) Microanatomy and function of the spleen. Adv Anat Embryol Cell Biol 151(III-IX):1–101

Steiniger B, Barth P, Hellinger A (2001) The perifollicular and marginal zones of the human splenic white pulp: do fibroblasts guide lymphocyte immigration? Am J Pathol 159:501–512

Strutz F, Muller GA (2000) Transdifferentiation comes of age. Nephrol Dial Transplant 15:1729–1731

Thierry J-P (2002) Epithelial-mesenchymal transitions in tumour progression. Nature Rev 2:442–454

Tomasek JJ, Gabbiani G, Hinz B, Chaponnier C, Browns RA (2002) Myofibroblasts and mechano-regulation of connective tissue remodelling. Nature Rev 3:349–362

Van den Hooff A (1986) Connective tissue as active participants in the process of malignant growth. Anticancer Res 6:775–780

Van den Hooff A (1988) Stromal involvement in malignant growth. In: Klein G, Weinberg S (eds) Advances in cancer research, vol 50. Academic Press, London, pp 159–196

Vescovi AL, Galli R, Gritti A (2001) The neural stem cells and their transdifferentiation capacity. Biomed Pharmacother 55:201–205

Woodbury D, Schwartz EJ, Prockop DJ, Black IB (2000) Adult rat and human bone marrow stromal cells differentiate into neurons. J Neurosci Res 61:364–370

Zhao Y, Glesne D, Huberman E (2003) A human peripheral blood monocyte-derived subset acts as pluripotent stem cells. Proc Natl Acad Sci USA 100:2426–2431

The Extracellular Matrix During Normal Development and Neoplastic Growth

J. Labat-Robert, L. Robert

1 Introduction

Few disciplines advanced so rapidly during the last decades of the 20th century as did the descriptive biochemistry of extracellular matrix (ECM) components and their role in normal and pathological processes. Therefore, even a succinct description of the involvement of ECM components in development and malignant growth represents a difficult challenge. Belonging to its four major components, collagens, elastin, glycosaminoglycans (GAGs), proteoglycans (PGs), and structural glycoproteins, we now recognize close to a hundred different macromolecules (Robert 1986; Comper 1994). These distinctions between tissue-born and circulating and within-ECM components became blurred by the recognition of GAG chains on some collagens (type IX) and the presence in tissues as well as in the blood of the same (or similar) glycoproteins (fibronectin, for instance). The term *structural glycoproteins* was proposed to distinguish glycoproteins synthesized in connective tissues from those circulating in the blood plasma, mostly of hepatic origin (Robert 1986). The descriptive biochemistry of ECM macromolecules is still progressing rapidly and renders therefore difficult the preparation of an exhaustive review monograph. For these reasons, we shall avoid an exhaustive description of all ECM components during development and tumor progression, and will concentrate instead on a limited number of examples. The choices are both subjective – work done in our laboratory – and objective – from the abundant recent literature. The chosen examples concern three of the four families of ECM components cited above, members of the structural glycoprotein family such as fibronectin and laminins, GAGs and PGs, and also elastin. The references give some orientation on the subjects omitted.

Before describing the involvement of these ECM components in normal development and cancer progression, we should look at the general principles

J. Labat-Robert, L. Robert (e-mail: lrobert5@wanadoo.fr)
Laboratoire de Recherche Ophtalmologique, Hôtel-Dieu, Université Paris 5, 1 place du parvis Notre Dame, 75181 Paris cedex 04, France

Progress in Molecular and Subcellular Biology
Alvaro Macieira-Coelho (Ed.)
Development Biology of Neoplastic Growth

of ECM biosynthesis. From the earliest stages of embryonic development, a mostly genetically orchestrated »program« of ECM biosynthesis starts in a large number of differentiating cells. This agenda comprises first the »choice« of genes coding for the macromolecules to be expressed in a given place at a given time, and then the regulation (up or down) of the level of the expression of the chosen genes. The »program« results in specific ECM composition and structure, variable from tissue to tissue and also changing continuously in time. ECM composition is not static; all of its components have measurable turnover, some very rapid as for instance glycoproteins, others slower as in some collagen fibers. This »program« of the relative rate of production of individual ECM components changes with age. For instance »old« collagens (cross-linked by the Maillard reaction) become indigestible by collagenases and will not be renewed. Another important discovery, which largely modified the study of the role of ECM during development and malignant growth, was the identification of cell receptors that »recognized« ECM components. These receptors, the integrins and some others such as elastin-laminin receptor (ELR), are instrumental in the cell-matrix interactions that play a major role in development and also in tumor progression. Tumor cells often interfere with the »normal program« of ECM biosynthesis and result in aberrant ECM production. This in turn may interfere (facilitate or slow down) tumor progression. Some examples will be discussed. Another principle that emerged during the last decades was the recognition of an (at least) double role of ECM components. First recognized was a structural role in the construction of a tissue-specific ECM, and second was a specific role in signaling pathways mediated by receptors. Several examples of this important principle will be discussed.

2 Extracellular Matrix During Development

Although the importance of »mesenchyme« was recognized by the early embryologists, a molecular approach to its role during embryonic development had to await the identification and detailed description of some of the ECM components. The two most studied ECM components in this respect were first fibronectin and soon after the laminins. The regain of interest in fibronectin (FN) started in the early 1970s (for detailed literature, see Hynes 1990). The first laminin was isolated from Engelbreth-Holm- Swarm (EHS) tumor (see Ekblom and Timpl 1996 for review), soon followed by the investigation of these glycoproteins in embryonic development. Already during gastrulation, after the third or fourth division cycle of the fertilized ovum some of the ECM components start to be synthesized. Among those we find basement membrane (BM) components, representative of the collagen type IV family and laminins. Soon after other components of BM appear, such as nidogen and perlecan, and also FN. The role of FN was shown to be important in directing cell movements during embryonic organogenesis. Immunochemical studies suggested that some of the cell movements take place over and might be guided by fibronectin fibrils laid down by surrounding cells. Let us first mention some of the observations concerning laminin isoforms in development.

These studies were carried out partly by local identification of laminins using immunohistochemistry, or of its message by in situ hybridization and Northern blotting. The reviews by Ekblom and Timpl (1996) as well as Yurchenko and Wadsworth (2004) give detailed information on this subject. Other experiments were carried out with reconstituted basement membrane, Matrigel, obtained from the EHS tumor of mice and rich among others in laminin (Kleinman et al. 1986). Some important observations concern the role of laminins and their integrin receptors in epithelial–mesenchymal interactions. Such interactions are important not only in early tissue differentiation but also in tumor–matrix interactions, as will be described later. One essential feature of epithelial cells is their polarization, yielding two lateral surfaces engaged in cell–cell interaction between epithelial cells, a basal surface anchored in the subepithelial BM and an apical surface turned to the lumen of ductal conduits in mammary glands or to the intestinal lumen (and other examples exist). The polarized nature of epithelial cells is a prerequisite for most of their differentiated functions, as for instance vectorial transport of nutriments from the intestinal lumen to the blood vessels in the subepithelial tissues, or the secretion of milk into the ductal lumen of mammary glands. It appeared that the subepithelial mesenchyme and especially laminin-1 plays a crucial role in epithelial cell polarization. As mentioned above, laminin can be detected in early mouse or chicken embryos from the two- to four-cell stages on, although the synthesis and excretion of the three composing laminin chains appear not to be completely synchronous at this stage. Chains β and γ appear before the α chain. Extracellular laminin is clearly detectable from the 16-cell stage, together with nidogen, another component of BMs. The $\alpha6\beta1$ integrin, a major laminin receptor, is present early as a maternal product in the oocytes (Sonnenberg et al. 1990). Even embryonic stem cell lines express integrin $\beta1$. During cell polarization, this integrin concentrates at the basal surface of epithelial cells. BM formation is a very early process during epithelial differentiation, as is also the expression of laminin-recognizing integrins that mediate the anchorage of epithelial cells in the BM. This sequential expression of laminin isoforms and corresponding integrins is similar in the primitive trophoectoderm formation, and later in the differentiation of intestinal epithelium from the endoderm. At this level, splice variants of the integrin $\alpha6$ chain (A and B) were observed to have discrete roles. This may have importance for differential cell responses.

An interesting example of epithelial–mesenchymal interactions is the developing kidney. Some of the mesenchymal cells of the kidney convert to epithelial cells during nephron morphogenesis. This is similar to mesenchymal–epithelial cell conversion during trophoectoderm formation. At the 10–11th gestational day of the mouse embryo, the kidney rudiment is composed of an epithelial bud surrounded by mesenchyme. Epithelial–mesenchymal interactions lead to the branching of the ureter bud leading to the formation of collecting ducts. This process is initiated by the conversion of mesenchyme to epithelium, accompanied by basement membrane formation, as observed already by Grobstein (1953, 1956, 1967) and Saxen et al. (1968). Some of the molecular processes involved in this conversion could be elucidated (Ekblom et al. 1996). In particular laminin-1 was shown to act as an autocrine enhancer of epithelial morphogenesis. At the start of conversion, this mesenchyme already expresses laminin β1 and γ1 chains. Expression of the α1-chain lags behind and increases only during epithelial polarization (Ekblom et al. 1990). Concomitantly, there is a local expression of E-cadherin and α6-integrin chains, apparently associated with the β1 subunit. It has been suggested that laminin-1 expression characterizes early epithelial differentiation. Interestingly, this was not the case during endothelial differentiation. During later development and adult life, there is mostly a sequential change in laminin and integrin isoform expression. Such studies are more difficult to perform on adults than on embryos, but the use of established cell lines was shown to be a satisfactory substitute. Studies conducted on the canine kidney-derived MDCK cell line, obtained from distal tubules (for a review, see Ekblom et al. 1996), gave some interesting results. Here, again, the production of laminin β1 and γ1 chains could be demonstrated, but only one out of several cell lines studied also expressed the α1 chain. Only in presence of all these chains of laminin-1 could its secretion in the extracellular space be demonstrated. Apparently, the presence of α1 chains was not an absolute prerequisite for the maintenance of epithelial cell polarity. Other experiments on the same and other cell lines confirmed the role of ECM components in general and of laminin-1 in particular in the maintenance and orientation of epithelial polarity. As shown by Wang et al. (1990), when cultured in a collagen-gel epithelial polarity is reversed, and β1 integrin is involved in this process. MDCK cells express a number of integrins, α2β1, α3β1, αvβ3, and α6β4, among others. Anti-β1 integrin antibodies efficiently inhibit binding MDCK cells to collagen or laminin-1. Comparable results concerning the role of laminins in branching processes contributes to the generalization as to the role of the ECM in early phases of embryonic morphogenesis. In a number of tissues epithelial cells are formed by such branching in the mesenchyme. A small epithelial rudiment, surrounded by mesenchymal cells, penetrates and proliferates by branching in the mesenchyme. The presence of mesenchyme for this process is also an absolute requirement (Grobstein 1956, 1967). The epithelial BM appears to orient the branching pattern of the proliferating epithelium. According to Bernfield et al. (1984), the PG components of BMs as well as the laminins play a special role in this respect. More details can be found in the above-cited review of Ekblom et al. (1996). The inverse of the above process, epithelial to mesenchymal transition also occurs during both normal development and tumor progression. Insulin-like GF-s (EGF-II in particular) and the AKT protein kinase family play an important role in these processes (Morali et al. 2001; Julien-Grille et al. 2003, 2004). These processes are of particular importance during cancer progression when epithelial cells become mobile and invasive. It has appeared recently that the same or similar mechanisms might be involved in triggering epithelial cell migration and conversion to mesenchyme during embryo development and tumor progression, including metastasis formation. Early tentative suggestions explained colon cancer formation by this principle. An international collaboration led by Robert Weinberg largely confirmed this hypothesis by showing that the tumor-regulating gene TWIST, which plays an important role in the development of mesoderm, is switched on in mouse breast cancer cells during metastasis formation. The inactivation of TWIST decreased metastasis formation without profoundly influencing the primary tumor. The transfection of TWIST in normal epithelial cells did trigger their transformation in mesodermal cells, E-cadherin being no longer expressed, a sign of dissociation of lateral cell–cell interactions (Yang et al. 2004). The roles of PG-s and GAGs in morphogenesis will be discussed in a later section. The purpose of this succinct description of the involvement of ECM components in general and of laminin isoforms in particular during embryonal cell differentiation and morphogenesis has been intended mainly to illustrate the dual role of ECM components in the formation of tissue structure, and in the accompanying signaling events. Similar processes have been shown to be involved in the modifications of tissue structure and loss of differentiated functions during tumor growth, as will be discussed in the following section.

2.1
Fibronectins in Development

As mentioned earlier, one of the important functions of FN appears to be in the mediation of cell adherence to ECM and also in the mediation of cell–cell interactions during cell migrations during early embryonic morphogenesis (for a review, see Hynes 1990). The first function, the mediation of cell-matrix interactions remains important well beyond embryonic development, as shown by a number of model experiments. Cells unable to attach to fibronectin matrices usually undergo cell death by apoptosis or necrosis (for a review, see Liotta and Kohn 2004). One of such model experiments performed in our laboratory by Péterszegi et al. (2002) consisted of exposing human skin fibroblasts in culture to increasing concentrations of ascorbic acid in order to study the effect of vitamin C on cell proliferation and matrix biosynthesis. Low concentrations ($<0.5\ \mu M$) of ascorbate stimulated cell proliferation and also matrix (collagen) biosynthesis. At a concentration of 1.0 mM or higher, however, ascorbate produced cell detachment and cell death. This could be attributed to the dose-dependent inhibition of FN biosynthesis by ascorbate. In this case, cell death was shown to be produced by lesions of the cell membrane-(necrosis), and could be inhibited by SOD and catalase but not by other ROS-scavenging agents. Regarding the guidance of cell migration by FN as observed during embryonic morphogenesis, besides focal adhesions and detachment (not to be discussed in this chapter), conformational modifications of FN are also involved. As shown by Hynes (1999) cell-matrix interaction and its modulation involve dynamic conformational changes in FN, and especially so, modulation of its elasticity. Besides its activity in orienting cell movements in the early embryo, FN appears to play a role in cell shape and cell migration through its integrin-mediated interaction with the cytoskeleton. Modulation of cell shape is important for tissue morphogenesis, comprising formation of cell sheets by cell-to-cell association, and stabilization of the structure of differentiated tissues. Integrin-mediated cell-to-cell signaling plays a key role in these processes (see Eble and Kühn 1997 for a review). The presence of significant quantities of FN reinforces the role of embryonal (and adult) BMs in the control of cell polarization and cell shape in tissues, as well as in its role in oriented cell movement. Among the specific issues examined (reviewed by Hynes 1990) are the formation of corneal epithelium over the Bowman membrane, limb bud development, chondrogenesis, osteogenesis, and myogenesis, as well as formation of adipose tissues and hematopoiesis. Besides ex vivo studies on embryonal tissues, a large number of experiments were carried out on model-cell cultures. Nearly without exception the stimulating role of FN in cell proliferation could be explained by its effect on cell adhesion, a prerequisite of initiation of the cell cycle. A specific role in cell proliferation could also be attributed to FN fragments, this topic will be discussed in a subsequent section.

3 Laminin Isoforms in Tumor Progression

As mentioned in the previous section, laminin isoform expression is regulated in space and time in a tissue- and age-specific manner (Ekblom and Timpl 1996). This scheme is modified during malignant transformation and tumor growth (for reviews, see Martin et al. 1996; Pattaroyo et al. 2002). BMs are the first structures influenced by tumor–matrix interactions. Invading tumor cells have to pass this barrier in order to penetrate new vessels, and again from the circulation into the tissues. This process involves proteolytic attack of BM components by endopeptidases expressed by transformed cells and activated by such cells or nearby non-transformed cells. We shall return to this process in a later section. Similar mechanisms prevail during the passage of epithelial–mesenchymal BMs by malignant cells. Using antibodies to laminin-1, discontinuities could be demonstrated in BM close to tumor cells, a sign of the above-mentioned proteolytic attack of BMs during tumor progression. Using specific monoclonal anti-laminin-chain antibodies, it could be shown that a number of tumor cells of epithelial origin (carcinomas, over 90% of all tumors) display a laminin-chain composition similar to the one seen in the surrounding original, non-transformed cells (Määttä et al. 2001). The laminin $\alpha1$, $\alpha2$, and $\alpha5$ chains are found in most differentiated epithelia. Expression of the α1 chain is restricted to kidney, prostate, breast, ovary, and lung carcinomas. The expression pattern of the β1 and γ1 chains is mostly linear, although discontinuities appear in some invasive tumor areas. Laminin α5, α3, β2, and γ2 chains are often present in carcinomas forming a more disrupted pattern with fragmentation. In contrast, laminin-5 staining appears continuous and even increased in some adenomas (Pyke et al. 1994, 1995). In invading malignant cells at the tip of progressing tumors γ2 chains are expressed; they are absent from the more distant cancer cells, away from the leading edge of the tumor (Pattaroyo et al. 2002, for review). The intracellularly localized γ2 laminin chain colocalizes with the receptor for urokinase plasminogen activator (Pyke et al. 1994, 1995). This relatively specific expression of γ2 laminin chains was confirmed also in melanomas but not in sarcomas. Therefore, several authors consider the expression of laminin γ2 chain as a marker of invasive carcinomas, even in the absence or much lower level of the α3 and β3 chains. Normal fibroblasts also express laminin chains. The expression of the α4 chain is increased in non-transformed tumor-stromal fibroblasts. Laminin chain production was studied also during in vitro cultivation of malignant cells. The deposition in culture of laminin-1, -5 and -10 and the »attachment« of the tumor cells appear to be mediated by α3β1, α6β1, and α6β4 integrins. In a model of invasion using penetration through Matrigel, expression of α6β4 integrin expression increased migration. The role of this integrin shifts apparently during tumor progression from its contact-stabilizing role in hemidesmosomes to promotion of migration (Mercurio and Rabinowitz 2001). Qualitative and quantitative differences in laminin (and integrin) expression were reported in some specific tumors. Integrin α7β1 is absent in normal melanocytes but strongly expressed in malignant melanoma cells. The expression of α3β1 integrin was claimed to be associated with tumor progression. As mentioned in the previous section, laminin-1 is specifically involved in the polarity of epithelial cells. This function cannot be fulfilled by other laminin isoforms, such as laminin-5 or -10/11. This function is disturbed in transformed cells. Overexpression of the laminin α1 chain in colon cancer cells is accompanied by an increase in tumor growth. This situation is quite different in glioblastoma cells, which synthesize the laminin-8 that promotes the migration of these cells, using the α3 and α6 integrins (Fujiwara et al. 2001). Human glial tumors were shown to overexpress α4 laminin chains (Ljubimova et al. 2001).

Laminin isoforms are also important in tumor-angiogenesis. Chains α4, α5, β1, and γ1 were demonstrated in tumor-localized blood vessels but also in normal vessels, laminin-8 to -10 are present in normal vascular BMs produced by the vascular endothelial cells. The 10-isoform of laminin appeared to be more efficient in promoting migration than the 8-isoform. Integrins α6 and β4 chains are expressed in capillaries as well as in their sprouts where α2 and αv integrin chains are increased. Matrigel, rich in laminin-1, was shown to facilitate endothelial tube formation (Schnaper et al. 1993). The adhesion of endothelial cells to laminin-1 is mediated by α6β1 and αvβ3 integrins as well as by $\alpha1\beta1$ and $\alpha2\beta1$ of endothelial origin. Aggressive melanoma cells express $\gamma2$ laminin chains together with several MMPs (Seftor et al. 2001). This was claimed to be necessary for vasculogenesis in aggressive tumors. Such laminins absent from BMs might well be involved in migration promotion of such malignant cells. Some of the tumor-associated laminins, for instance laminin-8, appear to be secreted by platelets associated with tumor cells (Honn et al. 1992). Tumor cells often interact with lymph nodes rich in so-called reticular fibers using especially $\alpha4$ and $\alpha5$ laminin chains (see review in Pattaroyo et al. 2002). This might well facilitate the interaction with metastatic melanoma cells (Bernston et al. 1994). In this model, integrin $\alpha3$ chains were shown to be involved in tumor enhancement and $\alpha7$ chains in inhibition of migration. Interaction of tumor cells with laminin also influences cell-pattern formation. As shown by Noel et al. (1988), when MCF-7 cells were plated on a BM model consisting of 85% laminin, cells formed clusters attached to the matrix, in contrast to cells plated on collagen I, where they grow as monolayer. Soluble laminin induced the same patchy pattern, in contrast to FN, which induced spreading. These authors suggested, therefore, that laminin and FN act in an antagonistic fashion on tumor-cell patterning.

3.1 Fibronectin in Malignant Growth

A surge of interest in FN in malignant growth has stemmed from the observation made by immunohistochemical methods showing the disappearance of FN from the cell surface of transformed cells (Gahmberg and Hakomori 1973; Hynes 1973). The observation by Yamada et al. (1976, 1977) that addition of FN to transformed cells could reestablish some of the normal features of transformed cells added a further argument to the designation of FN as a »transformation-sensitive« surface glycoprotein essential for cell shape and the normal cell phenotype. These observations have triggered, in fact, most of the studies on the role of FN in cell shape and migration, its interaction with the cytoskeleton and tissue differentiation. This phenomenon, the loss of membrane localized FN, was confirmed for most cases of viral transformation (DNA or RNA viruses), chemical transformation, and oncogene activation (nuclear or cytoplasmic; see Hynes 1990 for review). Among the exceptions are chondrocytes which normally hardly express FN on their membrane, but do express it after RSV transformation. Some SV-40-transformed keratinocytes were also reported to remain positive for FN staining. X-ray-transformed fibroblasts were also reported to keep their membrane-bound FN. The tumor-promoting phorbol esters were shown to release membrane bound FN, although this effect is inconsistent and reversible and can be inhibited by retinoids (Blumberg et al. 1976; Zerlauth and Wolf 1985). These observations were rapidly extended to in vivo tumors. Our studies concerned normal human breast tissue, mammary adenocarcinomas, and benign tumors

(Labat-Robert et al. 1981b, 1983; Labat-Robert 2002). In normal mammary tissue, FN is present in periglandular BMs and around glandular epithelial cells. In cancer tissues, the BMs are frequently interrupted and FN is absent from the cell-surrounding membranes. None of these modifications could be seen in benign mammary tumors. These observations were extended to a number of other solid human tumors, and all of them exhibited a comparable pattern (Birembaut et al. 1981; Labat-Robert et al. 1981b; Szendroi et al. 1984). Another interesting observation in this respect was the increased expression of FN in the tumor stroma. A further intriguing observation made during these studies was the slower age-dependent increase in plasma FN in cancer patients as compared to a normal population (Labat-Robert et al. 1981a, 1992; Labat-Robert 2002). A number of experiments (and also speculations) were devoted to the significance of the above-summarized in vitro and in vivo observations with respect to in vivo tumor growth and progression. A number of authors reported good correlations between in vitro loss of FN and in vivo tumor progression (reviewed by Hynes 1990), but exceptions were also reported. A better correlation was found, as could have been expected, between loss of FN and invasion and metastasis formation. Reduced cell adhesion by transformed cells lacking FN can facilitate migration. Loss of membrane-bound FN might well be involved in the previously mentioned phenotypic modifications of transformed cells, such as cytoskeletal organization, focal contact formation, and oriented migration, as observed during normal cell differentiation and tissue morphogenesis. Such FN loss also appears to be involved in the loss of the early discovered normal phenotypical traits, such as the feature known as »contact inhibition of movement« (Abercrombie et al. 1971). This property of normal cells is an essential feature of their typical growth pattern and morphology. Transformed cells lose this contact inhibition and present mostly an aberrant morpho-pattern in culture, and seen also in tissue section, which is one of the bases of their recognition during histopathological examinations. Closer observation by Bell (1977) showed that transformed cells underlap more often than overlap during migration, probably as a result of the weakened adhesion of transformed cells to the cell substratum. Another consequence of the decrease or loss of substrate adhesion is the capacity of most transformed cells to grow in suspension, for instance, in soft agar. This loss also explains the localized, focal growth of transformed cells on normal cell membranes (foci formation). An interesting, nearly extreme, example of this capacity of transformed cells is the acquisition by some of these cells to grow in suspension in the ascites fluid of inoculated animals (mostly mice). A systematic study of Georg and Eva Klein of this selective process showed that some tumor cells are able to acquire the capacity of anchorage-free proliferation in the ascites fluid. Still, many other tumor cells cannot (Klein and Klein 1955; Ringertz et al. 1957). No recent detailed studies have been conducted on this interesting model, at least to our best knowledge, in order to explore its relationship to integrin-mediated cell–matrix interactions. A more recent study suggested, however, a relationship between p53-mutations and the capacity to thrive in anchorage-independent suspension (Magnusson et al. 1998). All these phenotypic traits of transformed cells (anchorage independence) play an important role in their recognition and identification, as well as in their malignant behavior.

A further argument for the central role for FN in the definition of the transformed phenotype is the above-mentioned partial restoration of some of the lost properties by the addition of purified cell FN to transformed cells. Not all the phenotypic traits of transformed cells are thus corrected, only those which concern morphology and adhesion. Another intriguing set of reports concerns the restoration, with no effect on tumor growth, of FN to transformed cells by pharmacological agents, such as cAMP derivatives, butyrate, BrdU, DMSO, DMF, or dexamethazone to mention only some of the tested substances (reviewed in Hynes 1990). This confirms the regulation of FN expression as being only one aspect of the malignant transformation, and the addition of FN to transformed cells only very partially restores the normal phenotype. An often-cited argument in this respect is that for any of the above-discussed phenotypic modifications, and their reversals, there are always exceptions. This might be the result of the necessary simultaneous co-expression of FN-recognizing receptors (integrins) in order to reestablish FN-related functions. Another factor might be the continuous modulation of FN biosynthesis in transformed cells; some perhaps only decrease it, while other cells more or less completely lose it. In the absence of cell membrane anchorage on receptors, FN could be lost even if its synthesis is maintained. As a number of factors are involved in the transcriptional regulation of FN biosynthesis, the detailed mechanisms of its altered expression may also vary from one transformed cell to another. Modulation of the alternative splicing of FN pre-messenger is another instance of tumor-specific modulation of its expression, as is also the case in normal cells. Another of the post-translational modifications possibly involved in the change of FN phenotype on transformation is increased glycosylation with a modified pattern (Nichols et al. 1986). This modified glycosylation pattern may also modulate the proteolytic fragmentation of FN, a subject to which we shall return in the next section. The phosphorylation pattern of FN was also shown to be modified on transformation, but functional consequences here remain unclear.

4
Proteolytic Degradation of Matrix Components, Generation of Biologically Active Peptides: Matricryptins

Although the initial observations on this topic date from the 1980s, the frequency and biological importance of these findings have been reviewed only recently (Davies et al. 2000; Labat-Robert et al. 2000; Labat-Robert 2002, 2004). A large number of proteolytic enzymes are expressed during differentiated growth and also during malignant transformation. About 560 different proteolytic enzymes were described, many of them registered in the voluminous handbook edited by Barrett et al. (1998). Some of the enzymes, as for instance the MMP-family and some serine proteases, are of special interest for the subject of this review. Among the relevant observations, we should mention the age-dependent upregulation of several of these proteases (Robert and Hornebeck 1989; Hornebeck et al. 2002 for reviews). MMPs are upregulated in nearly all human cancers, some synthesized and excreted by the malignant cells (MMP-7), some others (MMP-2, MMP-9) by other, non-transformed stromal cells, fibroblasts, endothelial cells, or invading mononuclear cells (Egeblad and Werb 2002). A comparable upregulation of elastase-type endopeptidase activity could also be demonstrated during sequential passages of mesenchymal cells. Together with the age-dependent upregulation of FN expression (Labat-Robert et al. 1981a; Labat-Robert 2002, 2004), these observations, suggesting the possibility of age-dependent increase in FN fragmentation, were confirmed by immunoblot analysis of plasma samples from geriatric patients (Labat-Robert et al. 2000). FN, because of its subunit structure, was shown to be especially sensitive to proteolytic degradation (Labat-Robert 2002, 2004 for review). FN-derived peptides were shown to exhibit original activities absent from the parent molecule. Similar phenomena were described for several other matrix macromolecules. Upon changes of conformation, new sites (matricryptic sites) are uncovered and become accessible. Peptides derived by proteolytic attack, endowed with original biological properties, are termed matricryptins (Davies et al. 2000). Several of them might well be involved in the above-discussed modifications of cell phenotype during malignant transformation and tumor growth. Among the earliest observations, we should cite the interesting contributions of Barlati and his team (1986). These authors first showed that some FN fragments, present in the plasma of cancer patients, potentiate the RSV-induced transformation of chicken embryo fibroblasts. All together, these observations suggest a role for FN fragments in tumor progression and in the modulation of the transformed phenotype. Some of the sites attacked by proteases are not available in the native conformation but become sensitive on transformation. Other observations in this respect were published by the team of Keil-Dlouha (Emod et al. 1990; Imhoff et al. 1990). They were able to show that a proteolytic fragment of FN expresses proper proteolytic activity, especially in breaking down laminin. This finding suggests the possibility of a vicious circle of matrix degradation starting with proteolytic attack of FN. Also of interest is the demonstration of mitogenic effect exhibited by some FN fragments (Humphries and Ayad 1983) .

As tumor growth is frequently associated with local inflammation, the original observations of Homandberg are also of interest, although they concern cartilaginous tissues. They showed that FN fragments induced IL-1 formation and activation of collagenolytic enzymes (Xie and Homandberg 1993). Upregulation of tumor-associated proteases were also claimed to be involved in the above-discussed release of membrane-bound FN, and especially in its patchy disappearance and discontinuity on tumor-associated BMs. Transformation-induced proteases might well be involved in this process (Hynes 1974, 1976; Bax et al. 2004). This subject attracted the attention of several research groups. An interesting observation was the increased loss of FN during cocultivation of transformed and normal cells (Hynes 1976). Membrane-bound proteases could be involved, although a more recently described ROS-mediated process might also play a role. A special role in this respect was also claimed for plasminogen activator (Qingley 1976, 1979; Qingley et al. 1980; Pyke et al. 1995), which might act directly on FN without the previous activation of plasminogen. Membrane-bound MMPs appear also to be involved in this process. Mouse keratinocytes immortalized with the large T antigen acquire the capacity to secrete MMP-9. This process is mediated by $\alpha3\beta1$ integrin expression (DiPersio et al. 2000). Thrombospondin-1 was shown by Rodriguez-Manzaneque et al. 2001) to inhibit activation of MMP-9 and of VEGF, and thereby suppressing spontaneous tumor growth. Further details of previous and early experiments on this important topic were reviewed by Hynes (1990). It should be remembered in this respect that some of the above-mentioned proteases, and especially those of the MMP family, are also involved in embryonic morphogenesis. In fact, the first description of morphogenetically related proteolytic activity was given by Gross and Lapière (1962) and concerned the removal of the tadpole tail by MMP collagenase.

As for the biological significance of the above-described proteolytically generated fragments of ECM macromolecules and their harmful effects, the only direct evidence is the presence of FN-degradation products in the plasma of elderly patients examined with a variety of pathologies (Labat-Robert et al. 2000). In the same experiments, a limited series of plasma samples from a selection of centenarians, clinically in good health, was also examined. Run on the same immunoblot plates, these latter plasmas did not show FN degradation products as did the pathological samples.

5 Tumor Growth and Matrix Biosynthesis

It was mentioned previously that the tissue-specific structure and function of the ECM are the result of the concerted »program« of biosynthesis of ECM macromolecules. This »program« is under the control of yet-to-be-discovered gene-expression regulation of a considerable complexity. A first step in this direction was the discovery of operons regulating bacterial gene expression (Miller and Reznikoff 1978, for review). A second phase was the identification of transcription factors and their regulation by receptor-mediated pathways (Bradshow and Dennis 2004, for review). Recently modified microarray procedures have enabled the identification of gene sets specifically involved in the regulation of the expression of a large number of genes (Morley et al. 2004; Kondro 2004). A considerable body of experimental evidence shows the role of growth factors and cytokines in gene-expression regulation. Regarding ECM components, TGF-β was shown to play a major role in the regulation of the biosynthesis of several of its macromolecules (Javelaud and Mauviel 2004). TGF-β signaling in fibroblasts has been shown recently to modulate the oncogenic potential of adjacent epithelial cells.

The first indication of tumor-induced modifications of the »program« of ECM biosynthesis came from the histopathological investigations of solid human tumors. The distribution pattern of the fibrous components of tumor stroma, and the tumor-specific distribution of malignant cells within this stroma, clearly indicated a profound modification of ECM biosynthesis. A limited number of experiments confirmed some of the details of ECM production disrupted by tumor growth. Instead of reviewing all the literature regarding this still-to-be-explored field of tumor biology, we shall cite only one example studied in some detail in our laboratory. It concerns the increased elastin biosynthesis by some human breast tumors of the scirrhus type (Kadar et al. 2002 for review). This variety of breast tumor, has seldom been seen since the 1970s because of earlier detection and treatment of breast cancers. As the more or less selective stimulation of elastin biosynthesis in solid human tumors was specific for some human breast cancers (some other localizations were, however, also cited in the literature), a more detailed investigation was carried out (see Kadar et al. 2002). Among the intriguing results was the specific stimulation of elastin biosynthesis by tumor extracts obtained only from elastogenic breast cancers. Extracts from other tumors did not show stimulation of elastin biosynthesis but did stimulate collagen biosynthesis, as shown by ^{3}H-Pro incorporation in polymeric Hyp (unpubl. results). The identification of the (growth) factor(s) involved and secreted by elastogenic human tumors remains, however, to be done. An often-neglected factor in this respect is the role of dynamic rheological influences on cell behavior. Such factors and their effect on cell behavior and tumor growth have been recently reviewed by Macieira-Coelho (2002). Cellular biosynthetic processes are also controlled by the structure and rheological properties of the surrounding matrix, as shown by the effect of collagen gels on matrix biosynthesis by cells inside the gel. Such factors were also shown to influence the effect of growth factors such as EGF on mesenchymal cells (Paye et al. 1987; Colige et al. 1988, 1990).

6 Role of Receptor Signaling in Cell Matrix Interaction During Normal Development and Tumor Growth

It was mentioned in the previous sections that the specific role of ECM components in embryonic development and morphogenesis is tightly linked to the expression of cell receptors conveying the signals to the cell interior. The same is true for tumor cell behavior as modified by matrix components. Receptor expression in tumor cells deviates often from the normal pattern, at least as far as their temporo-spatial expression is concerned, and also by the very nature of receptors expressed. We shall briefly mention two types of receptors, the integrins and the elastin-laminin receptor (ELR). Regarding the integrins, after their identification a more static role was assigned to them in holding together cells within the matrix (for review see Hynes 1990; Eble and Kuhn 1997; Hynes and Zhao 2000; Travis et al. 2003). Only later did it become evident that their primary role, besides holding cells and ECM components together, is signal transduction from the matrix to the cells (outside-in) and from the cells to the matrix (inside-out; Hynes and Zhao 2000; Hynes 2002; Mostafavi-Pour et al. 2003). Their expression is qualitatively and quantitatively modified in most malignant cells. Suppression of some integrins, for instance β3 or β3/β5 chain expression (knockout mice), was shown to increase tumor growth (Taverna et al. 1998, 2004). Mouse keratinocytes immortalized with the large T antigen express α3β1 integrin-dependent MMP-9 secretion (DiPersio et al. 2000). The α5β1 integrin as well as ADAM-7 colocalize with migrating HeLa cells (Bax et al. 2004). Integrins also play an important role in tumor angiogenesis (Hynes et al. 2002). Their presence is required for cell survival during embryonic developmental processes (Yang et al. 1999; Haack and Hynes 2001). Absence of αv integrin chains is embryonically lethal (Bader et al. 1998), preceded by intensive vasculogenesis.

The situation is different for the ELR. From the first indication of an elastin-receptor's stimulation (inducement) of cell–elastin interactions (Hornebeck et al. 1986), it became evident that the primary role of this type of receptor was signaling (see Fülöp et al. 1998; Labat-Robert and Robert 2000 for reviews). For both types of receptors, integrins and ELR, the signaling pathways in several cell types were studied in detail (for ELR see Varga et al. 1989; Faury et al. 1998). Only the functional implications of these receptors are the main focus of this chapter; cell-matrix interactions in development and tumor growth will be only briefly mentioned. The role of integrins was already mentioned in some aspects of developmental processes in previous sections. For more details the monograph of Eble and Kuhn (1997) can be consulted. Studies on knockout mice with invalidated integrin genes confirmed their importance in development. This is especially true for the gene of the β1 integrin chain, its inactivation is lethal after the blastocyst stage, following implantation (Gullberg and Ekblom 1997 for review). Embryonic stem cells deficient in both β1 alleles are defective in cell migration and adhesion as well as in cell junction formation, among other anomalies. Some of the splice variants in the intracellular domain of β1-integrins (β1C) appear to interfere with the cell cycle. Such splice variants are sequentially expressed in some embryonic tissues, such as skeletal muscle. Inactivation of FN-recognizing integrin α chains (α5 for instance) produces embryonic lethality with major mesodermal defects, especially in the posterior part of the embryo. Inactivation of the α4 gene is also lethal, producing defects in placenta and heart. Integrins containing α4 recognize VCAM-1, and its inactivation results in a similar phenotype, showing again that ligand–receptor pairs are important for development and later function. Inactivation of individual FN-recognizing integrins does not result systematically in early embryonic lethality as is the case with FN-knockout mice, probably because of mutual compensation (at least eight different integrins react with FN). As an exception, we mentioned previously the case of αv-knockout mice. FAK-deficient embryos show a lethal phenotype similar to FN-knockout mice, suggesting an important role for this tyrosine kinase in FN-mediated signaling. Curiously, inactivation of laminin-binding integrins did not produce severe phenotypes as with FN inactivation. They did exhibit, however, anomalous BM development. The human equivalent of β4 integrin inactivation results in epidermolysis bullosa, and the absence of α7 in mice results in a muscular dystrophy syndrome, although a milder form than the one produced by dystrophin gene elimination. Elimination of α1 results also in a mild phenotype with deficiency in cell adhesion to collagen IV and to laminin-1 as well as of cell migration on these same ECM components. A number of »no-phenotype« results could be explained by compensation phenomena and »cross talk« between integrins (for more details see Gullberg and Ekblom 1997).

As for tumor growth, we mentioned above, as well as in previous sections, the role of some integrins. One interesting bit of information concerns α5β1 integrin, which was considered a tumor suppressor (Horwitz et al. 1985; Tamkun et al. 1986), as shown by studies on transfected cell lines. Its expression inactivates the immediate early genes of cell proliferation (Varner et al. 1995), c-fos, c-jun, and junB. This effect is, however, inhibited and cell proliferation reactivated when cells adhere to FN, confirming that α5β1 integrin mediates anchorage-dependent cell proliferation. Cell adhesion mediated by α5β1 also activates MMP expression (MMP-1, -3 and -9, involved in tumor invasion as well as in tissue remodeling (Werb et al. 1989; Hornebeck et al. 2002). This effect is counterbalanced by α4β1 integrin expression, its binding site to FN being different from that of α5β1 (the RGD binding site; Huhtala et al. 1995). These two integrins influence cell migration on FN differently (Beauvais et al. 1995).

6.1 The Elastin-Laminin Receptor

The functional subunit of the ELR is a 67-kDa loosely membrane-bound protein, apparently identical to the non-integrin laminin receptor (for reviews see Mecham and Hinek 1996; Fülöp et al. 1998; Lapis and Timar 2002). This subunit exhibits site-recognizing elastin-derived peptides such as the VGVAPG hexapeptide, but also shorter sequences such as VGV, and some others, acting as agonists (Faury et al. 1998). Another site on this same subunit reacts with oligo- and polysaccharides exhibiting a galactose end group, such substances acting as antagonists. Its cell-membrane anchorage and signal-transmission activity is probably dependent on its close association with two other membrane-bound subunits of 55 and 61 kDa. This receptor was shown to be present on most normal and several transformed cell lines (Timar et al. 1990, 1995). A number of physiologically and pathologically important reactions were shown to be mediated by this receptor, including chemotactic cell migration and release of lytic enzymes and free radicals, to mention only those important for the present review. In endothelial cells at least, ELR is coupled to NOS, and elastin peptides added to noradrenalin-precontracted rat aorta rings produce endothelium- and NO-dependent vasorelaxation (Faury et al. 1995). This effect decreases with age suggesting an »uncoupling« of this receptor from NOS and also from its normal transmission pathway. Laminin-1 was shown to inhibit this vasorelaxation triggered by elastin peptides (Faury et al. 1997). These observations suggest the possibility of peroxynitrite production triggered by the activation of the ELR as a result of the simultaneous release of superoxide (O_2^-) and NO•. This highly reactive anion ($ONOO^-$) may well be involved in the observed cell and tissue injury and cell death by necrosis and apoptosis (Péterszegi et al. 1999). Not much is known about the role of this receptor during embryonic development. Its role in the mediation of cell-elastin adhesion (Hornebeck et al. 1986; Perdomo et al. 1994; Groult et al. 1998) suggests, however, an important role during vascular development when smooth muscle and endothelial cells have to adhere to the freshly synthesized elastic fibers. Both the elastin- and laminin-binding properties of the 67-kDa subunit-containing receptors (ELR) were shown to be important for tumor–matrix interaction. Extensive discussion of these interactions and their role in tumor progression can be found in the reviews of Timar et al. (1995); Mecham and Hinek (1996); Fülöp and Larbi (2002); Kadar et al. (2002); Lapis and Timar (2002).

The first indication showing that cells can adhere to elastin fibers – long believed to be »inert« – came from our demonstration of inducible adhesion of fibroblasts and vascular smooth muscle cells to elastin fibers, as followed by time-lapse video-microscopy (Hornebeck et al. 1986; Groult et al. 1998). This adhesion is »inducible« by the addition of elastin peptides that react with the ELR and induce the synthesis and membrane localization of a glycoprotein complex, which in turn mediates the adhesion of the cell membrane to the elastin fiber. The receptor-induced membrane component mediating adhesion to elastin was shown to be a 140-kDa protein. It was labeled with ^{35}S-methionine, identified electrophoretically (PAGE) and designated elastonectin (Hornebeck et al. 1986). It appears highly improbable that the easily removable 67-kDa subunit could mediate such a strong cell-fiber association that it could not be dissociated by bacterial collagenase, and yet yielded to PMN-elastase. This was a convincing demonstration of strong elastin–cell interaction, inducible by ELR activation. A decade later Blood and Zetter (1999) demonstrated the adhesion of tumor cells to elastin, as well as the chemotactic action of elastin-peptides to tumor cells. Similar reports were presented during the following years (for a review see Mecham and Hinek 1996; Kadar et al. 2002). More recently a surprisingly important role of elastin fibers in the limitation of cell proliferation during embryonic development was demonstrated by Li et al. (1998a,b). In elastin-gene knockout mice ($e^{-/-}$) lethality in utero was shown to be due to closure of the developing blood vessels as a result of unlimited proliferation of vascular smooth muscle cells. In contrast, in the heterozygous $e^{+/-}$ mice, where only one allele was inactivated, the other could still direct elastin synthesis; a compensatory increase in concentric elastic fibers in the vessel wall was demonstrated with extra elastin lamellae. These modifications had important functional consequences as shown by the elegant experiments of Li et al. (1998b). A number of investigations further confirmed the functional importance of ELR in some tumors, such as colon carcinomas. Instead of exhaustively citing the literature (allotted space forbids), we wanted to illustrate the fact that elastin, hitherto considered as 'biologically inert', plays an active role in cell-matrix interactions, both during normal development and in malignant processes, as shown by the above examples.

7 Glycosaminoglycans and Proteoglycans in Normal Development and Tumor Progression

Glycosaminoglycans (GAGs) and proteoglycans (PGs) represent a heterogeneous family of ECM components, heterogeneous as far as composition, structure, and tissue distribution are concerned (Hascall 1986; Laurent 1998; Comper 1996). The only non-protein linked GAG in vertebrate tissues is hyaluronan, a high molecular weight ($\geq 10^6$ kDa) polysaccharide, extensively studied in a number of laboratories since its description by the team of K. Meyer (Laurent 1998) All the other GAGs, chondroitin-4 and -6 sulfate, dermatan sulfate, keratan sulfate, heparan sulfate, are linked to proteins, mostly by O-glycosidic linkages. Some of them are early expressed during embryonal development as hyaluronan (HA) and membrane-bound heparan sulfate (HS), linked to both developing BMs and cell membranes. Most other PGs start to be synthesized during tissue differentiation, chondroitin sulfate (CS), dermatan sulfate (DS) mainly in cartilaginous tissues, keratan sulfate (KS) in cornea and intervertebral disks, also in some other tissues as vessel walls, glial cells, and neurons. Several important biological roles can be ascribed to GAGs and PGs. The earliest recognized functions were related to the mechanical properties of tissues. The suppleness and elastic recoil of several tissues, from the umbilical cord to skin and vessel wall are at least partially (vessel wall) or totally (umbilical cord) attributable to their GAG content. Because of its negative charge and its »excluded volume« (Laurent 1970), hyaluronan controls tissue hydration and molecular traffic. Other GAGs, according to their protein components, play a similar role as HS-PG (perlecan) in BMs, controlling molecular traffic across the membrane. CS-PGs have also similar roles in cartilaginous tissues. Pericellular hyaluronan is an obstacle to cell movements. It is degraded at the start of cell proliferation, as in the primitive cornea when mesenchymal cells (future keratocytes) »invade« it to form the corneal stroma, as shown by Trelstad and the Coulombres. Similar conditions must be fulfilled during malignant cell migration. In this respect, regulation of cell proliferation and migration for both normal and transformed cells is to some extent dependent on membrane-bound HSPGs as co-receptors for GFs. For these functions, as well as for some others such as the regulation of blood coagulation, some specific sequences in HS were detected, and indeed used in pharmacology against deep vein thrombosis and limitation of smooth muscle cell (SMC) migration in the atherosclerotic wall (the so-called low MW heparin derivatives).

Another important member of this large family of ECM constituents is Decorin, a DS-PG, of the small leucine-rich PG (SLRP) family. It was shown to play an important role in collagen fiber morphogenesis in several tissues by electron microscopy (Scott 1988) and also by gene-disruption procedures (Reed and Iozzo 2003). Most importantly for the topic of this chapter, it was also shown by the Iozzo-team (Santra et al. 2002) to bind to a specific region of the EGF-receptor, distinct but overlapping with the EGF-binding site. As such, it was shown by Iozzo's team to play a role as a key regulator for tumor growth by acting as an agonist of the growth factor EDFR tyrosine kinase. It also was shown to suppress tumor cell-mediated angiogenesis (Grant et al. 2002). GAGs in general (Timar et al. 1995, 2002) and HS proteoglycans in particular were shown to be involved in tumor growth and progression. One important argument in this respect was the demonstration by the team of Vlodavsky of the stimulation of tumor angiogenesis and metastasis formation by the cell surface expression of an endo-β glucuronidase or heparanase (Goldschmidt et al. 2002, 2003). The Vlodavsky team explored in detail the mechanisms involved in HS- and heparanase-mediated (or inhibited) tumor growth interfering with penetration of malignant cells across continuous endothelial cell layers and basement membranes (Vlodavsky and Friedman 2001). The detailed analysis of the mode of action of cloned heparanases in dimer formation (Levy-Adam et al. 2003), as well as the identification of some of their type-specific inhibitors (as the eosinophilic, major basic protein, by Temkin et al. 2004) appears to be of special importance in this respect. Overexpression of human heparanase in mice led to important discoveries in normal embryonic morphogenesis as well as in tumor progression by the Vlodavsky team. Zchariaet al. (2004) showed that such heparanase-overexpressing mice exhibited an important deficit in HS-chain expression (chain length) in tissues, especially in BMs, but these mice still exhibited an apparently normal phenotype. They showed, however, an increased urinary excretion of proteins and creatinine, and reduced food consumption and body weight compared to control mice. Their mammary glands exhibited extensive branching and widening of ducts and enhanced neovascularization with disruption of epithelial BMs, a further confirmation of the importance of HS-PG (perlecan) in BM structure and function. A further interesting demonstration of the mediation of cell-adhesion by heparanase, independent of its enzyme activity, was described by Goldschmidt et al. (2002, 2003). They used non-adherent lymphoma cells, which became adherent to ECM and showed integrin-dependent cell spreading and tyrosine phosphorylation of paxillin, followed by cytoskeletal reorganization in the presence of heparanase. The presence of heparanase on the cell surface increased cell invasion through reconstituted BMs. The fact that these effects on cell adhesion, migration, and cytoskeletal organization were observed with active as well as inactive (mutated) heparanase confirms the dual role of these enzymes in normal and malignant cell behavior. This is an important example of the involvement of an ECM component-degrading enzyme in both normal embryonic development and in tumor progression, vascularization, and metastasis formation (Goldschmidt et al. 2002, 2003). A further interesting observation in this respect concerns the estrogen-dependent regulation of heparanase gene expression in breast cancer (Elkin et al. 2003). The estrogen-mediated upregulation of heparanase expression might well represent a key factor in the differential behavior of estrogen-dependent and -independent mammary cancers. As a result of the above finding, heparanase became an interesting molecular target for cancer chemotherapy (Simizu et al. 2004). The above-summarized recent developments in GAG-PG action in tumor progression further advance interest in the therapeutic use of substances derived from these glycoconjugates. One recent example, coming from the team of Iozzo (Bix et al. 2004) and comparable to the anti-angiogenic action of several collagen derivatives (endostatin and others), is the recently described endorepellin, a carboxy-terminal portion of HS-PG (perlecan), which is a potent inhibitor of angiogenesis. This novel biological activity is dependent on the interaction of a protein fragment of perlecan with the $\alpha 2\beta 1$-collagen-I-recognizing integrin followed by intracellular signaling events, increase in cAMP, activation of PKA and FAK, sequential activation of p38 MAPK and HSP 27, followed by their rapid downregulation, disassembly of stress fibers, and focal adhesion resulting in the arrest of endothelial cell migration and capillary formation. Proteolytic release of endorepellin from vascular or epithelial BMs might therefore represent a »natural« anti-tumor mechanism with foreseeable therapeutic applications.

8 Discussion and Conclusions

It appears from this rapid survey of cell-matrix interactions in developmental processes and in tumor progression that there are indeed some similarities and several major differences. Among the similarities, we can mention the importance of cell-matrix interactions in both processes. The biosynthesis of ECM components, especially of BM constituents such as laminins, nidogen, and collagen IV (as well as FN involved in cell adhesion to ECM) start very early during embryonal development. ECM biosynthesis is tightly gene regulated, although an increasing number of experiments suggest the importance of epigenetic (parental imprinting, for instance) and even post-translational processes. Similar conjectures were proposed for malignant cell formation and tumor progression. Transformation is widely recognized as a multistep process in which genetic and epigenetic factors play important roles (Klein 2002, 2004). It remains to be determined which steps during transformation are the most influenced by cell-matrix interactions. Selective (»Darwinian«) processes are involved both in embryonal cell survival (for instance, for differentiating neurons, about half of which are eliminated by apoptosis if not consolidated by synapses) as well as for tumor cell selection during growth, migration, and metastasis formation. It became evident during the last decades that for both processes matrix macromolecules play a dual role as structural elements (for instance in BM formation) and as signaling molecules. Several examples of this dual role of ECM components as well as of degrading enzymes were mentioned. As for differences between development and tumor progression – they outnumber the similarities. ECM construction and remodeling are orderly processes tightly regulated in space and time. Regulation also concerns the coordinated expression of ECM components during cell differentiation, migration, and tissue-pattern formation. Tumor cells or their secretion products (GFs, cytokines, enzymes) also produce profound matrix remodeling mostly by altering the normal »program« of matrix biosynthesis, as shown by some examples, for instance elastogenic breast carcinomas. For tumor cells also several matrix components represent signaling events, as shown among others by the above-discussed role of some laminin chains during invasion. These signals do not result, however, in an ordered matrix pattern, a prerequisite for normal histogenesis. Recent reports further confirm the essential role of a tumor matrix in malignant cell behavior in an unexpected manner. Chemical carcinogenesis (α-nitroso-methyl urea) treatment of the stroma, devoid of epithelial cells, was shown to induce transformation of normal epithelial cells into tumor cells, a result not obtained in this experiment when the epithelial cells were selectively treated with the carcinogen without the stroma (Maffini et al. 2004). Such observations can best be explained by altered signaling from the matrix to the cells. Altered signaling events might well play a major role in cancer progression. A further major difference between normal tissue patterning and tumor-induced modifications concern the post-translational processes triggered by the production of biologically active proteolytic fragments of matrix macromolecules. The conformational modifications uncovering matricryptic sites as well as the age- and pathology-dependent up(or down)regulation of the biosynthesis of target macromolecules (like FN) as well as the proteases underlying the auto-amplifying vicious circles that appear to be involved in aging and age-related pathologies (Robert 2000; Labat-Robert 2002, 2004). Such processes need time in order to be amplified to reach biologically relevant intensity and should therefore play an increasing role with aging but not during development. Remaining to be answered is the question as to gene- or environmentally triggered modifications in cell-matrix interactions during development with late effects on tumor progression. Some indications in favor of such processes were obtained during the discussion of the role of GAG components in both developmental processes and tumor progression, suggesting that such early events might occur. Further studies are clearly necessary to extend and confirm this possibility. As stated in the introduction, no extensive coverage of the topic proposed by the title of this chapter would be conceivable at present in the frame of a short chapter because of the exponentially increasing number of publications on both topics, cell-matrix interaction in normal development and in tumor progression. We had to limit our review to some matrix components only, with no or only brief mention of the role of other components in these processes. To be complete and to do justice to most of the relevant contributions, a large volume would be necessary. In fact, the role of the ever-increasing number of members of the collagen family had to be omitted. Let us hope that the examples given illustrate cell-matrix interactions in both normal development and tumor progression sufficiently to encourage scientists to become involved in this important field of research.

References

Abercrombie M, Heaysman J, Pegrum SM (1971) The locomotion of fibroblasts in culture. Exp Cell Res 65:359–367

Bader BL, Rayburn H, Crowley D, Hynes RO (1998) Extensive vasculogenesis, angiogenesis, and organogenesis precede lethality in mice lacking all αv integrins. Cell 95:507–519

Barlati S, Adamoli A, de Petro G(1986) Presence and role of fibronectin fragments in transformed cells. In: Labat-Robert J, Timpl R, Robert L (eds) Front matrix biology, vol 11. Karger, Basel, pp 174–182

Barrett AJ, Rawlings ND, Woessner JF (eds) (1998) Handbook of proteolytic enzymes. Academic Press, San Diego

Bax DV, Messen AJ, Tart J, van Hoang M, Kott J, Maciewicz RA, Humphries MJ (2004) Integrin α5β1 and ADAM-17 interact in vitro and co-localize in migrating HeLa cells. J Biol Chem 279:22377–22386

Beauvais A, Erickson CA, Goins T, Craig SE, Humphries MJ, Thiery JP, Dufour S (1995) Changes in the fibronectin-specific integrin expression pattern modify the migratory behavior of sarcoma S180 cells in vitro and in embryonic development. J Cell Biol 128:699–713

Bell PB (1977) Locomotory behavior, contact inhibition and pattern formation of 3T3 and polyoúa virus-transformed 3T3 cells in culture. J Cell Biol 74:963–982

Bernfield M, Banerjee SD, Koda J, Rapraeger A (1984) Remodeling of the basement membrane: morphogenesis and maturation. CIBA Found Symp 108:179–196

Bernston ED, Ramos DM, Kramer RH (1994) Metastatic melanoma cells interact with the reticular fibers of the lymph node. Melanoma Res 4:115–125

Birembaut P, Adnet JJ, Labat-Robert J, Mercantini F, Robert L (1981) Fibronectin in breast cancer. In: Hollman KH, de Brux J, Werley JM (eds) New frontiers in mammary pathology, vol II. Plenum Press, London, pp 175–184

Bix G, Fu J, Gonzales EM, Macro L, Barker A, Campbell S, Zutter MM, Santoro SA, Kim JK, Höök M, Reed CC, Iozzo RV (2004) Endorepellin causes endothelial cell disassembly of actin cytoskeleton and focal adhesions through α2β1 integrin. J Cell Biol 166:97–109

Blood CH, Zetter BR (1999) Membrane-bound PKC modulates receptor affinity and chemotactic responsiveness of Lewis lung carcinoma sublines to an elastin-derived peptide. J Biol Chem 264:10614–10620

Blumberg PM, Driedger PE, Rossow PW (1976) Effect of a phorbol ester on a transformation-sensitive surface protein of chick fibroblasts. Nature 264:446–447

Bradshaw RA, Dennis EA (2004) Handbook of cell signaling, vol I. Academic Press, New York

Colige A, Nusgens B, Lapière CM (1988) Effect of EGF on human skin fibroblasts is modulated by the extracellular matrix. Arch Dermatol Res 280 [Suppl]:S42–S46

Colige A, Nusgens B, Lapière CM (1990) Response to epidermal Growth Factor of skin fibroblasts from donors of varying age is modulated by the extracellular matrix. J Cell Physiol 145:450–457

Comper WD (ed) (1996) Extracellular matrix, vols 1–2. Harwood Academic Publ, Australia

Davies GE, Bayless KJ, Davis MJ, Leininger GA (2000) Regulation of tissue injury responses by the exposure of matricryptic sites within extracellular matrix molecules. Am J Pathol 156:1489–1498

DiPersio CM, Shao M, Di Costanzo L, Kreidberg JA, Hynes RO (2000) Mouse keratinocytes immortalized with large T antigen acquire alpha3beta1 integrin-dependent secretion of MMP-9/gelatinase. Br J Cell Sci 113:2909–2921

Eble JA, Kuhn K (1997) Integrin-ligand interactions. Springer-Landes, New York, Austin

Egeblad M, Werb Z (2002) New functions for the matrix metalloproteinases in cancer progression. Nature Rev Cancer 2:163–176

Ekblom P, Timpl R (eds) (1996) The laminins. Harwood Academic, Amsterdam

Ekblom P, Durbeej M, Ekblom M.(1996) Laminin isoforms in development. In: Ekblom P, Timpl R (eds) The laminins. Harwood Academic, Amsterdam, pp 185–216

Ekblom M, Klein G, Mugrauer G, Fecker L, Deutzmann R, Timpl R, Ekblom P (1990) Transient and locally restricted expression of laminin A chain mRNA by developing epithelial cells during kidney organogenesis. Cell 60:337–346

Elkin M, Cohen I, Zcharia E, Orgel A, Guatta-Rangini A, Peretz T, Voldavsky I, Kleinma, HK (2003) Regulation of heparanase gene expression by estrogen in breast cancer. Cancer Res 63:8821–8826

Emod I, Lafaye P, Planchenault T, Lambert Vidmar S, Imhoff JM, Keil-Dlouha V (1990) Potential proteolytic activity of fibronectin: fibronectin laminase and its substrate specificity. Biol Chem Hoppe-Seyler 371:129–135

Faury G, Ristori MT, Verdetti J, Jacob MP, Robert L (1995) Effect of elastin peptides on vascular tone. J Vasc Res 32:112–119

Faury G, Chabaud A, Ristori MT, Robert L, Verdetti J (1997) Effect of age on the vasodilatory action of elastin peptides. Mech Age Dev 95:31–42

Faury G, Usson Y, Robert-Nicoud M, Robert L, Verdetti J (1998) Nuclear and cytoplasmic free calcium level changes induced by elastin peptides in human endothelial cells. Proc Natl Acad Sci USA 95:2967–2972

Fujiwara H, Kikkawa Y, Sanzen N, Sekiguchi K (2001) Purification and characterization of human laminin-8. Laminin-8 stimulates cell adhesion and migration through $\alpha3\beta1$ and $\alpha6\beta1$ integrins. J Biol Chem 276:17550–17558

Fülöp T, Larbi A (2002) Putative role of 67 kDa elastin-laminin receptor in tumor invasion. Semin Cancer Biol 12:219–229

Fülöp T, Jacob MP, Khalil A, Wallach J, Robert L (1998) Biological effects of elastin peptides. Pathol Biol 46:497–506

Gahmberg CG, Hakomori SI (1973) Altered growth behavior of malignant cells associated with changes in externally labelled glycoprotein and glycolipid. Proc Natl Acad Sci USA 70:3329–3333

Goldschmidt O, Zcharia E, Abramovitch R, Metzger S, Aingorn H, Friedmann Y, Schirmacher V, Mitrani E, Vlodavsky I (2002) Cell surface expression and secretion of heparanase markedly promote tumor angiogenesis and metastasis. Proc Natl Acad Sci USA 99:10031–10036

Goldschmidt O, Zcharia E, Cohen M, Aingorn H, Cohen I, Nadav L, Katz BZ, Geiger B, Vlodavsky I (2003) Heparanase mediates cell adhesion independent of its enzymatic activity. FASEB J 17:1015–1025

Grant DS, Yenisey C, Wesley Rose R, Tootell M, Santra M, Iozzo RV (2002) Decorin suppresses tumor-cell-mediated angiogenesis. Oncogene 21:4765–4777

Grobstein C (1953) Epithelio-mesenchymal specificity in the morphogenesis of mouse submandibular rudiments in vitro. J Exp Zool 124:383–413

Grobstein C (1956) Inductive tissue interactions in development. Adv Cancer Res 4:187–236

Grobstein C (1967) Mechanism of organogenetic tissue interaction. Natl Cancer Inst Monogr 26:279–299

Gross J, Lapière CM (1962) Collagenolytic activity in amphibian tissues: a tissue culture assay. Proc Natl Acad Sci USA 48:1014–1022

Groult V, Hornebeck W. Robert L, Pouchelet M, Jacob MP (1998) Interactions of elastin fibers with fibroblasts – a time-lapse cinemicrographic study. Pathol Biol 46:507–516

Gullberg D, Ekblom P (1997) Integrins during development. In: Eble JA, Kühn K (eds) Integrin-ligand interaction. Springer, Berlin Heidelberg New York, pp 253–267

Haack H, Hynes RO (2001) Integrin receptors are required for cell survival and proliferation during development of the peripheral glial lineage. Dev Biol 233:38–55

Hascall V (ed) (1986) Functions of the proteoglycans. Wiley, Chichester

Honn KV, Tang DG, Chen YQ (1992) Platelets and cancer metastasis: more than an epiphenomenon. Semin Thromb Hemost 18:392–415

Hornebeck W, Tixier JM, Robert L (1986) Inducible adhesion of mesenchymal cells to elastic fibers: elastonectin. Proc Natl Acad Sci USA 83:5517–5520

Hornebeck W, Emonard H, Monboisse JC, Bellon G (2002) Matrix-directed regulation of pericellular proteolysis and tumor progression. Semin Cancer Biol 12:231–241

Horwitz A, Duggan K, Greggs R, Decker C, Buck C (1985) The cell substrate attachment (CSAT) antigen has properties of a receptor for laminin and fibronectin. J Cell Biol 101:2134–2144

Huhtala P, Humphries MJ, McCarthy JB, Tremble PM, Werb Z (1995) Cooperative signaling by $\alpha5\beta1$ and $\alpha4\beta1$ integrins regulates metalloproteinase gene expression in fibroblasts adhering to fibronectin. J Cell Biol 129:8857–879

Humphries MJ, Ayad SR (1983) Stimulation of DNA synthesis by cathepsin D digests of fibronectin. Nature 305:811–813

Hynes RO (1973) Alteration of cell-surface proteins by viral transformation and by proteolysis. Proc Natl Acad Sci USA 70:3170–3174

Hynes RO (1974) Role of surface alterations in cell transformation: the importance of proteases and surface proteins. Cell 1:147–158

Hynes RO (1976) Cell surface proteins and malignant transformation. Biochim Biophys Acta 458:73–107

Hynes RO (1990) Fibronectins. Springer, Berlin Heidelberg New York

Hynes RO (1999) The dynamic dialogue between cells and matrices: implication of fibronectin's elasticity. Proc Natl Acad Sci USA 96:2588–2590

Hynes RO (2002) Integrins: bidirectional, allosteric signaling machines. Cell 110:673–687

Hynes RO, Lively JC, McCarthy JH, Taverna D, Francis SE, Hodivala-Dilke K, Xiao Q (2002) The diverse roles of integrins and heir ligands in angiogenesis. Cold Spring Harbor Symp Quant Biol 67:143–153

Hynes RO, Zhao Q (2000) The evolution of cell adhesion. J Cell Biol 113:2909–2921

Imhoff JM, Blondeau X, Planchenault T, Emod I, Keil-Dlouha V (1990) Collagenase activation of latent matrix degrading proteinases from human plasma fibronectin. Biol Chem Hoppe-Seyler 371:137–144

Javelaud D, Mauviel A (2004) Mammalian transforming growth factor-βs: Smad signaling and physio-pathological roles. Int J Biochem Cell Biol 36:1161–1165

Julien-Grille S, Bellacosa A, Upson J, Klein-Szanto A, van Roy F, Lee-Kwon W, Donowitz M, Tsichlis PN, Larue L (2003) The protein kinase Akt induces epithelial mesenchymal transition and promotes enhanced motility and invasiveness of squamous cell carcinoma lines. Cancer Res 63:2172–2178

Julien-Grille S, Moore R, Denat L, Morali OG, Delmas V, Bellacosa A, Larue L (2004) The role of insulin-like growth factors in the epithelial to mesenchymal transition. In: Savagner P (ed) Rise and Fall of epithelial phenotype. Eurekah.com and Kluwer Academic/ Plenum, Dordtrecht, pp 1–21

Kadar A, Tökés AM, Kulka J, Robert L (2002) Extracellular matrix components in breast carcinomas. Semin Cancer Biol 12:243–257

Klein G (2002) Introduction: genetic and epigenetic contributions to tumor evolution. Semin Cancer Biol 12:327–330

Klein G (2004) Cancer, apoptosis and nonimmune surveillance. Cell Death Differ 11:13–17

Klein G, Klein E (1955) Variation in cell populations of transplanted tumors as indicated by studies on the ascites transformation. Exp Cell Res (Suppl 3):218–229

Kleinman HK, McGarvey ML, Hassel JR, Cannon VL, Laurie GW, Martin GR (1986) Basement membrane complexes with biological activity. Biochemistry 25:312–318

Kondro W (2004) Molecular biology. Consortium tackles mouse regulome. Science 304(5673):942

Labat-Robert J (2002) Fibronectin in malignancy. Effect of aging. Semin Cancer Biol 12:187–195

Labat-Robert J (2004) Cell-matrix interactions in aging: role of receptors and matricryptins. Ageing Res Rev 3:233–247

Labat-Robert J, Robert L (2000) Interaction between cells and extracellular matrix: signaling by integrins and the elastin-laminin receptor. In: Macieira-Coelho A (ed) Progress in molecular and subcellular biology, vol 25. Springer, Berlin Heidelberg New York, pp 57–70

Labat-Robert J, Potazman JP, Derouette JC, Robert L (1981a) Age-dependent increase of human plasma fibronectin. Cell Biol Int 5:969–973

Labat-Robert J, Birembaut P, Robert L, Adnet JJ (1981b) Modification of fibronectin distribution pattern in solid human tumors. Diagn Histopathol 4:299–306

Labat-Robert J, Birembaut P, Robert L, Adnet JJ (1983) Fibronectin distribution pattern in solid human tumors. In: Lapis K, Jeney A, Price MR (eds) Tumor progression and markers. Kugler Publ, Amsterdam, pp 387–392

Labat-Robert J, Kern P, Robert L (1992) Biomarkers of connective tissue aging: biosynthesis of fibronectin, collagen type III and elastase. Ann NY Acad Sci 673:16–22

Labat-Robert J, Marques MA, N'Doye S, Alperovitch A, Moulias R, Allard M, Robert L (2000) Plasma fibronectin in French centenarians. Arch Gerontol Geriatr 31:95–105

Lapis K, Timar J (2002) Role of elastin-matrix interactions in tumor progression. Semin Cancer Biol 12:209–217

Laurent TC (1970) Structure of hyaluronic acid. In: Balazs EA (ed) Chemistry and molecular biology of the intercellular matrix, vol 2. Academic Press, London, pp 703–742

Laurent TC (ed) (1998) The biology of hyaluronan. Wiley, Chichester

Levy-Adam F, Miao HQ, Heinrikson RL, Vlodavsky I, Ilan N (2003) Heterodimer formation is essential for heparanase enzymatic activity. Biochem Biophys Res Commun 308:885–891

Li DY, Brooke B, Davis EC, Mecham RP, Sorensen LK, Boak BB, Eichwald E, Keating MT (1998a) Elastin is an essential determinant of arterial morphogenesis. Nature 393:276–280

Li DY, Faury G, Taylor DG, Davis EC, Boyle WA, Mecham RP, Stenzei P, Boek B, Keating MT (1998b) Novel arterial pathology in mice and humans hemizygous for elastin. J Clin Invest 102:1783–1787

Liotta LA, Kohn E (2004) Cancer and the homeless cell. Nature 430:973–974

Ljubimova JY, Lakhter AJ, Loksh A, Yong WH, Riedinger MS, Miner JH, Sorokin LM, Ljubimov AV, Black KL (2001) Overexpression of α4 chain containing laminins in human glial tumors identified by gene microarray analysis. Cancer Res 61:5601–5610

Määttä M, Virtanen I, Burgeson R, Autio-Harmainen H (2001) Comparative analysis of the distribution of laminin chains in the basement membranes in some malignant epithelial tumors: the α1 chain of laminin sows a selected expression pattern in human carcinomas. J Histochem Cytochem 49:711–725

Macieira-Coelho A (2002) The biology of conformation in the regulation of the senescent and transformed cell phenotype. Semin Cancer Biol 12:165–171

Maffini MV, Soto AM, Calabro JM, Ucci AA, Sonnenschein C (2004) The stroma as a crucial target in rat mammary gland carcinogenesis. J Cell Sci 117:1495–1502

Magnusson KP, Satalino R, Qian W, Klein G, Wiman KG (1998) Is conversion of solid into more anoxic ascites tumors associated with p53 inactivation? Oncogene 17:2333–2337

Martin GR, Liotta LA, Kleinman HK (1996) Interaction of tumor cells with basement membranes and laminin-1. In: Ekblom P, Timpl R (eds) The laminins. Harwood Academic Publ, Amsterdam, pp 277–290

Mecham RP, Hinek A (1996) Non-integrin laminin receptors. In: Ekbom P, Timpl R (eds) The laminins. Harwood Academic, Amsterdam, pp 159–183

Mercurio AM, Rabinovitz I (2001) Towards a mechanistic understanding of tumor invasion-lessons from the α6β4 integrin. Semin Cancer Biol 11:129–141

Miller JH, Reznikoff WS (eds) (1978) The operon. Cold Spring Harbor Laboratory Press, Cold Spring Harbor

Morali OG, Delmas V, Moore R, Jeanney C, Thiery JP, Larue L (2001) IGF-II induces rapid β-catenin relocation to the nucleus during epithelium to mesenchyme transition. Oncogene 20:4942–4950

Morley M, Molony CM, Weber T, Devlin JL, Ewens KG, Spielman RS, Cheung VG (2004) Genetic analysis of genome-wide variation in human gene expression. Nature 430:743–747

Mostafavi-Pour Z, Askari JA, Parkinson SJ, Parker PJ, Ng TTC, Humphries MJ (2003) Integrin-specific signaling pathways controlling focal adhesion formation and cell migration. J Cell Biol 161:155–167

Nichols EJ, Fenderson BA, Carter WG, Hakomori SI (1986) Domain-specific distribution of carbohydrates in human fibronectin and the transformation-dependent translocation of branched type 2 cahin defined by monoclonal antibody C6. J Biol Chem 261:11295–11301

Noel A, Callé A, Emonard H, Nusgens B, Foidart JM, Lapiere CM (1988) Antagonistic effects of laminin and fibronectin in cell to cell and cell to matrix interactions in MCF-7 cultures. In Vitro Cell Dev Biol 24:373–380

Pattaroyo M, Tryggvason K, Virtanen I (2002) Laminin isoforms in tumor invasion, angiogenesis and metastasis. Semin Cancer Biol 12:197–207

Paye M, Nusgens BV, Lapière CM (1987) Modulation of cellular biosynthetic activity in the retracting collagen lattice. Eur J Cell Biol 45:44–50

Perdommo JJ, Gounon P, Schaeverbeke M, Schaeverbeke J, Groult V, Jacob MP, Robert L (1994) Interaction between cells and elastin fibers: an ultrastructural and immunocytochemical study. J Cell Physiol 158:451–458

Péterszegi G, Texier S, Robert L (1999) Cell death by overload of the elastin-laminin receptor on human activated lymphocytes: protection by lactose and melibiose. Eur J Clin Invest 29:166–172

Péterszegi G, Dagonet F, Labat-Robert J, Robert L (2002) Inhibition of cell proliferation and fibronectin biosynthesis by Na ascorbate. Eur J Clin Invest 32:372–380

Pyke C, Römer J, Kallunki P, Lund LR, Ralfkiär E, Danö K, Tryggvason K (1994) The γ2 chain of kalinin/laminin-5 is preferentially expressed in invading malignant cells in human cancers. Am J Pathol 145:782–791

Pyke C, Salo S, Ralfkiär E, Römer J, Danö K, Tryggvason K (1995) Laminin-5 is a marker of invading cells in some human carcinomas and is coexpressed with the receptor for urokinase plasminogen activator in budding cancer cells in colon adenocarcinomas. Cancer Res 55:4132–4139

Quigley JP (1976) association of a protease (plasminogen activator) with a specific membrane fraction from transformed cells. J Cell Biol 71:472–486

Quigley JP (1979) Phorbol ester-induced morphological changes in transformed chick fibroblasts: evidence for direct catalytic involvement of plasminogen activator. Cell 17:131–141

Quigley JP, Goldfarb RH, Scheiner C, O'Donnell-Tormey J, Yeo TK (1980) Plasminogen activator and the membrane of transformed cells. In: Hynes RO, Fox CF (eds) Tumor cell surfaces and malignancy. Liss, New York, pp 773–796

Reed CC, Iozzo RV (2003) The role of decorin in collagen fibrillogenesis and skin homeostasis. Glucoconjugate J 19:249–255

Ringertz N, Klein E, Klein G (1957) Histopathologic studies of peritoneal implantation and lung metastasis at different stages of the gradual transformation of the MC1 M mouse sarcoma into ascites form. J Natl Cancer Inst 18:173–199

Robert L (1986) Structural glycoproteins. In: Labat-Robert J, Timpl R, Robert L (eds) Structural glycoproteins in cell-matrix interactions. Front matrix biology, vol 11. Karger, Basel, pp 1–16

Robert L (2000) Cellular and molecular mechanisms of aging and age related diseases. Path Oncol Res 6:3–9

Robert L, Hornebeck W (1989) Elastin and elastases, vols I and II. CRC Press, Boca Raton

Robert L, Labat-Robert J (2000) Aging of connective tissues, from genetic to epigenetic mechanisms. Biogerontology 1:123–131

Rodriguez-Manzaneque JC, Lane TF, Ortega MA, Hynes RO, Lawler J, Iruela-Harispe ML (2001) Thrombospondin-1 suppresses spontaneous tumor growth and inhibits activation of matrix metalloproteinase-9 and mobilization of vascular endothelial growth factor. Proc Natl Acad Sci USA 98:12485–12490

Santra M, Reed CC, Iozzo RV (2002) Decorin binds to a narrow region of the epidermal growth factor (EGF) receptor, partially overlapping but distinct from the EGF-binding epitope. J Biol Chem 277:35671–35681

Saxen L, Koskimies O, Lahti A, Miettinen H, Rapola J, Wartiovaara J (1968) Differentiation of kidney mesenchyme in an experimental model system. Adv Morphogr 7:251–293

Scott JE (1988) Proteoglycan-fibrillar collagen interactions. Biochem J 252:313–323

Schnaper HW, Kleinman HK, Grant DS (1993) Role of laminin in endothelial cell recognition and differentiation. Kidney Int 43:20–25

Seftor REB, Seftor EA, Koshikawa N, Meltzer PS, Gardner EMG, Bilban M, Stetler-Stevenson WG, Quaranta V, Hendrix MJC (2001) Cooperative interaction of laminin-5 γ2 chain, matrix metalloproteinase-2, and membrane type-1-matrix/metalloproteinase are required for mimicry of embryonic vasculogenesis by aggressive melanoma. Cancer Res 61:6322–6327

Simizu S, Ishida K, Osada H (2004) Heparanase as a molecular target of cancer chemotherapy. Cancer Sci 95:553–558

Sonnenberg A, Linders CJT, Modderman PW, Damsky CH, Aumailley M, Timpl R (1990) Integrin recognition of different cell-binding fragments of laminin (P1, E3, E8) and evidence that α6β1 but not α6β4 functions as a major laminin receptor for fragment E8. J Cell Biol 110:2145–2155

Szendroi M, Lapis K, Zalatnai A, Robert L, Labat-Robert J (1984) Appearance of fibronectin in putative preneoplastic lesions and in hepato-cellular carcinoma during chemical hepatocarcinogenesis in rats and in human hepatomas. J Exp Pathol 1:189–199

Tamkun JW, DeSimone DW, Fonda D, Patel RS, Buck C, Horwitz AF, Hynes RO (1986) Structure of integrin, a glycoprotein involved in the transmembrane linkage between fibronectin and actin. Cell 46:271–282

Taverna D, Disatnik MH, Rayburn H, Bronson RT, Yang J, Raando TA, Hynes RO (1998) Dystrophic muscle in mice chimeric for expression of alpha5 integrin. J Cell Biol 143:849–859

Taverna D, Moher H, Crowley D,Borsig L, Varki A, Hynes RO (2004) Increased primary tumor growth in mice null for beta3- or beta3/beta5-integrins or selectins. Proc Natl Acad Sci USA 101:763–768

Temkin V, Aingorn H, Puxeddu H, Goldschmidt O, Zcharia E, Gleich GJ, Vlodavsky I, Levi-Schaffer F (2004) Eosinophil major basic protein: first identified natural heparanase-inhibiting protein. J Allergy Clin Immunol 113:703–709

Timar J, Tixier JM, Robert L, Hornebeck W, Lapis K (1990) Metastatic phenotype-dependent interactions between elastin and tumor cells. Clin Exp Metastasis 8:50

Timar J, Diczasi Cs, Ladanyi A, Raso E, Hornebeck W, Robert L, Lapis K (1995) Interaction of tumour cells with elastin and the metastatic phenotype. In: Robert L (ed) The molecular biology and pathology of elastic tissues. Wiley, Chichester, pp 321–337

Timar J, Lapis K, Dudas J, Sebestyen A, Kopper L, Kovalszky I (2002) Proteoglycans and tumor progression: Janus-faced molecules with contradictory functions in cancer. Semin Cancer Biol 12:173–186

Travis MA, Humphries JD, Humphries MJ (2003) An unraveling tale of how integrins are activated from within. Trends Pharmacol Sci 24:192–197

Varga Z, Jacob MP, Robert L, Fülöp T Jr (1989) Identification and signal transduction mechanism of elastin peptide receptor in human leukocytes. FEBS Lett 258:5–8

Varner JA, Emerson DA, Juliano RL (1995) Integrin $\alpha 5\beta 1$ expression negatively regulates cell growth:reversal by attachment to fibronectin. Mol Biol Cell 6:725–740

Vlodavsky I, Friedmann Y (2001) Molecular properties and involvement of heparanase in cancer metastasis and angiogenesis. J Clin Invest 108:341–347

Wang AZ, Ojakian GK, Nelson WJ (1990) Steps in the morphogenesis of polarized epithelium. Disassembly and assembly of plasma membrane domains during reversal of epithelia cell polarity in multicellular epithelial (MDCK) cysts. J Cell Sci 95:153–165

Werb Z, Tremble PM, Behrendtsen O, Crowley E, Damsky CH (1989) Signal transduction through the fibronectin receptor induces collagenase and stromelysin gene expression. J Cell Biol 109:877–889

Xie DL, Homandberg GA (1993) Fibronectin fragments bind to and penetrate cartilage tissue resulting in proteinase expression and cartilage damage. Biochim Biophys Acta 1182:189–196

Yamada KM, Yamada SS, Pastan I (1976) Cell surface protein partially restores morphology, adhesiveness and contact inhibition of movement to transformed fibroblasts. Proc Natl Acad Sci USA 73:1217–1221

Yamada KM, Schlessinger DH, Kennedy DW, Pastan I (1977) Characterisation of a major fibroblast cell surface glycoprotein. Biochemistry 16:5552–5559

Yang JT, Bader BL, Kreidberg JA, Ullman-Cullere M, Trevithick JE, Hynes RO (1999) Overlapping and independent functions of fibronectin receptor integrins in early mesodermal development. Dev Biol 215:264–277

Yang J, Mani SA, Donaher JL, Ramaswamy S, Itzykson RA, Come C, Savagner P, Gitelman I, Richardson A, Weinberg RA (2004) Twist, a master regulator of morphogenesis, plays an essential role in tumor metastasis. Cell 117:927–939

Yurchenco PD, Wadsworth WG (2004) Assembly and tissue function of early embryonic laminins and netrins. Curr Opin Cell Biol 16:572–579

Zcharia E, Metzger S, Chakek-Shaul T, Aingorn H, Elkin M, Friedmann Y, Weinstein T, Li JP, Lindahl U, Vlodavsky I (2004) Transgenic expression of mammalian heparanase uncovers physiological functions of heparan sulfate in tissue morphogenesis, vascularization, and feeding behavior. FASEB J 18:252–263

Zerlauth G, Wolf G (1985) Studies on the tumour promoter-induced release of fibronectin from human lung fibroblasts, and its counteraction by retinoic acid. Carcinogenesis 6:531–534

Wilms' Tumor: Starting Off the Kidney All Over Again?

M. Little

1 Wilms' Tumor and the WT1 Gene

Wilms' tumor or nephroblastoma, while comprising only 6% of all renal tumors, is one of the most common solid tumors of childhood, occurring in around 1 in 10,000 live births worldwide with no evidence of demographic clustering or environmental etiological factors (Beckwith 1983). This condition most frequently arises in a sporadic fashion (95% of presentations) within the first three years of life, although it can also present in adulthood (Kozman et al. 1989; Terenziani et al. 2004) or in utero (Beckwith 1999). Usually, only one kidney is affected (unilateral), but bilateral cases represent approximately 5% of cases and sometimes occur in association with genitourinary anomalies, including double ureter, horseshoe kidney, cryptorchidism, micropenis, and hypospadias (Barakat et al. 1974; Miller et al. 1964). Less than 1% of cases present in a familial fashion (Matsunaga 1981).

Current treatment regimes for Wilms' tumor favor combination therapy with chemotherapy prior to resection, sometimes also including radiotherapy. The result is a very high patient survival rate (>85%; McLorie 2001). While reduction of morbidity and long-term complications of the therapy remain an issue, the primary interest in Wilms' tumor for the last 15 years has been as a paradigm for the two-hit model of tumorigenesis. This led to the isolation of the Wilms' tumor suppressor gene, WT1 (Call et al. 1990; Gessler et al. 1990). Indeed, it was the 1% of Wilms' tumor patients that presented with Wilms' tumor as part of the *WAGR* syndrome (Wilms' tumor, Aniridia, Genitourinary anomalies, mental Retardation; Riccardi et al. 1978) that were critical in the identification of the WT1 gene. WAGR patients were shown to have constitutional deletions of the short arm of human chromosome 11 (11p13) (Francke et al. 1979), allowing the positional cloning of WT1 and Pax6. The latter was shown to be the gene in which mutation resulted in the aniridia component

M. Little
Institute for Molecular Bioscience, Level 4 North, Queensland Bioscience Precinct, The University of Queensland, St. Lucia, 4072, Qld, Australia

Progress in Molecular and Subcellular Biology
Alvaro Macieira-Coelho (Ed.)
Development Biology of Neoplastic Growth

of the syndrome (Ton et al. 1991). Both the Wilms' tumor and genitourinary anomalies result from loss of the WT1 gene (Bruening et al. 1992; Pelletier et al. 1991b).

The isolation of WT1 in 1990 (Call et al. 1990; Gessler et al. 1990) was heralded as the 'answer' for Wilms' tumor. This gene was shown to encode a putative nuclear transcription factor containing four tandem C-terminal C_2H_2 zinc finger motifs similar to those present in the transcriptional activator EGR1 (Haber et al. 1990; Rauscher et al. 1990). In fact, WT1 proteins were shown to bind EGR1-like DNA binding sites. Unlike the Rb1 *tumor suppressor* that had at that time been recently identified as the site of mutations resulting in retinoblastoma (Dunn et al. 1988), expression of the WT1 gene was not ubiquitous but highly tissue-specific. WT1 mRNA is expressed in both humans and mice in the developing gonads, mesonephros, and *metanephric mesenchyme*, but not the Wolffian duct (Armstrong et al. 1992; Pritchard-Jones et al. 1990). WT1 expression increases as the mesenchyme undergoes a mesenchyme-to-epithelial transition to form the nephrons, and expression persists post-natally in the *podocytes* of the glomeruli. WT1 is expressed in the indeterminate gonad, after which expression in seen in the Sertoli cells of the testis and granulosa cells of the ovary. This expression pattern supported the concept of this protein having a role in gonad as well as in kidney development (van Heyningen et al. 1990; Pelletier et al. 1991b). Intragenic mutations in WT1 were subsequently detected in the sporadic form of Wilms' tumor (Haber et al. 1990; Pelletier et al. 1991b; Little et al. 1992), demonstrating that this gene was involved not only in the syndromic presentation of Wilms' tumor. Patients with the rare *Denys-Drash syndrome* (DDS) were also shown to display missense point mutations within the DNA-binding domain of WT1 (Pelletier et al. 1991a; Little et al. 1993). DDS patients again show defects in both the kidney and the gonads, with patients displaying XY pseudohermaphroditism, early renal failure due to mesangial sclerosis, and predisposition to Wilms' tumor and gonadoblastoma (Denys et al. 1967; Drash et al. 1970). However, subsequent analyses of a large number of Wilms' tumors revealed that the prevalence of intragenic mutation in this gene was as low as 10% (Varanasi et al. 1994; reviewed in Little and Wells 1997). Hence, while two hits in the WT1 gene were sufficient for tumorigenesis, this was not the only way in which Wilms' tumor arose, leaving the reality that Wilms' tumor was not genetically uniform but likely to result from mutations in more than one causative gene with individual tumors representing phenocopies.

The decade to follow saw a large body of literature proposing target genes for the WT1 gene, primarily using the techniques of promoter analysis to find EGR1-like potential binding sites followed by reporter analysis in cell lines (reviewed in Little et al. 1999). In this way a number of other transcription factors (EGR1, Pax2, c-myc, c-myb, Bcl2, myoD, p21, Sry, Dax1) including WT1 itself, growth factors (Igf2, PDGF-A, TGFβ, CSF-1, CTGF, amphiregulin, midkine, inhibin α) and their receptors (IGF1R, EGFR, RARα, insulin R, vitamin D receptor), extracellular matrix and cell–cell communication proteins (syndecan, thrombospondin 1, NovH, E cadherin), signaling molecules (sprouty 1), and other genes have been proposed to be regulated by WT1 (S.B. Lee et al. 1999; reviewed in Little et al. 1999 and Scharnhorst et al. 2001; see also Gross et al. 2003). None of the proposed WT1 target genes have been shown to contain mutations in Wilms' tumor. Indeed, the results of expression profiling to verify existing target genes and identify new target genes have been quite varied (Thate et al. 1998; S.B. Lee et al. 1999; T.H. Lee et al. 1999; Lee and Pelletier 2001). WT1 has been shown to have indirect effects on a number of pathways elicited, at least in some instances, via protein-protein interactions. For example, WT1 affects the expression of anti-Mullerian hormone (AMH/MIS) and aromatase P450 (P450arom) during gonad development by competing with Dax1 for protein binding to SF1 (Natchigal et al. 1998; Gurates et al. 2003). We have also proposed that an interaction between steroid-response-element-binding proteins (SREBPs) results in the repression of the mevalonate pathway of cholesterol biosynthesis, potentially affecting protein prenylation and lipoprotein production (Rae et al. 2004). None of these data coherently explain why a disruption in WT1 leads to Wilms' tumor, however, there are clearly multiple roles for WT1 in kidney development (see Table 1). The fact that a total knockout of the gene results in renal agenesis suggests a role in survival of the metanephric mesenchyme (Kreidberg et al. 1993). The histology of Wilms' tumors suggests a role in nephron formation and the continued podocyte expression suggests a role in ongoing renal function for WT1. However, there are clearly other genes involved in Wilms' tumor etiology.

Table 1. Evidence for multiple roles for WT1 in kidney development and renal function

Functional role	Evidence
Maintenance of the survival of the metanephric mesenchyme	- Renal and gonadal agenesis - Metanephric mesenchyme arises in knockout mice, then apoptosis (Kreidberg et al. 1993)
Regulation of the mesenchyme to epithelial transition required for nephron induction	- Triphasic histology of Wilms' tumors - Evidence of indirect activation of Wnt4 (Sim et al. 2002) - Knockdown of WT1 in explants (Davies et al. 2004) - Loss of caudal mesonephric tubules (Sainio et al. 1997)
Maintenance of podocyte function in the postnatal period	- Development of diffuse mesangial sclerosis and early onset of renal failure in Denys-Drash syndrome

2 A Role for Other Genes and Chromosomal Regions in Wilms' Tumor

Even without the results of WT1 mutation analysis, there were known to be multiple chromosomal regions implicated in Wilms' tumor. Original loss of heterozygosity studies in sporadic Wilms' tumors prior to the isolation of WT1 identified two distinct regions of the short arm of chromosome 11 involved in Wilms' tumor; 11p13 and 11p15. The latter region contains the gene insulin-like growth factor 2 (Igf2) and is a region that is linked to the fetal overgrowth syndrome, *Beckwith Wiedemann syndrome* (BWS) (Ping et al. 1989). BWS patients have a predisposition to Wilms' tumor and other embryonal tumors such as hepatoblastoma and rhabdomyosarcoma. Subsequent studies have revealed that the 11p15 region contains an imprinting center such that a large number of genes in the area are paternally expressed, including Igf2. The adjacent H19 gene is also imprinted such that it is the maternally inherited copy that is transcriptionally active (Lewis and Murrell 2004). Disruption to the imprinting of this region resulting in BWS and Wilms' tumor can result from trisomy (two paternal copies), uniparental disomy for the paternal allele, preferential maternal allele loss, translocation to disrupt the imprinting center, and reactivation of the maternal allele. Ogawa et al. (1993) showed that some sporadic Wilms' tumors have reactivation of maternal 11p15 genes. However, the actual 11p15 causative gene for Wilms' tumor has not been positively identified. Wilms' tumor can also occur as a part of the overgrowth syndrome *Simpson-Golabi-Behmel syndrome* (SGBS). The gene responsible for SGBS1 is glypican 3 (GPC3) on chromosome Xq26. Mutations in GPC3 have now been detected in sporadic Wilms' tumors (White et al. 2002).

The Wnt4 gene was shown to be indirectly activated by WT1 using nylon array analysis (Sim et al. 2002). Wnt4 is a critical secreted factor in normal kidney development, being involved in the mesenchyme-to-epithelial transition events which lead to the formation of the nephrons from the metanephric mesenchyme (Stark et al. 1994). The downstream effector of the canonical Wnt signaling pathway, beta-catenin, has been shown to be mutated in 15% of Wilms' tumors, resulting in the loss of key phosphorylation sites and the consequent activation of beta-catenin (Koesters et al. 1999). While this might suggest a pathway linking WT1 and beta-catenin, Maiti et al. (2000) have shown a very high correlation between the presence of WT1 and beta-catenin mutations in Wilms' tumors and suggest that these two genes actually regulate different pathways both critical for normal kidney development and both needing to be disrupted to form a tumor.

Although rarely presenting in a familial fashion, a number of large familial Wilms' tumor pedigrees have been described, and analysis of these has ruled out the involvement of chromosome 11p13 or 11p15 (Grundy et al. 1988; Huff et al. 1988; Schwartz et al. 1991). Rahmen et al. (1996) subsequently identified a familial Wilms' tumor gene (FWT1) on chromosome 17q21 (1996), but the gene has not been identified. Other known Wilms' tumor pedigrees do not show linkage to this region. Chromosomal anomalies in sporadic Wilms' tumor also include sites on chromosomes 1p36, 1q21, 7p (Miozzo et al. 1996; Grundy et al. 1998), 16q (Maw et al. 1992), and 18 (Olson et al. 1995; reviewed in Slater and Mannens 1992), but no causative genes have been identified at these sites.

This takes us back to the drawing board in terms of what we know about the molecular basis of Wilms' tumor. This chapter will re-examine what we know about Wilms' tumor ontology based upon its histology, presentation, and genetic expression profile, and investigate the proposal that a Wilms' tumor represents normal kidney development gone wrong. Indeed, the case for Wilms' tumor representing arrested development is convincing and will be discussed further, not only for what normal development can tell us about tumorigenesis, but also what Wilms' tumor can tell us about development and regeneration.

3
Setting the Stage – Normal Kidney Development

The permanent kidney, the metanephros, is regarded as arising from two distinct mesodermal derivatives in the mid-gestation fetus; the ureteric bud (UB) and the metanephric mesenchyme (MM). The UB is a side branch of the mesonephric (Wolffian) duct, whereas the MM is a specific region of the intermediate mesoderm that becomes committed to a renal fate. A series of reciprocal inductive events, involving both long- and short-range signals, occurs between the UB and the MM. The MM signals to the adjacent Wolffian duct in order to give rise to the UB, which then grows towards the MM (Saxen 1987; see Fig. 1). This process is driven at least in part by the production of the glial-derived neurotrophic factor (GDNF) by the MM with the signal being transduced by the UB via the c-ret receptor tyrosine kinase and another GDNF receptor, GDNFRα (Durbec et al. 1996; Jing et al. 1996; Moore et al. 1996; Pichel et al. 1996; Sanchez et al. 1996). BMP4 expression along the Wolffian duct and the trunk of the UB is known to limit the formation of additional ureteric buds (Miyazaki et al. 2003). Having reached the adjacent MM, the UB is induced to undergo dichotomous branching until it forms an extensive 'tree' that will serve as the collecting duct system of the kidney. In turn, morphogens produced by the tips of the advancing UB signal to the adjacent MM to induce the formation of aggregates that subsequently undergo a mesenchyme-to-epithelial transition (MET). These epithelial vesicles elongate to form comma- and then S-shaped bodies. Endothelial and mesangial cells migrate into the cleft of each S-shaped body to form the glomerular tufts, while the rest of the S-shaped body elongates to form the proximal tubules, distal tubules, and loops of Henle (reviewed in Saxen 1987; Davies and Bard 1998; see Figure 1). Until the completion of nephrogenesis (36 weeks of gestation in

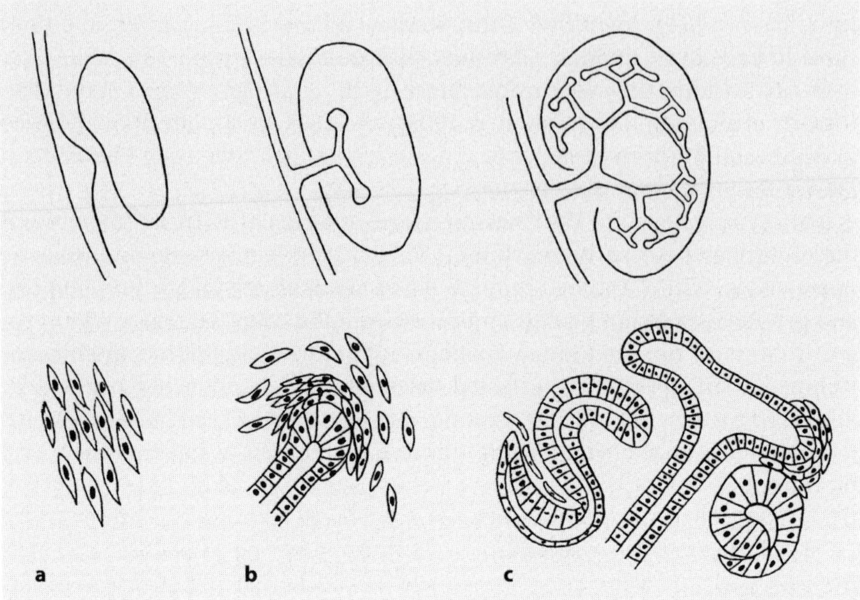

Fig. 1. Normal development of the metanephros. For each diagram the tissue structure is shown at the top and the cellular organization shown directly below. **a** The permanent kidney or metanephros arises from around embryonic day 10 in the mouse, when the ureteric bud (UB) starts as an outgrowth of the mesonephric (Wolffian) duct and grows toward the committed but uninduced metanephric mesenchyme (MM). As the UB reaches the MM, the MM condenses around the tip of the bud to form the cap mesenchyme. **b** At embryonic day 11.5 in the mouse, the UB has branched once (T-stage) and induces the surrounding MM to condense at its tips. **c** Condensed MM gives rise to aggregates at the edge of the cap mesenchyme and just away from the tip of the UB. These pre-tubular aggregates undergo a mesenchyme-to-epithelial transition (MET) to form a comma-shaped body (seen on right of UB). This unwinds to form an S-shaped body. Endothelial cells migrate into the cleft of the lower limb of the S-shaped body to form the glomerular capillaries. The lower limb cells facing the cleft give rise to the visceral epithelial cells (podocytes) while the cells on the other side of the lower limb give rise to the parietal epithelial cells which surround the Bowman's space. Each MET event represents the induction of a new nephron

the human), the periphery of the kidney contains a region of undifferentiated MM referred to as the nephrogenic zone (NZ).

Understanding the molecular basis of kidney development is a work in progress that, until recently, has been investigated gene by gene using knockout and transgenic mice together with embryonic kidney explant culture. Most of the well-known growth factor families (e.g., TGFs, IGFs, Wnts, FGFs, BMPs, PDGFs, LIF, EGF, hedgehogs) and transcription factors (Pax, Hox, forkhead, and winged helix proteins) involved in many organ systems are critical for kidney development (reviewed in Schedl and Hastie 2000). In early kidney development, the MM produces a number of factors that both modulate and repress nephron formation and/or regulate UB branching, including BMP7, FGF2, midkine, IGF2, LIF, TGFβ, HGF, and FGF2. The UB also produces morphogens, including Wnt11, EGF, and TGFα. The process of MET is known to require autocrine factors such as Wnt4 (Stark et al. 1994). A number of key transcriptional regulators are known to be key to kidney morphogenesis right from the point of commitment of the MM. These include Eya1, Emx2, and Lim1, which mark the intermediate mesoderm, and WT1 and Pax2, which are first expressed in the MM prior to the generation of the UB. Many of these genes go on to play additional roles in UB branching versus elongation, and in the maturation of the nephron. For example, Pax2 and WT1 overlap in expression in the MM, upregulate in the S-shaped bodies, but become spatially restricted as the S-shaped body unwinds. WT1 eventually shows specific expression in the presumptive visceral epithelium of the glomerulus (the podocytes), whereas Pax2 is expressed in the developing tubules until maturation. There is another component of the kidney, referred to in the literature as the stroma. This tissue expresses the winged helix transcription factor Foxd1 (previously called BF-2) and is found around the periphery of the developing kidney. The loss of Foxd1 expression results in small, dysgenic kidneys, suggesting a critical role for the stroma in normal kidney development (Hatini et al. 1996). Whether the stroma comes from the MM itself or an adjacent mesodermal source has not been clarified.

The final kidney is known to contain over 20 distinct cellular subtypes, including many distinct tubular cells marked by the expression of specific transporters essential to renal function (Al-Awqati and Oliver 2002). Global expression profiling of kidney development has demonstrated distinct stages of kidney development as the renal structures mature (Stuart et al. 2001; Schwab et al. 2003, 2004; Challen et al. 2004). Early stages predominantly express DNA- and chromatin-binding proteins as well as cell-cycle regulators. Most of these genes are not restricted to the kidney and play morphogenetic roles in other organs. Later stages of development show the predominant expression of extracellular matrix proteins (laminins, nidogen, perlecan, agrin), transporters, cell–cell communication proteins (e.g., integrins, cadherins) and specialized structural proteins including nephrin in the slit diaphragms of the developing podocytes (Stuart et al. 2001; our unpubl. work).

Several key features of kidney development are of importance in understanding Wilms' tumor. Firstly, the oldest nephrons are the most central nephrons. Secondly, the Foxd1-expressing mesenchymal compartment of the developing kidney referred to as the stroma is not necessarily synonymous with the tissue termed stroma in a Wilms' tumor. There are mesenchymal elements in the renal interstitium that are MM-derived but do not express Foxd1. Finally, the MM is a highly plastic tissue, thought to give rise to almost all of the >20 distinct cell types of the final kidney, including most sections of the nephron and the intervening interstitial cells. Indeed, there is evidence that endogenous vasculogenesis occurs in the developing kidney with the MM also giving rise to endothelial cells.

4
A Case for Disrupted Development

4.1
Histology Suggesting a Recapitulation of Development

Wilms' tumor has classically been referred to as one of the group of childhood malignancies termed embryonal tumors due to the histological appearance of persistent metanephric mesenchyme or blastema. Classical Wilms' tumors display what is termed triphasic histology, composed of blastemal elements, epithelial elements, and stromal elements. Each of these elements may dominate in a given tumor, so that tumors can be predominantly blastemal, predominantly stromal, or predominantly epithelial. Nevertheless, usually all three elements are present. The prognosis of a given tumor does not appear to correlate with the histological elements prevalent. Where do the classical elements of Wilms' tumor come from? The blastemal elements are thought to represent undifferentiated metanephric mesenchyme that has proliferated rather than differentiated. Analysis of the epithelial elements of Wilms' tumors suggests that these represent regions of metanephric blastema that has attempted to undergo a normal process of mesenchyme-to-epithelial transition but failed to proceed to nephron formation. This can result in the formation of what are called glomeruloid bodies. The stromal elements resemble the interstitial tissue of the early metanephros. Together these observations supports the notion that the metanephros has arrested during development. But how do they reach their typical size?

4.2
Nephrogenic Rests

The existence of *nephrogenic rests* (NR) in normal kidneys has long been interpreted as residual embryonal elements that have failed to differentiate normally. *Nephroblastomatosis* is a condition in which there is a large number of nephrogenic rests. Beckwith et al. (1990) subdivided nephrogenic rests into perilobar (PLNR) and intralobar (ILNR) rests based upon their location within the kidney. Nephroblastomatosis can present with only PLNR, only ILNR, both ILNR and PLNR, or widely spread, diffuse NRs. The very high association between the presence of rests in the kidneys of Wilms' tumor patients (between 28 and 40% of sporadic Wilms' tumor kidneys are rest positive) has raised the possibility that such rests act as precursor lesions for the tumor itself. This again argues that Wilms' tumor is a case of development gone awry. The combination of location, rest histology, age at diagnosis, and associated developmental defects suggests that PLNR represent a later event than ILNR (see Table 2), reflecting the fact that the earliest nephrons are the most central nephrons. Additionally, the tumors found in association with PLNR tend to

Table 2. Features of nephrogenic rests

Rest type	Location	Histology	Age at diagnosis	Associated defects	Associated tumor histology
PLNR	Peripheral	Predominantly blastemal	Late	Sporadic WT, BWS	Tubular and glomeruloid
ILNR	Central	Prominent stroma, some blastemal and epithelial elements	Early	DDS, WAGR	Stroma, blastema and epithelial elements, often presence of heterologous elements

be more tubular/glomeruloid while the tumors associated with ILNR display a greater spectrum of histological features. Of note, Park et al. (1993) showed that WT1 mutations can occur within nephrogenic rests prior to the development of a tumor, suggesting that the WT1 mutation was an early event with an additional event required for formation of a Wilms' tumor.

4.3 Presentation and Location

While almost always occurring in the kidney, there are cases of extra-renal Wilms' tumor. The location of these tumors is almost always retroperitoneal, including sites along the genital tract (testis, uterus, cervix, inguinal canal, ovotestis; Akhtar et al. 1977; Aterman 1989). These locations have been proposed to indicate that extrarenal Wilms' tumors arise from mesonephric remnants rather than metanephric remnants (Pritchard-Jones and Hastie 1990), reflecting the close relationship between the mesonephros and the metanephros and supporting the notion that Wilms' tumor is development gone awry. The fact that Wilms' tumor can arise in association with several fetal overgrowths (BWS, SGBS, Perlman syndrome) and congenital anomaly syndromes (DDS, Frasier syndrome), these latter usually involving the urogenital tract, also demonstrates a link between the genesis of this tumor type and abnormal development.

4.4 The Potential Role of WT1 Gene in MET

The functions of WT1 in normal kidney development are complex and still unclear. However, given that WT1 mutation is sufficient for the formation of a Wilms' tumor, it is important to consider what we do know in order to understand Wilms' tumor etiology. Due to its plasticity, the MM is regarded as the progenitor population of the kidney (Herzlinger et al. 1992). A knockout of WT1 in mice results in total renal and gonadal agenesis, together with no initial induction of the UB (Kreidberg et al. 1993). The MM forms, but undergoes subsequent apoptosis without differentiating. Hence, WT1 appears to be initially involved in the maintenance of this progenitor identity, and possibly in quiescence, in advance of the next inducing signal. Many Wilms' tumors overexpress WT1. This may result in the persistence of metanephric blastema with a subsequent block in nephron formation or may merely mark the origin of the blastema via continued WT1 expression.

While the knockout mouse indicates a role for WT1 in MM commitment and survival, it tells us little about its subsequent role in kidney development. WT1 expression is upregulated as the MM undergoes a MET event and here WT1 may promote differentiation as opposed to proliferation. The observation of clusters of epithelial cells within Wilms' tumor has been suggested to represent attempts at MET that have failed. All these observations support the concept that Wilms' tumor represents arrested development.

4.5 Continued Expression of Early Markers of Kidney Development

Not only WT1 is overexpressed in Wilms' tumor. Many other markers of early kidney development continue to show high levels of expression in these tumors. Insulin-like growth factor 2 was the first such developmental gene shown to be highly expressed in Wilms' tumor (Reeve et al. 1985; Scott et al. 1985), but the literature documents many other examples (see Table 3).

The dilemma here for all of these genes is the question of whether overexpression is causative or indicative of the embryological nature of the tumor. Both WT1 and EGR1 are overexpressed in the blastemal elements of nephroblastoma, reflecting their original expression in the developing kidney. It has been claimed that there is a prognostic relationship between WT1/EGR1 expression in tumors and clinical progression (Ghanem et al. 2000; Scharnhorst et al. 2001). Scharnhorst proposes that the proliferative effect of EGR1 is an important factor in tumor growth. Table 3 shows that in many cases the genes overexpressed are those normally lost in the adult, and their expression in Wilms' tumors reflects the structures in which they are normally expressed during development. Additionally, mutations have not been detected in any of these genes other than WT1 in Wilms' tumors, suggesting that their expression here simply indicates expression at the stage of developmental arrest represented in the tumor. The most definitive analysis of gene expression in Wilms' tumors was that reported by Li et al. (2002) in which they compared the expression patterns of Wilms' tumors with the temporal analysis of kidney development in the rat reported by Stuart et al. (2001). They defined a

Table 3. Developmental genes with maintained expression in Wilms' tumors

Gene	Embryonic expression	Adult expression	Expression in Wilms' tumors	References
CD24	Mesenchyme and developing nephrons	Distal tubules, collecting duct, loops of Henle, parietal epithelial cells	Blastema and epithelial elements	Droz et al. (1990)
Igf2	High expression	No expression	High expression	Haltia et al. (1997); Reeve et al. (1985); Scott et al. (1985)
c-myc; N-myc	Renal mesenchyme		Blastema	Nisen et al. (1986); Udtha et al. (2003)
EGR1	Broad expression	Broad expression	Expressed in all three cell types of classical Wilms' tumor	Ghanem et al. (2000); Udtha et al. (2003)
WT1	MM, commas, and S-shaped bodies, developing glomeruli in presumptive podocytes	Podocytes	Up expressed in blastemal and epithelial elements; high in immature and mature glomeruloid bodies	Pritchard-Jones and Fleming (1991); Grubb et al. (1994); Ghanem et al. (2000); Kanemoto et al. (2003)
Pax2	UB and developing nephrons; expression restricted as S-shaped body matures away from presumptive glomeruli	Little or no expression; some weak residual expression detectable by Northern	Upregulated in blastema and immature glomeruloid bodies	Haltia et al. (1997); Kanemoto et al. (2003)
MK	High	Not present	High expression	Haltia et al. (1997)
synaptopodin	Developing podocytes from the time of S-shaped body	Podocytes	Immature glomeruloid bodies	Kanemoto et al. (2003)
podocalyxin	Developing podocytes from the time of S-shaped body	Podocytes	Immature glomeruloid bodies	Kanemoto et al. (2003)

Table 3. *Continued.*

Gene	Embryonic expression	Adult expression	Expression in Wilms' tumors	References
nephrin	Developing podocytes from the time of S-shaped body	Podocytes	Mature glomeruloid bodies	Kanemoto et al. (2003)
Bcl-2	S-shaped bodies; developing parietal cells	Undetectable	Blastema, mature glomeruloid bodies	Kanemoto et al. (2003)
p27	Developing podocytes from the time of S-shaped body	Podocytes	Mature glomeruloid bodies	Kanemoto et al. (2003)
Cyto-keratin	S-shaped bodies; developing parietal cells	Undetectable	Blastema; Immature and mature glomeruloid bodies	Droz et al. (1990); Kanemoto et al. (2003)
vimentin	Mesenchyme, S-shaped bodies; developing parietal and visceral epithelial cells	Podocytes, endothelium	Immature and mature glomeruloid bodies	Kanemoto et al. (2003)
villin	Not expressed	Brush border of proximal tubules	Tubular elements	Droz et al. (1990)
CD26	Not expressed	Brush border of proximal tubules	Tubular elements	Droz et al. (1990)
Glypican 3	Renal mesenchyme	Not expressed	Blastema; upregulated with IGFII	Saikali and Sinnett (2000)
CD117 (c-kit)	Diffuse staining	Not described	Strong diffuse staining in some tumors	Smithey et al. (2002); Pan et al. (2004)
BA-1	UB and MM	Blastemal and distal epithelial cells	Loop of Henle only	Ishii et al. (1987)

Table 3. *Continued.*

Gene	Embryonic expression	Adult expression	Expression in Wilms' tumors	References
BA-2	Ureteric bud	Distal epithelial structures	Collecting ducts and distal tubules	Ishii et al. (1987)
CALLA (CD10)	Developing glomeruli and proximal tubules	Glomerular epithelium (podocytes) and proximal tubules	Proximal epithelial component and glomeruloid bodies	Ishii et al. (1987); Droz et al. (1990)

set of genes that were preferentially expressed in Wilms' tumor with respect to fetal kidney and other unrelated fetal tissues. This analysis strongly suggests that Wilms' tumors represent arrest at the point of undifferentiated MM, with overexpression of genes of the committed but uninduced MM, including members of the *Pax, Six, Eya, Mox,* and *Sall* families of transcription factors, the Wnt receptor *FZD7*, and regulators of proteoglycan decoration such as the carbohydrate sulfotransferase gene, *CHST1*. Genes such as Foxd1 (*Bf2*) varied, depending upon the degree of stromal tissue present.

What other occasions show re-expression of developmental genes? There is a growing literature suggesting that not only Wilms' tumors show recapitulation of developmental genes, but that renal injury can also lead to the reinitiation of developmental gene networks. Devarajan et al. (2003) have reviewed the expression profiling of the kidney after renal ischemia/reperfusion injury, highlighting the rapid re-expression of genes involved in development during the post-ischemic period. They propose that this ability to recapitulate development is critical in the process of regeneration. What about other renal tumors? Do they re-express developmental pathways? In an analysis of the expression profiles of a spectrum of kidney cancer types, Takahashi et al. (2003) demonstrated a clear clustering of clear cell carcinoma versus other kidney tumors, with Wilms' tumor clustering alongside papillary renal cell carcinoma (RCC), oncocytoma, and transitional cell carcinoma. The fact that the five Wilms' tumors grouped together within the larger cluster supports the disruption of a common pathway in their etiology, despite the fact that WT1 itself is only mutated in 10% of sporadic Wilms' tumors. So while Wilms' tumors do not appear to have a uniform primary mutation, they are more genetically related than individual granular cell RCCs, for example. However, there was no obvious correlation between the co-expression of developmental markers and this cluster of tumor types, arguing that the re-expression of genes involved in development is a hallmark of the Wilms' tumor.

5 Lessons to be Learned

Having presented the evidence that Wilms' tumor does result from disrupted development, what can Wilms' tumor tell us about development and vice versa? What does it tell us about the nature of embryonal tumors as opposed to adult tumors, and what can we draw from this about the nature of childhood? What lessons are there to be drawn relating to progenitor populations and stem cells?

5.1 Teratomatous Behavior in Wilms' Tumor

If Wilms' tumor represents disrupted development, how do we interpret the teratomatous nature of some Wilms' tumors? A teratoma is a tumor formed of a multiplicity of component tissues apparently derived from tissues foreign to the site of the tumor (Bolande 1976). On rare occasions, Wilms' tumor presents as a teratoid form in which >50% of the tumor is made up of teratomatous elements (Variend et al. 1984). While this more extreme teratomatous form is rare, it is not uncommon for Wilms' tumor to contain elements other than blastema, stroma, and epithelium. These teratomatous elements most commonly include bone, cartilage, and skeletal muscle. Normal embryological dogma suggests a mesodermal origin for all kidney elements, including both the MM and the UB. Hence it may not seem surprising to see the disrupted MM adopting other mesodermal alternatives such as bone, cartilage, and muscle, or even fat, which is seen on occasion (Fernbach et al. 1988; Bakshi et al. 2003; Park et al. 2003). However, tissues of apparently non-mesodermal origin can also be present, including squamous epithelium, sometimes with keratinization and formation of hair follicles and sebaceous glands (Karaca et al. 2000; Otani et al. 2001). There is also one case of the presence of a developing tooth (Govender and Hadley 1999). It is likely that in some of these cases, the diagnosis of Wilms' tumor is incorrect and other terms such as mature cystic teratoma have been used (Otani et al. 2001). However, there have also been scattered reports of the presence of neural elements in what are regarded as Wilms' tumors (Masson 1980; reviewed in Magee et al. 1987). As a result, Masson (1980) proposed an alternative neuroepithelial origin for Wilms' tumor, with the suggestion of the involvement of neural crest elements in kidney development. The arguments against this have ranged from the general lack of occurrence of ectopic Wilms' tumors, given the diverse locations of neural crest cells, to the misidentification of the neural elements in the tumors. There is a growing list of genes that are expressed in the developing kidney together with the nervous system, including three of the most important genes in kidney development, Pax2, GDNF, and c-ret. Indeed, the WT1 gene is also expressed in parts of the brain and in the interneurons of the developing spinal cord. The expression of apparently neural markers may simply reflect the diversity of roles for such molecules rather than challenge the embryological origin of Wilms' tumor. Magee et al. (1987) investigated the expression of three neural proteins (S100, GFAP, and NSE) in five cases of Wilms' tumor, ranging in histological type from blastemal to teratomatous, and again identified the presence of neuroepithelial rosettes and glial bodies. While S100 is not specific to such structures, they claimed to co-expression of all three proteins; and the observation of such structures in a Wilms' tumor, with no other heterogeneous structures, argues strongly that their identity is neural. The alternative interpretation is that the MM itself is more multipotent than we expected. Placed in the wrong environment or sufficiently disrupted, the MM may contain a mesodermal progenitor/stem cell population so that it can revert to a more pluripotent phenotype, allowing the development of structures with an apparent ectodermal or endodermal origin (see Fig. 2).

5.2 Persistence of a Stem Cell or Multipotential Progenitor?

The concept of stem cell involvement in Wilms' tumor etiology is not new. The literature has stated for over a decade that Wilms' tumor arises from *renal stem cells* or nephroblasts (Hastie et al. 1990). All organ systems in the body are derived from progenitor cells that were once derived from a primitive germ layer, and before that time from a pluripotent inner cell mass. However, in most organs it has been assumed that such progenitor cells are transient in that they themselves become more and more restricted in potential during the process of maturation into the mature organ or tissue type. By definition, a stem cell must undergo self-renewal. There is no proof that there are cells able to both self renew and differentiate within the kidney. Hence, is it more prudent to say that Wilms' tumor arises from renal progenitors? Recent years have seen a complete rethink in terms of the existence within adult tissues of self-renewing cells with wide developmental potential. While multipotent stem cells are critical for the ongoing turnover of the skin, bone marrow, stomach, intestine, and cornea, the ability to derive cells with pluripotency has now been shown for the brain, bone marrow, skin, and fat (Azizi 2000; Mezey et al. 2000; Rietze et al. 2001; Toma et al. 2001; Zuk et al. 2001). From these observations, it has been hypothesized that there may be 'stem cells', or at least cells with an ability to return to a pluripotent state, in all adult tissues. Al-Awqati (Al-Awqati and Oliver 2002) suggests that this will also be the case for the kidney and that it is simply a matter of knowing what to look for.

Nephrogenesis has long been regarded as complete by week 36 of gestation as no new nephrons are 'born' after this time. At this point in time, it is assumed that the peripheral nephrogenic zone is exhausted. So where would a nephroblast lie? The presence of heterologous tissues in Wilms' tumor would argue that either a renal stem cell does exist and persist and that this cell has

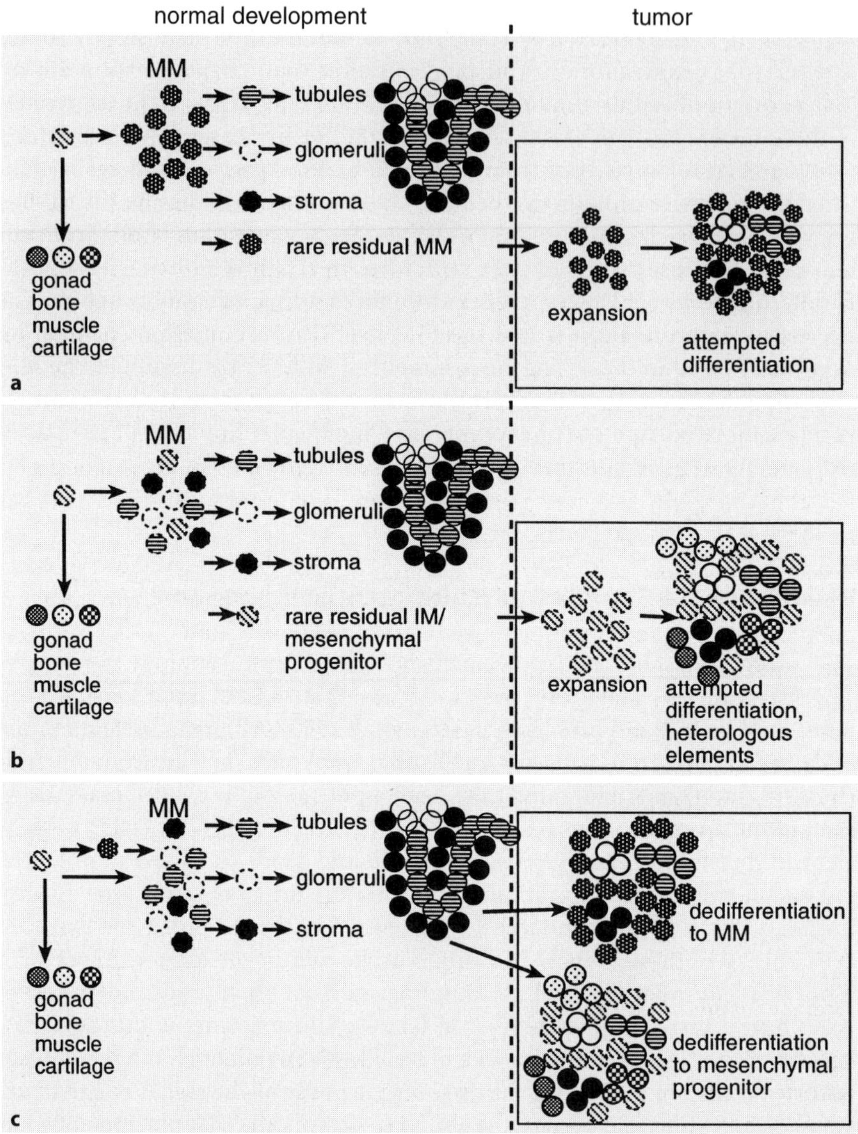

Fig. 2. Modeling the origin of Wilms' tumor based upon different hypotheses of normal kidney development from a progenitor metanephric mesenchyme (*MM*). Three possibilities have been presented for the development of the various differentiated cells types arising from the committed MM. A common mesenchymal progenitor with the potential to give rise to bone, cartilage, gonad, and MM exists in all models. **a** MM at the time of commitment consists of a homogeneous population of renal progenitor cells distinct from more primitive mesodermal progenitors, but with the potential to form all MM derivatives. This common MM progenitor gives rise to different cell types with more restricted stromal, tubular, and glomerular potential that go on to form the various MM-derived cell types of the kidney. Such derivative intermediates, as in the hematopoietic system, are committed to that sublineage. A rare population of renal progenitors persists from the committed time point. It is the aberrant expansion and subsequent attempted differentiation of this

pan-mesodermal potential, or that there are cells within the kidney that are able to dedifferentiate into a pre-renal mesodermal progenitor (see Fig. 2). The existence of PLNR and ILNR suggests that there is a progression from multipotential metanephric mesenchyme to differentiated renal cell which can be halted at various points in time and that in the case of ILNR, if arrested early enough, such cells can be redirected to other mesodermal fates. The very rare occurrence of skin or neural elements in such Wilms' tumors evokes an even wider potential, but was this inherent or did it result from de-differentiation? Indeed, it is not known whether the region of the intermediate mesoderm that gives rise to the metanephros is a homogeneous population or a mixed population of cells with varying potential. The origin of the metanephric Foxd1-positive stromal population has also not been clarified, and it is possible that not only do endothelial cells migrate into the kidney but also neural crest cells. Or does ILNR indicate the location of a persistent population of more highly potent progenitors within the core of the kidney, and is this the resting site of a 'renal stem cell'?

Using expression profiling of the murine metanephric mesenchyme at E10.5 compared to E10.5 rostral intermediate mesoderm (includes mesonephros and indeterminate gonad), we have defined the expression profile of the renal progenitor time point (Challen et al. 2004). In this way, it was hoped that we could find marker combinations that may identify the existence of a residual postnatal population of cells in the kidney (Challen et al. 2004). Of note, two genes that mark this renal progenitor time point are CD24 and cadherin 11. While both of these genes are strongly expressed in the MM at E10.5, each goes on to mark a distinct subpopulation of the devel-

◂

population that result in a Wilms' tumor. This model suggests the continued existence of renal progenitors that may be of value in regeneration. Without evoking de-differentiation, this model would not explain the occurrence of heterologous elements. Model A agrees with that presented by Li et al. (2002). **b** The committed MM contains a variety of lineage-restricted precursor cell types together with a minor population of primitive mesenchymal stem cells. The latter may act to resupply precursor populations and/or be involved in regeneration or repair. In this case, a tumor may arise via the expansion and inappropriate differentiation of this population of cells that, being primitive mesenchyme, can give rise to heterologous elements. **a** and **b** may represent different subgroups of Wilms' tumor. **C** The intermediate mesenchyme (IM) within the region of the developing kidney either differentiates into a homogeneous MM (renal progenitor), which can then form distinct lineage progenitors, or simultaneously differentiates into a heterogeneous population of different progenitor types, each restricted in potential and only able to differentiate into a limited number of renal cell types. These represent the entire potential of the kidney, and development proceeds until these progenitors are exhausted. Tumor formation can, therefore, only arise by the de-differentiation of such cells to either an MM or IM mesodermal progenitor, which then can give rise to a number of aberrant tissue types. De-differentiation may occur via the loss of an inhibitor of proliferation/differentiation signal or the re-activation of an embryologically active proliferation/inhibitor of differentiation signal oping and adult kidney. This divergence has taken place by E11.5. Both CD24 and cadherin 11 are re-expressed in Wilms' tumors. The expression of CD24 in Wilms' tumors has been used as evidence for recapitulation of development. The question remains, were these genes co-expressed on a common renal progenitor ($CD24^{+}cadherin11^{+}$) that has then differentiated into distinct nephron ($CD24^{+}cadherin11^{-}$) and stromal ($CD24^{-}cadherin11^{+}$) progenitors, or is the MM already a mixed population at E10.5? Are residual intermediate mesoderm (IM) cells retained in the developing metanephros to represent the origin of heterologous elements in Wilms' tumor, or do these result from de-differentiation? Alternatively, is commitment to metanephros simply a process of inhibition of other IM alternatives? In this case, do heterologous elements result from the loss of such suppression (see Fig. 2)? Miyagawa et al. (1994) propose that one of the roles of WT1 is to suppress a myogenic fate in the MM. Their evidence for this was the correlation between the presence of WT1 mutations and skeletal muscle elements in those tumors. If they are correct, Wilms' tumor, rather than being a recapitulation of kidney development or a case of disrupted normal development, may represent a process of de-differentiation. The molecular definition of the mesenchyme at the point of 'commitment' to metanephrogenesis will at least enable us to answer the question of the homogeneity of this timepoint.

5.3 Evidence for Renal Stem Cells in the Adult Kidney

If Wilms' tumor does represent evidence for the existence of a renal stem cell, shouldn't the postnatal kidney show some capacity for regeneration and repair? The entire complement of nephrons is laid down by gestational week 36 in humans, implying that the kidney ceases to have any regenerative capacity after birth. In fact, in response to acute damage the kidney can undergo significant remodeling. Unilateral ureteric obstruction can result in the near destruction of the medulla, with significant apoptosis and necrosis as well as the denuding of tubular basement membranes. However, once the obstruction is removed, there is a rapid process of reconstruction and repair that will regenerate the tubules of the medulla without forming new nephrons. The cells eliciting such repair may come from interstitial cell transdifferentiation, tubular cell de-differentiation and migration into the areas of damage prior to re-differentiation, the recruitment of stem cells from the bone marrow, or the generation of new tubular cells from an endogenous renal stem cell population. Maeshima et al. (2003), using BrdU pulse-chase labeling, identified putative renal stem cells in the renal tubules. This approach is based upon the premise that stem cells cycle slowly and should therefore retain BrdU label for a long time. This study did not isolate the label-retaining cells to verify their ability to self-renew and differentiate. Hence, while it is now accepted that there are stem cell populations in all adult organs, including the kidney, these may not be them.

5.4 Lack of Clonogenicity of Wilms' Tumor

One argument against Wilms' tumor arising from stem cells might be its lack of clonogenicity. Anyone doing research in the field of Wilms' tumor will be aware of the paucity of Wilms' tumor cell lines. There is even controversy about the origin of some of these lines. Attempts to derive tumor cell lines from primary Wilms' tumor are generally unsuccessful. If Wilms' tumors arise from residual stem/progenitor cells, why are they so difficult to grow? Other tumor types thought to represent tumors of stem cells, including glioblastoma, are notoriously metastatic, very hard to control and easy to culture as cell lines. Indeed, the propensity for Wilms' tumors to give rise to highly differentiated tissues such as muscle and bone does not seem to suggest that the bulk of the tumor represents a self-renewing cell type. In the adult, a stem cell is traditionally regarded as a slow dividing cell that will divide in an asymmetric fashion so as not to increase the size of the stem cell pool. Stem cells are also frequently associated with niches in which adjacent cells regulate this process of controlled self-renewal. It is possible that there is a stem cell component within a Wilms' tumor that is maintaining asymmetric division at a higher than usual pace, but that the bulk of the tumor is the immediate offspring of this stem cell and that the stem cell population does not expand. Hence, if the expanding elements of the tumor are separated from the stem cell or the stem cell loses its niche and is extinguished upon primary culture, there will be no resulting cell line. Why then do some Wilms' tumors metastasize and invade other tissues? One possibility evokes an additional mutational event resulting in metastatic capability or immortality for other elements of the tumor. The other possibility is that it is the stem cell population that moves and, if it finds another niche that will support it, it will recreate the same sort of tumor elsewhere.

5.5 A Window of Opportunity?

Wilms' tumor is a malignancy of children. The mean age of onset for Wilms' tumor is 3 years of age. While there have been reports of children born with Wilms' tumor or Wilms' tumor detected in utero (Beckwith 1999), it is essentially a tumor of childhood. However, there are occasional patients presenting with Wilms' tumors as adults. It was thought that these showed a poorer prognosis, but this is likely to be due simply to a lack of early diagnosis and appropriate treatment rather than adult Wilms' tumor representing a different entity (Terenziani et al. 2004). Adult Wilms' tumors respond best to the same chemotherapeutic regimes as children, and molecular analyses indicate involvement of the short arm of chromosome 11 (Kozman et al. 1989). This begs the question, if many of us are born with nephrogenic rests (1% of fetuses show evidence of nephrogenic rests), why does the oncogenic event only occur extremely rarely in adults? Wilms' tumor is not alone in being a disease restricted to children. There are a range of leukemias which affect children preferentially, yet the literature is silent on why this would be the case. Clearly, children are actively growing and many of the proliferative processes involved in growth to maturity are likely to differ from the 'resting' adult state. Does this tell us that the kidney retains a pool of progenitors until such time as adulthood is reached and that after that 'window of opportunity' has closed, there is no longer a stem cell population or a population that can undergo de-differentiation present? This may imply that the child, like the premetamorphic insect, has unique properties including a distinctly different regenerative potential. This is an area that needs considerable investigation with respect to our understanding of adult stem cells.

In closing, Wilms' tumor is a fascinating malignancy in terms of what it can tell us about normal development. The evidence that it represents a case of disrupted development is compelling, but perhaps more fascinating is what the presentation and histology of Wilms' tumor can teach us about lineage commitment and cell plasticity.

References

Akhtar M, Kott E, Brooks B (1977) Extrarenal Wilms' tumor: report of a case and review of the literature. Cancer 40:3087–3091

Al-Awqati Q, Oliver JA (2002) Stem cells in the kidney. Kidney Int 61:387–395

Armstrong J, Pritchard-Jones K, Bickmore WA, Hastie ND, Bard JBL (1992) The expression of the Wilms' tumour gene, WT1, in the developing mammalian embryo. Mech Dev 40:85–97

Aterman K (1989) Extrarenal nephroblastomas. J Cancer Res Clin Oncol 115:409–417

Azizi SA (2000) Exploiting nonneural cells to rebuild the nervous system: from bone marrow to brain. Neuroscientist 6:353–361

Bakshi N, Mansoor I, Venkataramu NK, Katariya S (2003) An unusual renal malignancy of childhood: unilateral teratoid Wilms's tumor. Pediatr Pathol Mol Med 22:435–441

Barakat AY, Papadopoulou ZL, Chandra RS, Hollerman CE, Calcagno PL (1974) Pseudohermaphroditism, nephron disorder and Wilms' tumor: a unifying concept. Pediatrics 54:366–369

Beckwith JB (1983) Wilms' tumor and other renal tumors of childhood: a selective review from the National Wilms' Tumor Study Pathology Center. Hum Pathol 14:481–492

Beckwith JB (1999) Prenatal detection of a Wilms' tumor. Pediatr Radiol 29:64

Beckwith JB, Kiviat NB, Bonadio J F (1990) Nephrogenic rests, nephroblastomatosis, and the pathogenesis of Wilms' tumor. Pediatr Pathol 10:1–36

Bolande RP (1976) Neoplasia of early life and its relationships to teratogenesis. Perspect Pediatr Pathol 3:145–183

Bruening W, Bardeesy N, Silverman BL, Cohn RA, Machin GA, Aronson AJ, Housman D, Pelletier J (1992) Germline intronic and exonic mutations in the Wilms' tumour gene (WT1) affecting urogenital development. Nat Genet 1:144–148

Call KM, Glaser T, Ito CY, Buckler AJ, Pelletier J, Haber DA, Rose EA, Kral A, Yeger H, Lewis WH et al (1990) Isolation and characterization of a zinc finger polypeptide gene at the human chromosome 11 Wilms' tumor locus. Cell 60:509–520

Challen GA, Martinez G, Davis MJ, Taylor DF, Crowe M, Teasdale RD, Grimmond SM, Little MH (2004) Identifying the molecular phenotype of renal progenitor cells. J Am Soc Nephrol 15:2344–2357

Davies JA, Bard JBL (1998) The development of the kidney. Curr Top Dev Biol 39:245–301

Davies JA, Ladomery M, Hohenstein P, Michael L, Shafe A, Spraggon L, Hastie N (2004) Development of an siRNA-based method for repressing specific genes in renal organ culture and its use to show that the Wt1 tumour suppressor is required for nephron differentiation. Hum Mol Genet 13:235–246

Denys P, Malvaux P, van den Berghe H, Tanghe W, Proesmans W (1967) Association of an anatomo-pathological syndrome of male pseudohermaphroditism, Wilms' tumor, parenchymatous nephropathy and XX/XY mosaicism. Arch Fr Pediatr 24:729–739

Devarajan P, Mishra J, Supavekin S, Patterson LT, Steven Potter S (2003) Gene expression in early ischemic renal injury: clues towards pathogenesis, biomarker discovery, and novel therapeutics. Mol Genet Metab 80:365–376

Drash A, Sherman F, Hartmann WH, Blizzard RM (1970) A syndrome of pseudohermaphroditism, Wilms' tumor, hypertension, and degenerative renal disease. J Pediatr 76:585–593

Droz D, Rousseau-Merck MF, Jaubert F, Diebold N, Nezelof C, Adafer E, Mouly H (1990) Cell differentiation in Wilms' tumor (nephroblastoma): an immunohistochemical study. Hum Pathol 21:536–544

Dunn JM, Phillips RA, Becker AJ, Gallie BL (1988) Identification of germline and somatic mutations affecting the retinoblastoma gene. Science 241:1797–1800

Durbec, P, Marcos-Gutierrez CV, Kilkenny C, Grigoriou M, Wartiowaara K, Suvanto P, Smith D, Ponder B, Costantini F, Saarma M et al (1996) GDNF signalling through the Ret receptor tyrosine kinase. Nature 381:789–793

Fernbach SK, Donaldson JS, Gonzalezcrussi F, Sherman JO (1988) Fatty Wilms tumor simulating teratoma – occurrence in a child with horseshoe kidney. Pediatr Radiol 18:424–426

Francke U, Holmes LB, Atkins L, Riccardi VM (1979) Aniridia-Wilms Tumor Association – evidence for specific deletion of 11p13. Cytogenet Cell Genet 24:185–192

Gessler M, Poustka A, Cavenee W, Neve RL, Orkin SH, Bruns GA (1990) Homozygous deletion in Wilms' tumours of a zinc-finger gene identified by chromosome jumping. Nature 343:774–778

Ghanem MA, van der Kwast TH, Den Hollander JC, Sudaryo MK, Oomen MH, Noordzij MA, van den Heuvel MM, Nassef SM, Nijman RM, van Steenbrugge GJ (2000) Expression and prognostic value of Wilms' tumor 1 and early growth response 1 proteins in nephroblastoma. Clin Cancer Res 6:4265–4271

Govender D, Hadley GP (1999) Recurrent Wilms' tumour or retroperitoneal teratoma? A challenging case. Virchows Arch Int J Pathol 435:67–70

Gross I, Morrison DJ, Hyink DP, Georgas K, English MA, Mericskay M, Hosono S, Sassoon D, Wilson PD, Little M, Licht JD (2003) The receptor tyrosine kinase regulator Sprouty1 is a target of the tumor suppressor WT1 and important for kidney development. J Biol Chem 278:41420–41430

Grubb GR, Yun K, Williams BR, Eccles MR, Reeve AE (1994) Expression of WT1 protein in fetal kidneys and Wilms tumors. Lab Invest 71:472–479

Grundy P, Koufos A, Morgan K, Li FP, Meadows AT, Cavenee WK (1988) Familial predisposition to Wilms' tumour does not map to the short arm of chromosome 11. Nature 336:374–376

Grundy RG, Pritchard J, Scambler P, Cowell JK (1998) Loss of heterozygosity for the short arm of chromosome 7 in sporadic Wilms tumour. Oncogene 17:395–400

Gurates B, Amsterdam A, Tamura M, Yang S, Zhou J, Fang Z, Amin S, Sebastian S, Bulun SE (2003) WT1 and DAX-1 regulate SF-1-mediated human P450arom gene expression in gonadal cells. Mol Cell Endocrinol 208:61–75

Haber DA, Buckler AJ, Glaser T, Call KM, Pelletier J, Sohn RL, Douglass EC, Housman DE (1990) An internal deletion within an 11p13 zinc finger gene contributes to the development of Wilms' tumor. Cell 61:1257–1269

Haltia A, Solin ML, Muramatsu T, Jalanko H, Holmberg C, Miettinen A, Holthofer H (1997) Expression of nine developmental stage-specific genes in human kidney and cultured renal cells. Exp Nephrol 5:457–464

Hastie ND, Bickmore W, Pritchard-Jones K, Porteous DJ, van Heyningen V (1990) Wilms' tumour: a developmental anomaly. In: Knudsnon AG, Stanbridge EJ, Sugimura M, Terada M, Watanabe S (eds) Genetic basis for carcinogenesis: tumour suppressor genes and oncogenes, Taylor and Francis, London, pp 145–150

Hatini V, Huh SO, Herzlinger D, Soares VC, Lai E (1996) Essential role of stromal mesenchyme in kidney morphogenesis revealed by targeted disruption of Winged Helix transcription factor BF-2. Genes Dev 10:1467–1478

Herzlinger D, Koseki C, Mikawa T, al-Awqati Q (1992) Metanephric mesenchyme contains multipotent stem cells whose fate is restricted after induction. Development 114:565–572

Huff V, Compton DA, Chao LY, Strong LC, Geiser CF, Saunders GF (1988) Lack of linkage of familial Wilms' tumour to chromosomal band 11p13. Nature 336:377–378

Ishii E, Fujimoto J, Tanaka S, Hata J (1987) Immunohistochemical analysis on normal nephrogenesis and Wilms' tumour using monoclonal antibodies reactive with lymphohaemopoietic antigens. Virchows Arch A Pathol Anat Histopathol 411:315–322

Jing S, Wen D, Yu Y, Holst PL, Luo Y, Fang M, Tamir R, Antonio L, Hu Z, Cupples R, Louis JC, Hu S, Altrock BW, Fox GM (1996) GDNF-induced activation of the ret protein tyrosine kinase is mediated by GDNFR-alpha, a novel receptor for GDNF. Cell 85:1113–1124

Kanemoto K, Takahashi S, Shu Y, Usui J, Tomari S, Yan K, Hamazaki Y, Nagata M (2003) Variable expression of podocyte-related markers in the glomeruloid bodies in Wilms tumor. Pathol Int 53:596–601

Karaca I, Sencan A, Ortac R, Bostanci-Sencan A, Mir E (2000) Teratoid Wilms' tumor: a case report. Turk J Pediatr 42:242–245

Koesters R, Ridder R, Kopp-Schneider A, Betts D, Adams V, Niggli F, Briner J, von Knebel Doeberitz M (1999) Mutational activation of the beta-catenin proto-oncogene is a common event in the development of Wilms' tumors. Cancer Res 59:3880–3882

Kozman HM, Clarke JM, Little MH, Smith PJ (1989) Molecular genetic evidence for common pathogenesis of childhood and adult Wilms' tumor. Cancer Genet Cytogenet 38:121–125

Kreidberg JA, Sariola H, Loring JM, Maeda M, Pelletier J, Housman D, Jaenisch R (1993) WT-1 is required for early kidney development. Cell 74:679–691

Lee SB, Huang K, Palmer R, Truong VB, Herzlinger D, Kolquist KA, Wong J, Paulding C, Yoon, SK, Gerald W, Oliner JD, Haber DA (1999a) The Wilms tumor suppressor WT1 encodes a transcriptional activator of amphiregulin. Cell 98:663–673

Lee TH, Pelletier J (2001) Functional characterization of WT1 binding sites within the human vitamin D receptor gene promoter. Physiol Genom 7:187–200

Lee TH, Moffett P, Pelletier J (1999b) The Wilms' tumor suppressor gene (wt1) product represses different functional classes of transcriptional activation domains. Nucleic Acids Res 27:2889–2897

Lewis A, Murrell A (2004) Genomic imprinting: CTCF protects the boundaries. Curr Biol 14:R284–R286

Li CM, Guo M, Borczuk A, Powell CA, Wei M, Thaker HM, Friedman R, Klein U, Tycko B (2002) Gene expression in Wilms' tumor mimics the earliest committed stage in the metanephric mesenchymal-epithelial transition. Am J Pathol 160(6):2181–2190

Little M, Wells C (1997) A clinical overview of WT1 gene mutations. Hum Mutat 9:209–225

Little M, Holmes G, Walsh P (1999) WT1: what has the last decade told us? Bioessays 21:191–202

Little M, Prosser J, Condie A, Smith PJ, van Heyningen V, Hastie ND (1992) Zinc finger point mutations within the WT1 gene in Wilms tumor patients. Proc Natl Acad Sci USA 89:4791–4795

Little M, Williamson KA, Mannens M, Kelsey A, Gosden C, Hastie ND, van Heyningen V (1993) Evidence that WT1 mutations in Denys-Drash syndrome patients may act in a dominant-negative fashion. Hum Mol Genet 2:259–264

Maeshima A, Yamashita S, Nojima Y(2003) Identification of renal progenitor-like tubular cells that participate in the regeneration processes of the kidney. J Am Soc Nephrol 14:3138–3146

Magee F, Mah RG, Taylor GP, Dimmick JE (1987) Neural differentiation in Wilms' tumor. Hum Pathol 18:33–37

Maiti S, Alam R, Amos CI, Huff V (2000) Frequent association of beta-catenin and WT1 mutations in Wilms tumors. Cancer Res 60:6288–6292

Masson P (1980) Wilms' tumors in human tumors. Wayne State University Press, Detroit

Matsunaga E (1981) Genetics of Wilms' tumor. Hum Genet 57:231–246

Maw MA, Grundy PE, Millow LJ, Eccles MR, Dunn RS, Smith PJ, Feinberg AP, Law DJ, Paterson MC, Telzerow PE et al (1992) A third Wilms' tumor locus on chromosome 16q. Cancer Res 52:3094–3098

McLorie GA (2001) Wilms' tumor (nephroblastoma). Curr Opin Urol 11:567–570

Mezey E, Chandross KJ, Harta G, Maki RA, McKercher SR (2000) Turning blood into brain: cells bearing neuronal antigens generated in vivo from bone marrow. Science 290:1779–1782

Miller RW, Fraumeni JF Jr, Manning MD (1964) Association of Wilms's tumor with aniridia, hemihypertrophy and other congenital malformations. N Engl J Med 270:922–927

Miozzo M, Perotti D, Minoletti F, Mondini P, Pilotti S, Luksch R, Fossati-Bellani F, Pierotti MA, Sozzi G, Radice P (1996) Mapping of a putative tumor suppressor locus to proximal 7p in Wilms tumors. Genomics 37:310–315

Miyagawa K, Kent J, Schedl A, van Heyningen V, Hastie ND (1994) Wilms' tumour – a case of disrupted development. J Cell Sci Suppl 18:1–5

Miyazaki Y, Oshima K, Fogo A, Ichikawa I (2003) Evidence that bone morphogenetic protein 4 has multiple biological functions during kidney and urinary tract development. Kidney Int 63:835–844

Moore MW, Klein RD, Farinas I, Sauer H, Armanini M, Phillips H, Reichardt LF, Ryan AM, Carver-Moore K, Rosenthal A (1996) Renal and neuronal abnormalities in mice lacking GDNF. Nature 382:76–79

Nachtigal MW, Hirokawa Y, Enyeart-VanHouten DL, Flanagan JN, Hammer GD, Ingraham HA (1998) Wilms' tumor 1 and Dax-1 modulate the orphan nuclear receptor SF-1 in sex-specific gene expression. Cell 93:445–454

Nisen PD, Zimmerman KA, Cotter SV, Gilbert F, Alt FW (1986) Enhanced expression of the N-myc gene in Wilms' tumors. Cancer Res 46:6217–6222

Ogawa O, Eccles MR, Yun K, Mueller RF, Holdaway MD, Reeve AE (1993) A novel insertional mutation at the third zinc finger coding region of the WT1 gene in Denys-Drash syndrome. Hum Mol Genet 2:203–204

Olson JM, Hamilton A, Breslow NE (1995) Non-11p constitutional chromosome abnormalities in Wilms' tumor patients. Med Pediatr Oncol 24:305–309
Otani M, Tsujimoto S, Miura M, Nagashima Y (2001) Intrarenal mature cystic teratoma associated with renal dysplasia: case report and literature review. Pathol Int 51:560–564
Pan CC, Chen PC, Chiang H (2004) Overexpression of KIT (CD117) in chromophobe renal cell carcinoma and renal oncocytoma. Am J Clin Pathol 121:878–883
Park S, Bernard A, Bove KE, Sens DA, Hazen-Martin DJ, Garvin AJ, Haber DA (1993) Inactivation of WT1 in nephrogenic rests, genetic precursors to Wilms' tumour. Nat Genet 5:363–367
Park CM, Kim WS, Cheon JE, Kim CJ, Kim WY, Kim IO, Shin HY, Yeon KM (2003) Teratoid Wilms tumor in childhood: CT and ultrasonographic appearances. Abdom Imaging 28:440–443
Pelletier J, Bruening W, Kashtan CE, Mauer SM, Manivel JC, Striegel JE, Houghton DC, Junien C, Habib R, Fouser L et al (1991a) Germline mutations in the Wilms' tumor suppressor gene are associated with abnormal urogenital development in Denys-Drash syndrome. Cell 67:437–447
Pelletier J, Bruening W, Li FP, Haber DA, Glaser T, Housman DE (1991b) WT1 mutations contribute to abnormal genital system development and hereditary Wilms' tumour. Nature 353:431–434
Pichel JG, Shen L, Sheng HZ, Granholm AC, Drago J, Grinberg A, Lee EJ, Huang SP, Saarma M, Hoffer BJ, Sariola H, Westphal H (1996) Defects in enteric innervation and kidney development in mice lacking GDNF. Nature 382:73–76
Ping AJ, Reeve AE, Law DJ, Young MR, Boehnke M, Feinberg AP (1989) Genetic linkage of Beckwith-Wiedemann syndrome to 11p15. Am J Hum Genet 44:720–723
Pritchard-Jones K, Fleming S (1991) Cell types expressing the Wilms' tumour gene (WT1) in Wilms' tumours: implications for tumour histogenesis. Oncogene 6:2211–2220
Pritchard-Jones K, Hastie ND (1990) Wilms' tumour as a paradigm for the relationship of cancer to development. Cancer Surv 9:555–578
Pritchard-Jones K, Fleming S, Davidson D, Bickmore W, Porteous D. Gosden C, Bard J, Buckler A, Pelletier J, Housman D et al (1990) The candidate Wilms' tumour gene is involved in genitourinary development. Nature 346:194–197
Rae FK, Martinez G, Gillinder KR, Smith A, Shooter G, Forrest AR, Grimmond SM, Little MH (2004) Analysis of complementary expression profiles following WT1 induction versus repression reveals the cholesterol/fatty acid synthetic pathways as a possible major target of WT1. Oncogene 23:3067–3079
Rahman N, Arbour L, Tonin P, Renshaw J, Pelletier J, Baruchel S, Pritchard-Jones K, Stratton MR, Narod SA (1996) Evidence for a familial Wilms' tumour gene (FWT1) on chromosome 17q12-q21. Nat Genet 13:461–463
Rauscher FJ 3rd, Morris JF, Tournay OE, Cook DM, Curran T (1990) Binding of the Wilms' tumor locus zinc finger protein to the EGR-1 consensus sequence. Science 250:1259–1262
Reeve AE, Eccles MR, Wilkins RJ, Bell GI, Millow LJ (1985) Expression of insulin-like growth factor-II transcripts in Wilms' tumour. Nature 317:258–260
Riccardi VM, Sujansky E, Smith AC, Francke U (1978) Chromosomal imbalance in the Aniridia-Wilms' tumor association: 11p interstitial deletion. Pediatrics 61:604–610
Rietze RL, Valcanis H, Brooker GF, Thomas T, Voss AK, Bartlett PF (2001) Purification of a pluripotent neural stem cell from the adult mouse brain. Nature 412:736–739
Saikali Z, Sinnett D (2000) Expression of glypican 3 (GPC3) in embryonal tumors. Int J Cancer 89:418–422
Sainio K, Hellstedt P, Kreidberg JA, Saxen L, Sariola H (1997) Differential regulation of two sets of mesonephric tubules by WT-1. Development 124:1293–1299

Sanchez MP, Silos-Santiago I, Frisen J, He B, Lira SA, Barbacid M (1996) Renal agenesis and the absence of enteric neurons in mice lacking GDNF. Nature 382:70–73

Saxen L (1987) Organogenesis of the kidney. Cambridge Univ Press, Cambridge

Scharnhorst V, van der Eb AJ, Jochemsen AG (2001) WT1 proteins: functions in growth and differentiation. Gene 273:141–161

Schedl A, Hastie ND (2000) Cross-talk in kidney development. Curr Opin Genet Dev 10:543–549

Schwab K, Patterson LT, Aronow BJ, Luckas R, Liang HC and Potter SS (2003) A catalogue of gene expression in the developing kidney. Kidney Int 64:1588–1604

Schwab K, Witte D, Aronow B, Devarajan P, Potter S, Patterson L (2004) Microarray analysis of focal segmental glomerulosclerosis. Am J Nephrol 24:438–447

Schwartz CE, Haber DA, Stanton VP, Strong LC, Skolnick MH, Housman DE (1991) Familial predisposition to Wilms tumor does not segregate with the WT1 gene. Genomics 10:927–930

Scott J, Cowell J, Robertson ME, Priestley LM, Wadey R, Hopkins B, Pritchard J, Bell GI, Rall LB, Graham CF et al (1985) Insulin-like growth factor-II gene expression in Wilms' tumour and embryonic tissues. Nature 317:260–262

Sim EU, Smith A, Szilagi E, Rae F, Ioannou P, Lindsay MH, Little MH (2002) Wnt-4 regulation by the Wilms' tumour suppressor gene, WT1. Oncogene 21:2948–2960

Smithey BE, Pappo AS, Hill DA (2002) C-kit expression in pediatric solid tumors: a comparative immunohistochemical study. Am J Surg Pathol 26:486–492

Slater RM, Mannens MM (1992) Cytogenetics and molecular genetics of Wilms' tumor of childhood. Cancer Genet Cytogenet 61:111–121

Stark K, Vainio S, Vassileva G, McMahon AP (1994) Epithelial transformation of metanephric mesenchyme in the developing kidney regulated by Wnt-4. Nature 372:679–683

Stuart RO, Bush K, Nigam SK (2001) Changes in global gene expression patterns during development and maturation of the rat kidney. Proc Natl Acad Sci USA 98:5649–5654

Takahashi M, Sugimura J, Yang X, Vogelzang N, Teh BS, Furge K, Teh BT (2003) Gene expression profiling of renal cell carcinoma and its implications in diagnosis, prognosis, and therapeutics. Adv Cancer Res 89:157–181

Terenziani M, Spreafico F, Collini P, Piva L, Perotti D, Podda M, Gandola L, Massimino M, Cereda S, Cefalo G, Luksch R, Casanova M, Ferrari A, Polastri D, Valagussa P, Fossati-Bellani F (2004) Adult Wilms' tumor: a monoinstitutional experience and a review of the literature. Cancer 101:289–293

Thate C, Englert C, Gessler M (1998) Analysis of WT1 target gene expression in stably transfected cell lines. Oncogene 17:1287–1294

Toma JG, Akhavan M, Fernandes KJ, Barnabe-Heider F, Sadikot A, Kaplan DR, Miller FD (2001) Isolation of multipotent adult stem cells from the dermis of mammalian skin. Nat Cell Biol 3:778–784

Ton CC, Hirvonen H, Miwa H, Weil MM, Monaghan P, Jordan T, van Heyningen V, Hastie ND, Meijers-Heijboer H, Drechsler M et al. (1991) Positional cloning and characterization of a paired box- and homeobox-containing gene from the aniridia region. Cell 67:1059–1074

Udtha M, Lee SJ, Alam R, Coombes K, Huff V (2003) Upregulation of c-MYC in WT1-mutant tumors: assessment of WT1 putative transcriptional targets using cDNA microarray expression profiling of genetically defined Wilms' tumors. Oncogene 22:3821–3826

van Heyningen V, Bickmore WA, Seawright A, Fletcher JM, Maule J, Fekete G, Gessler M, Bruns GA, Huerre-Jeanpierre C, Junien C et al (1990) Role for the Wilms tumor gene in genital development? Proc Natl Acad Sci USA 87:5383–5386

Varanasi R, Bardeesy N, Ghahremani M, Petruzzi MJ, Nowak N, Adam MA, Grundy P, Shows TB, Pelletier J (1994) Fine structure analysis of the WT1 gene in sporadic Wilms tumors. Proc Natl Acad Sci USA 91:3554–3558

Variend S, Spicer RD, Mackinnon AE (1984) Teratoid Wilms' tumor. Cancer 53:1936–1942

White GR, Kelsey AM, Varley JM, Birch JM (2002) Somatic glypican 3 (GPC3) mutations in Wilms' tumour. Br J Cancer 86:1920–1922

Zuk PA, Zhu M, Mizuno H, Huang J, Futrell JW, Katz AJ, Benhaim P, Lorenz HP, Hedrick MH (2001) Multilineage cells from human adipose tissue: implications for cell-based therapies. Tissue Eng 7:211–228

Testicular Teratomas: Back to Pluripotent Stem Cells

Tohru Kimura, Kazushige Murayama, Toshinobu Nakamura, Shoko Watanabe, Hiroki Umehara, Maya Tomooka, Toru Nakano

1 Introduction

Mammals are made up of as many as 200 differentiated cell types, all of which originate from a single totipotent cell, the fertilized egg. The germline is the specialized cell lineage that transmits new combinations of genetic and epigenetic information from generation to generation.

Germ cells are differentiated from a group of pluripotent epiblasts that migrate into the extra-embryonic region during gastrulation at embryonic day 7.5 (E7.5) in mice (Rugh 1968; Wylie 2000; McLaren 2003). The founder germ cells, which are called primordial germ cells (PGCs), produce all the germ lineage cells. PGCs migrate from the extra-embryonic mesoderm through the hindgut and dorsal mesentery, and populate the genital ridges up to E11.5. In the gonads, PGCs undergo a series of sexual differentiation steps that are under the control of intrinsic programs, and that are influenced by the neighboring somatic cells. In males, PGCs enter mitotic arrest at E13.5. Shortly after birth, the spermatogonia resume mitosis, and spermatogenesis is completed at puberty. In females, PGCs start meiosis at E13.5, and oocytes reach the quiescent (dictyate) stage at prophase of the first meiosis, about three days after birth. The follicles are matured in sequence, leading to ovulation.

PGCs ultimately differentiate into gametes, i.e., oocytes and sperm, and totipotency is restored at fertilization. However, PGCs never differentiate into

T. Kimura, K. Murayama, T. Nakamura, S. Watanabe, H. Umehara, M. Tomooka, T. Nakano
Department of Pathology, Osaka University Medical School

T. Nakamura, S. Watanabe, H. Umehara, M. Tomooka, T. Nakano
Graduate School of Frontier Biosciences, Osaka University, 2–2 Yamada-oka, Suita, Osaka, Japan 565–0871

T. Kimura, T. Nakano
Department of Molecular Cell Biology, Research Institute for Microbial Diseases, Osaka University, 3–1 Yamada-oka, Suita, Osaka, Japan 565–0871

Progress in Molecular and Subcellular Biology
Alvaro Macieira-Coelho (Ed.)
Development Biology of Neoplastic Growth

other cell types during normal development. In addition, PGCs do not lead to the development of chimeric mice when injected into blastocysts (Labosky et al. 1994b). These facts suggest that PGCs are already restricted to germline development and cannot exhibit pluripotency under physiological conditions.

Nonetheless, PGCs retain the potential to recover pluripotency (Fig. 1). First, PGCs are the originators of testicular teratomas, tumors that contain collections of various differentiated and undifferentiated cells (Stevens 1967b, 1980). Teratomas contain undifferentiated embryonal carcinoma (EC) cells that can differentiate into a variety of cell types in vitro and in vivo. The successful generation of EC cell-derived chimeric mice has prompted efforts to establish embryonic stem (ES) cell lines from the inner cell mass (Mintz and Illmensee 1975; Martin 1981; Stewart and Mintz 1981; Bradley et al. 1984). Second, pluripotent stem cell lines, which are called embryonic germ (EG) cells, can be established from PGCs that are cultured in the presence of a growth factor cocktail (Matsui et al. 1992; Resnick et al. 1992). The developmental potency of EG cells resembles that of ES cells. When injected into blastocysts, EG cells are incorporated into the developmental cycle and differentiate into all three germ layers as well as germ cells (Stewart et al. 1994). Thus, PGCs are

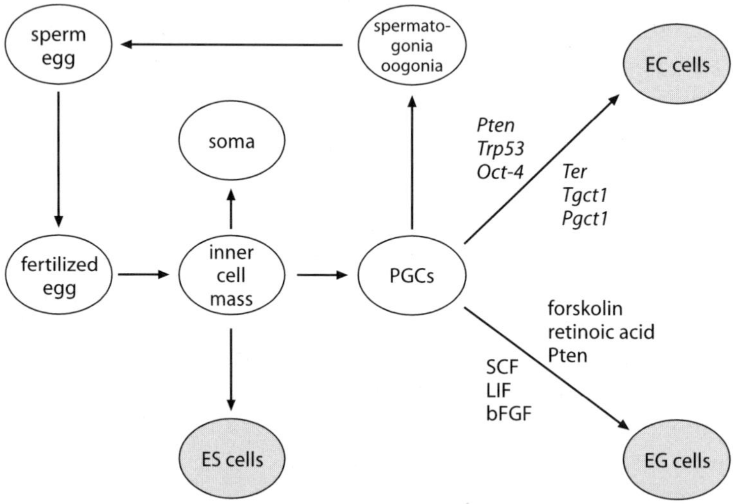

Fig. 1. Three pluripotent stem cell lines: ES, EG, and EC cells. The cells of the inner cell mass and PGCs can produce the pluripotent cell ES and EG cell lines, respectively. Both of these cell lines can differentiate into three germ layers and germ cells. The EC cells in teratomas also originate from PGCs and have a pluripotency generally more limited than that of EG cells. Thus, ES cells are derived from pluripotent cells, whereas EG and EC cells are from the restricted cell lineage. Growth factors, such as SCF, LIF, and bFGF, promote the derivation of EG cells. The PKA activator forskolin and retinoic acid are also implicated in conversion to EG cells. The *Pten* deficiency, which involves activation of the PI3 K/Akt signaling pathway, enhances both EG and EC cell formation. EC cell formation is influenced by other genetic factors, such as *Trp53, Oct-4, Ter, Tgct1,* and *Pgct1*

unique from the perspective of stem cell biology, in that they appear to "de-differentiate" from germ lineage cells to pluripotent stem cells under both pathological and experimental conditions. In this review, we summarize our current understanding of the cellular and molecular basis for the formation of teratomas and EG cells in mice, and discuss the relevance of these studies to stem cell biology, as well as to germline development and testicular cancer in humans.

2
Cellular Basis of Teratoma Formation

2.1
Teratomas in the Mouse Model

2.1.1
The 129/Sv Mouse

Steven's pioneering studies of testicular teratomas and EC cells were conducted using the inbred 129/Sv mouse strain (Stevens 1967b, 1980). Testicular teratomas arise in 3-week-old 129/Sv males at an incidence of 1–10%, which is significantly higher than the incidence of 0.01–0.1% for other mouse strains (Stevens and Hummel 1957). The first teratomatous foci appear within the seminiferous tubules of E15.5 testes, and are composed of cells that resemble the inner cell masses of blastocysts and epiblasts at the egg cylinder stages. The EC cell mass proliferates, escapes from the tubules to the interstitial region, and differentiates into various cell types and tissue structures in a disorganized fashion. The teratomas at 5 days of age contain both differentiated and undifferentiated cells, and the tumors differentiate completely in most adult mice. In some cases, EC cell lines can be established from teratocarcinomas and maintained in ascites or in culture (Stevens 1958). EC cells lead to the development of chimeric mice when injected into blastocysts (Mintz and Illmensee 1975; Stewart and Mintz 1981). The variable cell types that compose the tumors and the differentiation capacity of the established EC cell lines indicate that the EC cells in teratomas are pluripotent.

2.1.2
Origin of Teratomas

Testicular teratomas can be induced experimentally by transplanting the genital ridges into adult testes. This experimental system has been used to identify the developmental stages of teratoma onset and the origin of teratomas (Stevens 1964). When E12.5 testes of 129/Sv mice were transplanted, approximately 80% of the grafts developed teratomas, whereas the tumor incidence decreased to 8% when E13.5 testes were grafted. Furthermore, examination of the sizes of the spontaneous tumors at various stages allowed the extrapolation of tumor onset as E12.5. The results of both experimental and spontaneous tumors suggest that tumorigenesis onset is at approximately E12.5 in 129/Sv mice.

The *Steel* locus encodes a growth factor called stem cell factor (SCF), which transmits signals via the receptor tyrosine kinase c-kit (Flanagan and Leder 1990; Huang et al. 1990). Homozygous mutant (*Sl/Sl*) mice show defects in the development of hematopoietic cells, melanocytes, and germ cells. The number of PGCs is reduced dramatically in the migratory and post-migratory stages. In addition, *W (White-spotting)* mice that harbor mutations in the c-kit locus exhibit essentially the same phenotype (Chabot et al. 1988; Geissler et al. 1988). Therefore, SCF/c-kit signaling plays a critical role in PGC development. In order to examine the origin of teratomas, the *Steel* (Sl) mutant locus was introduced into the 129/Sv background (Stevens 1967a). When the testes from the E12.5 *Sl/Sl* mice in the 129/Sv background were transplanted, very few grafts developed teratomas. This result clearly shows that testicular teratomas originate from PGCs.

2.2 Human Testicular Cancers

In humans, testicular cancers can be classified into three categories, based on their epidemiological, histological, and clinical features (Looijenga and Oosterhuis 1999). The first group includes teratomas and yolk sac tumors, which occur in fetuses and infants. The second group consists of adult testicular cancer, which occurs in the second to fourth decades of human life and comprises both seminomas and nonseminomas. The third class of testicular cancer is spermatocytic cancer, which occurs in elderly males. The mouse model described above may be valuable for studying the human teratomas and yolk sac tumors in infantile testes, given the strong similarities between these tumors in both species, i.e., the tumors arise from PGCs, are classified histologically as nonseminomas, and appear to be karyotypically normal.

It is a matter of debate as to whether the mouse model is valid for adult testicular cancer, which is the major form of testicular cancer in humans. The human adult testicular tumors encompass a mixture of seminomas and nonseminomas. Chromosomal abnormalities, such as iso-12p, are observed consistently in these cancers (von Eyben 2004). It is generally agreed that adult testicular cancer originates from carcinoma in situ (CIS), which corresponds to germ cell neoplasia in seminiferous tubules (Skakkebaek 1972; Rajpert-de Meyts et al. 2003). Although it is still unclear when CIS occurs, it is believed to arise early in development, since CIS cells are similar in character to PGCs. Therefore, common mechanisms may operate in the generation of EC cells and CIS cells. The spermatocytic cancers comprise seminomas. Since this type of cancer shows characteristics of spermatogonia stage B and is not accompanied by CIS, it is generally accepted that spermatocytic cancer originates from a later stage of germ cell development than do teratomas or adult testicular cancer (Rosenberg et al. 1998).

2.3 Environmental Influences on Teratomas

Teratoma formation is influenced by a number of environmental factors. For example, in mice teratomas can be induced at an incidence of 80% when E12.5 genital ridges are grafted into adult testes. However, when E12.5 genital ridges are transplanted into the spleen, the incidence of teratoma drops to 9%, and the tissue composition is much simpler (Stevens 1964). Furthermore, teratomas occur more frequently in the left than in the right testes of 129/Sv mice (Stevens 1967b). In humans, estrogen exposure during pregnancy may be a causative factor for testicular cancers (Looijenga and Oosterhuis 1999). It has been demonstrated that endogenous and environmental estrogens increase c-kit expression in the somatic cells of genital ridges in vitro, which promotes PGC proliferation and EG cell formation (Moe-Behrens et al. 2003). In addition, gonadal dysgenesis, cryptorchidism, and childhood hernia are also associated with testicular germ cell tumors in humans (Looijenga and Oosterhuis 1999).

3 EG Cells

3.1 Behavior of PGCs in Culture

Mouse PGCs can be cultured on feeder cells, such as STO cells, in the short term (Donovan et al. 1986). When migratory phase PGCs at E8.5 are cultured in vitro, the majority show elongated and polar morphology with pseudopods. These cells proliferate actively for several days, but eventually stop cell division. In contrast, E13.5 PGCs have round morphology and do not proliferate in vitro. Some PGCs can enter the early meiotic cycle in vitro, regardless of sex (Chuma and Nakatsuji 2001), which supports the notion that female development is the default in mammals. In normal development, the PGCs move to the genital ridges until E11.5, and enter mitotic and meiotic arrest at E13.5 in males and females, respectively. Thus, it appears that PGCs execute their intrinsic developmental programs to some degree in vitro.

3.2 PGC Growth Factors

A number of studies have identified the factors that support the proliferation and survival of PGCs in vitro. *W* and *Sl* mutant mice, with mutations in the *c-kit* and *SCF* loci, respectively, show PGC deficiency (Chabot et al. 1988; Geissler et al. 1988; Flanagan and Leder 1990; Huang et al. 1990). Consistent with this phenotype, the addition of SCF in culture promotes the proliferation and survival of PGCs (de Felici and Dolci 1991; Dolci et al. 1991; Godin et al. 1991; Matsui et al. 1991; Pesce et al. 1993). Leukemia inhibitory factor (LIF) also enhances PGC proliferation and survival in vitro, although LIF-deficient mice do not show any abnormalities in terms of PGC development (Matsui et al. 1991; Pesce et al. 1993; Cheng et al. 1994). A synergistic effect on proliferation and survival is observed for the combination of SCF and LIF. Furthermore, it has been reported that oncostatin M, tumor necrosis factor (TNF)-α, interleukin (IL)-4, PACAP (pituitary adenylate cyclase-activating polypeptide), neuregulin β, and retinoic acid act as mitogens and anti-apoptotic factors (Kawase et al. 1994; Koshimizu et al. 1995; Cooke et al. 1996; Pesce et al. 1996; Hara et al. 1998; Toyoda-Ohno et al. 1999). In contrast, transforming growth factor (TGF)-β1 and activin inhibit the proliferation, and TGF-β1 promotes the migration of PGCs (Godin and Wylie 1991; Richards et al. 1999).

3.3 Generation of EG Cells

The culturing of PGCs in the presence of SCF, LIF, and basic fibroblast growth factor (bFGF) dramatically enhances PGC proliferation and survival, and leads to the formation of EG cell colonies (Fig. 2; Matsui et al. 1992; Resnick et al. 1992). The morphology of the EG cell colonies resembles that of ES cells; the colonies are multi-layered and composed of compact, round cells, whereas PGCs cultured in the presence of SCF and LIF generate scattered colonies of cells with elongated morphology. Intriguingly, EG cells have pluripotency despite their origin; they can differentiate into all three germ layers and germ cells in chimeric mice, as is the case with ES cells (Stewart et al. 1994). Stimulation with LIF, SCF, and bFGF is required for the establishment of EG cells; in the secondary culture and thereafter only LIF is required for maintenance (Matsui et al. 1992). Stimulation with bFGF can be replaced by the addition of forskolin, which activates the cyclic-AMP (cAMP)-dependent protein kinase (PKA) by increasing the levels of intracellular cAMP, and the addition of retinoic acid (Koshimizu et al. 1996).

The formation of EG cells is influenced by the developmental stage of PGCs: EG cells can be established efficiently from E8.5 PGCs, but the capacity decreases gradually, and is eventually lost at E13.5 (Matsui et al. 1992). Meanwhile, EG cells can be established from various inbred and outbred mouse strains other than 129/Sv, although testicular teratomas are rarely found in these strains (Stevens 1967b).

3.4
Human EG Cells

Shortly after the report on the establishment of human ES cells appeared (Thomson et al. 1998), human EG cell lines were generated from mesenteries and genital ridges of 5- to 9-week post-fertilization human embryos (Sham-

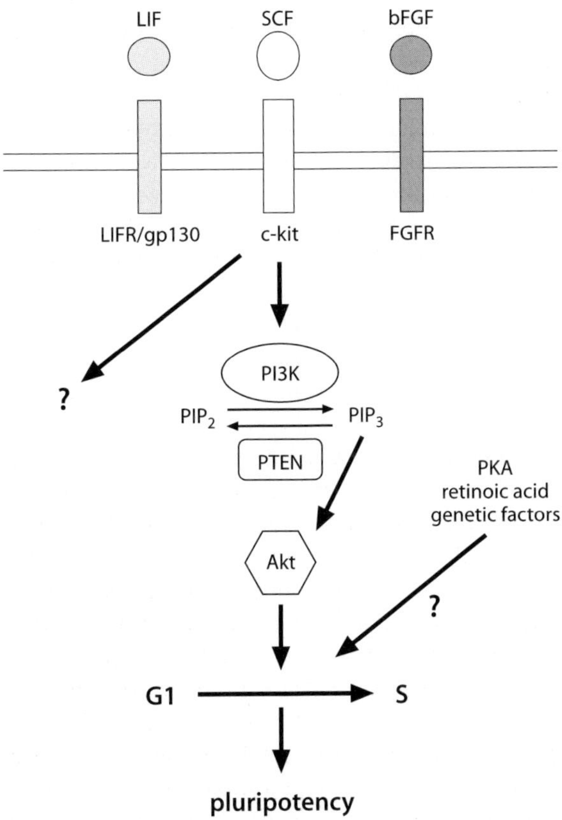

Fig. 2. Signaling pathways that regulate the formation of EG cells and EC cells. SCF, LIF, and bFGF activate a variety of signaling pathways, including the PI3 K signaling cascade. PGC-specific, *Pten*-deficient mice have been used to reveal that enhanced PI3 K/Akt signaling plays a pivotal role in the generation of both EG and EC cells. PI3 K/Akt signaling promotes G1 to S-phase transition by activating cyclin/CDK complexes through phosphorylation of various downstream molecules. Enhanced cell cycle progression may be involved in the conversion of PGCs to EG and EC cells. The mechanisms by which PKA, retinoic acid, and other genetic factors control conversion to stem cells remain to be elucidated

blott et al. 1998; Turnpenny et al. 2003). The derivation was essentially the same as that for mouse EG cells, i.e., PGCs were cultured on STO feeder cells that expressed SCF, in medium that contained human LIF, bFGF, and forskolin. EG cells are karyotypically normal and stable, and have been shown to differentiate into all three germ layers in embryoid body formation assays and transplantation assays in mice. Thus, human EG cells represent valuable material for studies of human embryogenesis and cell therapy, although there may be some obstacles to clinical applications, as discussed in the next section.

3.5 Genomic Imprinting in PGCs and EG Cells

Genomic imprinting is an epigenetic modification that leads to preferential expression of a specific parental allele in somatic cells (Surani 2001). To ensure parental-specific monoallelic expression in the next generation, imprinting must be reprogrammed in the germ line. Reprogramming involves erasure of the epigenetic marks and the establishment of new marks. Nuclear transplantation cloning experiments in mice have revealed that erasure in PGCs begins at E10.5 to E11.5, and that the timing of erasure varies among the genes (Lee et al. 2002; Yamazaki et al. 2003). Investigations of the methylation patterns in the imprinting genes of PGCs have also shown that erasure occurs at E10.5 to E13.5 (Hajkova et al. 2002; Szabo et al. 2002).

When the epigenetic marks of the EG cells derived from E8.0 and E8.5 PGCs were examined, about half of the cells retained imprinting marks to variable degrees and another half of cells showed complete loss of imprinting (Labosky et al. 1994a). The EG cells from E11.5 and E12.5 PGCs had lost the parental-specific epigenetic marks, and they did not reacquire monoallelic gene expression during differentiation (Labosky et al. 1994a; Tada et al. 1998). In addition, epigenetic instability has been observed in mouse ES cells, although they are derived from inner cell masses, which have normal parental-specific marks (Dean et al. 1998; Humpherys et al. 2001). The imprinting erasure patterns and abnormalities that have been observed in mouse ES and EG cells may represent an impediment to the clinical use of tissues that are derived from human ES and EG cells.

On the other hand, it has been reported that human mesenchymal fibroblast-like cells differentiated from EG cells display monoallelic expression of imprinted genes (Onyango et al. 2002). In addition, these authors have demonstrated that mouse EG cells established from E8.5 PGCs show biallelic expression of imprinted genes, whereas differentiated cells derived from the same EG cells reacquire monoallelic expression. Although these results suggest that epigenetic erasure in EG cells may not be an obstacle to the transplantation of EG cell-derived differentiated cells, further investigations into the biology of ES and EG cells are needed to evaluate the risks and benefits of the clinical application of these cell types.

4
Genetic and Molecular Bases of Teratoma Formation

4.1
Genetic Traits and Mendelian Inheritance Factors

The 129/Sv mice spontaneously develop testicular teratomas at high rates, while the incidence of this cancer is very low in other strains. Since the incidence of testicular teratoma is low in inter-strain hybrids with 129/Sv, susceptibility to this type of cancer appears to involve a genetic trait. It is estimated that as many as 10 to 15 susceptibility genes are present in the 129/Sv genome (Matin et al. 1999; Jiang and Nadeau 2001). Although no susceptibility gene has been identified as yet, linkages to genetic markers have been reported.

Analyses of naturally occurring alleles and gene knockout mice have identified the Mendelian factors that control susceptibility to testicular teratomas in 129/Sv mice. These factors include: *Ter (Teratoma)*, *Trp53* and *Tgct-1* (*testicular germ cell tumor 1*) (Noguchi and Noguchi 1985; Harvey et al. 1993; Matin et al. 1999). In addition, it has been demonstrated recently that phosphoinositide-3 kinase (PI3K) signaling regulates cancer susceptibility in mixed backgrounds (Kimura et al. 2003).

4.2
Genetic Control of Teratomas in Mice

4.2.1
Ter Locus

Ter (Teratoma) is a recessive mutation that enhances dramatically testicular teratoma formation in the 129/Sv background, in that 75 and 19% of homozygous 129/Sv-*Ter* male mice develop bilateral and unilateral teratomas, respectively (Stevens 1973; Noguchi and Noguchi 1985). PGCs enter mitotic arrest at E13.5 in normal mice, whereas the PGCs of 129/Sv-*Ter/Ter* mice retain mitotic activity even after E14.5 (Noguchi and Stevens 1982). However, the majority of the mutant mice display germ cell deficiency at E16.5, presumably due to apoptosis. It appears that most mutant PGCs undergo apoptosis, while low numbers of survivors continue to proliferate and form the teratomas.

Ter mutant mice in another genetic background, such as C57BL/6, also show germ cell deficiency, but they do not develop testicular teratomas, which indicates that genetic factors in the 129/Sv mice affect tumorigenesis (Noguchi et al. 1996). Exchange co-culture experiments using normal PGCs and *Ter/Ter* somatic cells suggest that the deficiency in somatic cells may cause germ cell deficiency in *Ter* mutant mice (Takabayashi et al. 2001a,b). The *Ter* locus has been mapped to chromosome 18 (Asada et al. 1994; Sakurai et al. 1994).

4.2.2
PI3 Kinase Signaling Pathway

Many growth factors, including SCF, LIF, and bFGF, activate directly and/or indirectly the PI3K signaling pathway (Fig. 2). PI3 K phosphorylates PIP2 (phosphatidylinositol 4,5-bisphosphate) to PIP3 (phosphatidylinositol 3,4,5-triphosphate; Cantley 2002). PIP3 then transmits the signals to downstream effector molecules, such as Akt/PKB, Rac, and Cdc42, and thereby regulates various physiological and pathological processes, such as cell growth, proliferation, survival, migration, differentiation, and tumorigenesis. On the other hand, *PTEN* (phosphatase and tensin homologue deleted on chromosome 10) is a tumor suppressor gene that is mutated in human sporadic cancers and in hereditary cancer syndromes, such as Cowden disease and Bannayan-Zonana syndrome (Kishimoto et al. 2003; Sulis and Parsons 2003; Stiles et al. 2004). *PTEN* encodes a phosphatase that catalyzes the reverse reaction of PI3K, thereby antagonizing the PI3K-mediated processes. Thus, in the absence of PTEN, PI3 K signaling is accelerated, leading to the uncontrolled proliferation and survival of cancer cells.

PGC-specific, *Pten*-deficient mice show similar phenotypes to 129/Sv-*Ter/Ter* mice (Kimura et al. 2003). The *Pten* mutant mice develop bilateral testicular teratomas. High levels of mitotic activity and apoptosis are observed in PGCs after E14.5. However, there are some differences between *Pten*-deficient mice and 129/Sv-*Ter/Ter* mice. First, the deficiency exists, presumably, in the somatic cells of 129/Sv-*Ter/Ter* mice (Takabayashi et al. 2001a), whereas it appears to be in the PGCs of *Pten*-deficient mice. In culture, the proliferation and survival rates for PGCs purified from 129/Sv-*Ter/Ter* mice are normal (Takabayashi et al. 2001a,b). In contrast, the proliferation, survival, and EG cell generation rates are enhanced in *Pten*-deficient PGCs. Second, the *Ter* mutation produces teratomas exclusively in mice with the pure 129/Sv background and not in other mouse strains (Noguchi et al. 1996). In addition, the incidence of teratoma is very low, even in the F1 generation of hybrids between the 129/Sv strain and other strains. In contrast, teratomas arise in *Pten*-deficient mice in a mixed background. Since the *Pten*-deficient mice analyzed have a mixed background (partial 129 background), it is possible that *Pten* deficiency cooperates with factors present in the 129/Sv genetic background or that *Pten* itself is one of the susceptibility factors in 129/Sv mice. Third, female 129/Sv-*Ter/Ter* mice display germ cell deficiency but do not develop ovarian teratomas, which are generally believed to originate from parthenogenetically activated oocytes (Eppig et al. 1977). Some female *Pten*-deficient mice develop ovarian teratomas, the origin of which is under investigation.

In vitro experiments with *Pten*-deficient PGCs have demonstrated that the enhancement of PI3K signaling downstream of SCF, LIF, and bFGF promotes the proliferation and survival of PGCs and EG cell production (Kimura et al. 2003). In addition, it has been shown that PI3 K/Akt signaling downstream of SCF regulates the proliferation and survival of PGCs in vitro (de Miguel et al. 2002). On the other hand, a knockin mouse study using c-kit mutation has shown that although the PI3 K signal is dispensable for PGC development, it is essential for spermatogenesis (Blume-Jensen et al. 2000; Kissel et al. 2000). SCF, LIF, and bFGF are expressed in the migratory pathways of PGCs and genital ridges, and their receptors are found in PGCs (Matsui et al. 1990; Keshet et al. 1991; Cheng et al. 1994; Resnick et al. 1998). Therefore, the potency of the PI3K signaling downstream of these growth factors may control germ line commitment and conversion to EG and EC cells.

4.2.3 *Tumor Suppressor Trp53*

Trp53 is a tumor suppressor gene that is involved in a broad spectrum of hereditary and sporadic cancers. Malignant lymphomas develop in *Trp53*-deficient mice with the 129/Sv, C57BL/6, and (129/Sv × C57BL/6) mixed backgrounds at high incidence (60–80%; Harvey et al. 1993), which suggests that lymphomas are the direct result of *Trp53* loss rather than the result of a particular genetic background. In contrast, teratomas arise more frequently in the 129/Sv background (35%) than in the mixed (9%) and C57BL/6 (<2%) backgrounds, which indicates that the effect of *Trp53* deficiency on this tumor type is indirect.

4.2.4 *Oct-4*

Oct-4 (also known as Oct-3 and POU5f1) is a transcription factor with a POU-type homeodomain, the expression of which is restricted to pluripotent cells of early embryos and germ cell lineages. In *Oct-4*-deficient blastocysts, the inner cell mass differentiates into the trophectoderm (Nichols et al. 1998). Conditional deletion of *Oct-4* in ES cell lines also promotes differentiation into the trophoectoderm lineage (Niwa et al. 2000). Conversely, increased Oct-4 expression in ES cell lines induces differentiation into mesoderm and endoderm. Thus, the expression level of Oct-4 controls the pluripotency and direction of differentiation in the inner cell mass and ES cells.

Teratomas can be generated by transplanting ES cells into syngeneic mice. Teratoma formation by ES cell lines that contain variable levels of Oct-4 has been examined in this system (Gidekel et al. 2003). The potential for tumor formation has been correlated with the expression level of Oct-4 in ES cells, and the accumulation of malignant cells is also associated with the amount of Oct-4 present. In humans, Oct-4 overexpression has been found in various types of testicular tumors, which include teratomas and adult testicular cancers (Gidekel et al. 2003; Looijenga et al. 2003). Therefore, Oct-4 may regulate not only pluripotency, but also the tumorigenicity of germ cells.

4.2.5
Genetic Interactions: Identification of the Tgct1 and Pgct1 Loci

Genetic analysis is beginning to reveal the genetic traits of the 129/Sv genome and the interplay of susceptibility genes. Analysis of testicular teratoma-bearing progeny from backcrosses between 129/SV-*Ter*/+ and MOLF/Ei strains has provided evidence that chromosome 19 of MOLF/Ei enhances the development of bilateral teratomas (Matin et al. 1999; Youngren et al. 2003). An autosomal chromosome substitution strain (consomic strain) has been generated, in which chromosome 19 of 129/Sv is replaced by its MOLF-derived homologue. Even in the absence of the *Ter* allele, homosomic 129.MOLF-Chr19 mice develop tumors with an incidence of 80%, and heterosomic males show a 20% incidence of tumors. Linkage analysis has identified the *Tgct1 (testicular germ cell tumor 1)* locus. Although the susceptibility locus has not been identified, the locus lies close to the *Pten* locus.

The 129/Sv genetic background contains loci that cooperate with the *Trp53* deficiency to engender susceptibility to teratomas, as discussed above. An intercross-backcross strategy has been used to identify the susceptibility locus *Pgct1 (primordial germ cell tumor 1*; Muller et al. 2000). The *Pgct1* locus on the 129/Sv genome increases the tumor incidence in *Trp53*-deficient mice with the C57BL/6 genetic background, and this locus has been mapped to chromosome 13.

4.3
Genetic Control of Testicular Cancer in Humans

The genetic control of human testicular cancers is largely unknown. Although *Trp53* mutations are found in about half of the tumor types that occur in humans, they are rarely found in human testicular cancers (Levine 1997). *Trp53* deficiency may act indirectly in testicular cancer formation, as observed in the mouse model. Furthermore, the loss of *PTEN* function is frequently noted in a wide variety of human cancers, but it is rare in human testicular cancers (von Eyben 2004). In contrast, Oct-4 may be involved in human tumors, since high-level expression of this factor is observed in these cancers (Gidekel et al. 2003; Looijenga et al. 2003). The region of mouse chromosome 13 that contains *Pgct1* is syntenic to a portion of human chromosome 5q, which itself is implicated in human susceptibility to testicular cancer (Muller et al. 2000). Genetic studies in humans have identified a susceptibility gene for bilateral testicular cancers, and the gene has been mapped to Xq27 (Rapley et al. 2000).

5
Perspectives: Implications for Stem Cell Biology

ES cells are established from pluripotent cells. In contrast, EG cells and EC cells are pluripotent cells derived from germline-committed PGCs. Therefore, EG cells and EC cells provide a unique system for enhancing our understanding of the pluripotency of stem cells in mammals. What mechanisms underlie the conversion of restricted cell lineages to pluripotent stem cells?

Stimulation with SCF and LIF, in combination with bFGF, forskolin, or retinoic acid drives the proliferation of PGCs, leading to EG cell formation. In *Pten*-deficient PGCs, in vitro proliferation is further enhanced and is accompanied by an increase in EG cell production (Kimura et al. 2003). In addition, mitotically active PGCs are present after E14.5 in teratoma-prone mice, such as *Pten*-deficient mice and 129/Sv-*Ter* mice (Noguchi and Stevens 1982; Kimura et al. 2003). Therefore, continued mitosis is probably a critical event for conversion to pluripotent stem cells. It is noteworthy that mitotic activation of adult neural stem cells by growth factors in vitro confers the developmental potential to other cell lineages, although the pluripotency of these cells is limited (Clarke et al. 2000).

PI3 K/Akt signaling promotes the G1 to S phase transition (Fig. 2; Brazil et al. 2004). The cyclin-dependent kinase (CDK) inhibitors $p21^{Cip1}$ and $p27^{Kip1}$ are blocked by PI3 K/Akt signaling, which causes the activation of CyclinE/CDK2. PI3 K/Akt increases the translation of *CyclinD* mRNA through the inhibition of glycogen synthase kinase β (GSK3β), which leads to the activation of CyclinD/CDK4/6. Consistent with the idea that mitotic activation may be implicated in conversion to stem cells, knockin mice carrying the mutant cdk4 gene insensitive to INK4 CDK inhibitors develop testicular teratomas at an incidence of 10% in the absence of *Trp53* (Sotillo et al. 2001). The process that links cell cycle progression to the regulation of pluripotency remains to be determined.

The genome of the 129/Sv mouse contains 10–15 loci that are involved in susceptibility to teratomas. An intercross-backcross strategy with subsequent generation of chromosome substitution strains has successfully identified these genetic traits (Matin et al. 1999; Muller et al. 2000). These strategies and ENU mutagenesis, in combination with genome informatics, represent powerful genetic tools for unveiling the genetic regulation of PGC conversion to pluripotent stem cells, as well as germ line development.

PI3K signaling promotes the conversion of PGCs to EG and EC cells (Kimura et al. 2003). In addition, self-renewal and the number of neural stem cells are enhanced in brain-specific *Pten*-deficient mice (Groszer et al. 2001). From these results, it appears that PI3 K signaling pathway is one of the general regulators of the stem cell system. Thus, studies of EG and EC cell formation have an impact on our understanding of the regulation, plasticity, and reprogramming of various stem cell systems.

Acknowledgements. We thank Ms. M. Ikeuchi, Y. Fujita, N. Asada, and A. Mizokami for their continuous assistance. This work was supported in part by grants from the Ministry of Education, Science, Sports and Culture; Senri Life Science Foundation; and Uehara Memorial Foundation.

References

Asada Y, Varnum DS, Frankel WN, Nadeau JH (1994) A mutation in the *Ter* gene causing increased susceptibility to testicular teratomas maps to mouse chromosome 18. Nat Genet 6:363–368

Blume-Jensen P, Jiang G, Hyman R, Lee KF, O'Gorman S, Hunter T (2000) Kit/stem cell factor receptor-induced activation of phosphatidylinositol 3′-kinase is essential for male fertility. Nat Genet 24:157–162

Bradley A, Evans M, Kaufman MH, Robertson E (1984) Formation of germ-line chimaeras from embryo-derived teratocarcinoma cell lines. Nature 309:255–256

Brazil DP, Yang ZZ, Hemmings BA (2004) Advances in protein kinase B signalling: AKTion on multiple fronts. Trends Biochem Sci 29:233–242

Cantley LC (2002) The phosphoinositide 3-kinase pathway. Science 296:1655–1657

Chabot B, Stephenson DA, Chapman VM, Besmer P, Bernstein A (1988) The proto-oncogene c-kit encoding a transmembrane tyrosine kinase receptor maps to the mouse W locus. Nature 335:88–89

Cheng L, Gearing DP, White LS, Compton DL, Schooley K, Donovan PJ (1994) Role of leukemia inhibitory factor and its receptor in mouse primordial germ cell growth. Development 120:3145–3153

Chuma S, Nakatsuji N (2001) Autonomous transition into meiosis of mouse fetal germ cells in vitro and its inhibition by gp130-mediated signaling. Dev Biol 229:468–479

Clarke DL, Johansson CB, Wilbertz J, Veress B, Nilsson E, Karlstrom H, Lendahl U, Frisen J (2000) Generalized potential of adult neural stem cells. Science 288:1660–1663

Cooke JE, Heasman J, Wylie CC (1996) The role of interleukin-4 in the regulation of mouse primordial germ cell numbers. Dev Biol 174:14–21

De Felici M, Dolci S (1991) Leukemia inhibitory factor sustains the survival of mouse primordial germ cells cultured on TM4 feeder layers. Dev Biol 147:281–284

De Miguel MP, Cheng L, Holland EC, Federspiel MJ, Donovan PJ (2002) Dissection of the c-Kit signaling pathway in mouse primordial germ cells by retroviral-mediated gene transfer. Proc Natl Acad Sci USA 99:10458–10463. Epub 12002 Jul 10424

Dean W, Bowden L, Aitchison A, Klose J, Moore T, Meneses JJ, Reik W, Feil R (1998) Altered imprinted gene methylation and expression in completely ES cell-derived mouse fetuses: association with aberrant phenotypes. Development 125:2273–2282

Dolci S, Williams DE, Ernst MK, Resnick JL, Brannan CI, Lock LF, Lyman SD, Boswell HS, Donovan PJ (1991) Requirement for mast cell growth factor for primordial germ cell survival in culture. Nature 352:809–811

Donovan PJ, Stott D, Cairns LA, Heasman J, Wylie CC (1986) Migratory and postmigratory mouse primordial germ cells behave differently in culture. Cell 44:831–838

Eppig JJ, Kozak LP, Eicher EM, Stevens LC (1977) Ovarian teratomas in mice are derived from oocytes that have completed the first meiotic division. Nature 269:517–518

Flanagan JG, Leder P (1990) The kit ligand: a cell surface molecule altered in steel mutant fibroblasts. Cell 63:185–194

Geissler EN, Ryan MA, Housman DE (1988) The dominant-white spotting (W) locus of the mouse encodes the c-kit proto-oncogene. Cell 55:185–192

Gidekel S, Pizov G, Bergman Y, Pikarsky E (2003) Oct-3/4 is a dose-dependent oncogenic fate determinant. Cancer Cell 4:361–370

Godin I, Wylie CC (1991) TGF beta 1 inhibits proliferation and has a chemotropic effect on mouse primordial germ cells in culture. Development 113:1451–1457

Godin I, Deed R, Cooke J, Zsebo K, Dexter M, Wylie CC (1991) Effects of the steel gene product on mouse primordial germ cells in culture. Nature 352:807–809

Groszer M, Erickson R, Scripture-Adams DD, Lesche R, Trumpp A, Zack JA, Kornblum HI, Liu X, Wu H (2001) Negative regulation of neural stem/progenitor cell proliferation by the Pten tumor suppressor gene in vivo. Science 294:2186–2189. Epub 2001 Nov 2181

Hajkova P, Erhardt S, Lane N, Haaf T, El-Maarri O, Reik W, Walter J, Surani MA (2002) Epigenetic reprogramming in mouse primordial germ cells. Mech Dev 117:15–23

Hara T, Tamura K, de Miguel MP, Mukouyama Y, Kim H, Kogo H, Donovan PJ, Miyajima A (1998) Distinct roles of oncostatin M and leukemia inhibitory factor in the development of primordial germ cells and sertoli cells in mice. Dev Biol 201:144–153

Harvey M, McArthur MJ, Montgomery CA Jr, Bradley A, Donehower LA (1993) Genetic background alters the spectrum of tumors that develop in p53-deficient mice. FASEB J 7:938–943

Huang E, Nocka K, Beier DR, Chu TY, Buck J, Lahm HW, Wellner D, Leder P, Besmer P (1990) The hematopoietic growth factor KL is encoded by the Sl locus and is the ligand of the c-kit receptor, the gene product of the W locus. Cell 63:225–233

Humpherys D, Eggan K, Akutsu H, Hochedlinger K, Rideout WM, 3rd, Biniszkiewicz D, Yanagimachi R, Jaenisch R (2001) Epigenetic instability in ES cells and cloned mice. Science 293:95–97

Jiang LI, Nadeau JH (2001) 129/Sv mice – a model system for studying germ cell biology and testicular cancer. Mamm Genome 12:89–94

Kawase E, Yamamoto H, Hashimoto K, Nakatsuji N (1994) Tumor necrosis factor-alpha (TNF-alpha) stimulates proliferation of mouse primordial germ cells in culture. Dev Biol 161:91–95

Keshet E, Lyman SD, Williams DE, Anderson DM, Jenkins NA, Copeland NG, Parada LF (1991) Embryonic RNA expression patterns of the c-kit receptor and its cognate ligand suggest multiple functional roles in mouse development. EMBO J 10:2425–2435

Kimura T, Suzuki A, Fujita Y, Yomogida K, Lomeli H, Asada N, Ikeuchi M, Nagy A, Mak TW, Nakano T (2003) Conditional loss of PTEN leads to testicular teratoma and enhances embryonic germ cell production. Development 130:1691–1700

Kishimoto H, Hamada K, Saunders M, Backman S, Sasaki T, Nakano T, Mak TW, Suzuki A (2003) Physiological functions of Pten in mouse tissues. Cell Struct Funct 28:11–21

Kissel H, Timokhina I, Hardy MP, Rothschild G, Tajima Y, Soares V, Angeles M, Whitlow SR, Manova K, Besmer P (2000) Point mutation in kit receptor tyrosine kinase reveals essential roles for kit signaling in spermatogenesis and oogenesis without affecting other kit responses. EMBO J 19:1312–1326

Koshimizu U, Watanabe M, Nakatsuji N (1995) Retinoic acid is a potent growth activator of mouse primordial germ cells in vitro. Dev Biol 168:683–685

Koshimizu U, Taga T, Watanabe M, Saito M, Shirayoshi Y, Kishimoto T, Nakatsuji N (1996) Functional requirement of gp130-mediated signaling for growth and survival of mouse primordial germ cells in vitro and derivation of embryonic germ (EG) cells. Development 122:1235–1242

Labosky PA, Barlow DP, Hogan BL (1994a) Mouse embryonic germ (EG) cell lines: transmission through the germline and differences in the methylation imprint of insulin-like growth factor 2 receptor (Igf2r) gene compared with embryonic stem (ES) cell lines. Development 120:3197–3204

Labosky PA, Barlow DP, Hogan BL (1994b) Embryonic germ cell lines and their derivation from mouse primordial germ cells. Ciba Found Symp 182:157–168; discussion 168–178

Lee J, Inoue K, Ono R, Ogonuki N, Kohda T, Kaneko-Ishino T, Ogura A, Ishino F (2002) Erasing genomic imprinting memory in mouse clone embryos produced from day 11.5 primordial germ cells. Development 129:1807–1817

Levine AJ (1997) p53, the cellular gatekeeper for growth and division. Cell 88:323–331

Looijenga LH, Oosterhuis JW (1999) Pathogenesis of testicular germ cell tumours. Rev Reprod 4:90–100

Looijenga LH, Stoop H, de Leeuw HP, de Gouveia Brazao CA, Gillis AJ, van Roozendaal KE, van Zoelen EJ, Weber RF, Wolffenbuttel KP, van Dekken H, Honecker F, Bokemeyer C, Perlman EJ, Schneider DT, Kononen J, Sauter G, Oosterhuis JW (2003) POU5F1 (OCT3/4) identifies cells with pluripotent potential in human germ cell tumors. Cancer Res 63:2244–2250

Martin GR (1981) Isolation of a pluripotent cell line from early mouse embryos cultured in medium conditioned by teratocarcinoma stem cells. Proc Natl Acad Sci USA 78:7634–7638

Matin A, Collin GB, Asada Y, Varnum D, Nadeau JH (1999) Susceptibility to testicular germ-cell tumours in a 129.MOLF-Chr 19 chromosome substitution strain. Nat Genet 23:237–240

Matsui Y, Zsebo KM, Hogan BL (1990) Embryonic expression of a haematopoietic growth factor encoded by the Sl locus and the ligand for c-kit. Nature 347:667–669

Matsui Y, Toksoz D, Nishikawa S, Williams D, Zsebo K, Hogan BL (1991) Effect of Steel factor and leukaemia inhibitory factor on murine primordial germ cells in culture. Nature 353:750–752

Matsui Y, Zsebo K, Hogan BL (1992) Derivation of pluripotential embryonic stem cells from murine primordial germ cells in culture. Cell 70:841–847

McLaren A (2003) Primordial germ cells in the mouse. Dev Biol 262:1–15

Mintz B, Illmensee K (1975) Normal genetically mosaic mice produced from malignant teratocarcinoma cells. Proc Natl Acad Sci USA 72:3585–3589

Moe-Behrens GH, Klinger FG, Eskild W, Grotmol T, Haugen TB, De Felici M (2003) Akt/PTEN signaling mediates estrogen-dependent proliferation of primordial germ cells in vitro. Mol Endocrinol 17:2630–2638. Epub 2003 Oct 2632

Muller AJ, Teresky AK, Levine AJ (2000) A male germ cell tumor-susceptibility-determining locus, pgct1, identified on murine chromosome 13. Proc Natl Acad Sci USA 97:8421–8426

Nichols J, Zevnik B, Anastassiadis K, Niwa H, Klewe-Nebenius D, Chambers I, Scholer H, Smith A (1998) Formation of pluripotent stem cells in the mammalian embryo depends on the POU transcription factor Oct4. Cell 95:379–391

Niwa H, Miyazaki J, Smith AG (2000) Quantitative expression of Oct-3/4 defines differentiation, dedifferentiation or self-renewal of ES cells. Nat Genet 24:372–376

Noguchi T, Stevens LC (1982) Primordial germ cell proliferation in fetal testes in mouse strains with high and low incidences of congenital testicular teratomas. J Natl Cancer Inst 69:907–913

Noguchi T, Noguchi M (1985) A recessive mutation (*ter*) causing germ cell deficiency and a high incidence of congenital testicular teratomas in 129/Sv-ter mice. J Natl Cancer Inst 75:385–392

Noguchi M, Watanabe C, Kobayashi T, Kuwashima M, Sakurai T, Katoh H, Moriwaki K (1996) The *ter* mutation responsible form germ cell deficiency but not testicular nor ovarian teratocarcinogenesis in *ter/ter* congenic mice. Dev Growth Differ 38:59–69

Onyango P, Jiang S, Uejima H, Shamblott MJ, Gearhart JD, Cui H, Feinberg AP (2002) Monoallelic expression and methylation of imprinted genes in human and mouse embryonic germ cell lineages. Proc Natl Acad Sci USA 99:10599–10604. Epub 12002 Jul 10511

Pesce M, Farrace MG, Piacentini M, Dolci S, de Felici M (1993) Stem cell factor and leukemia inhibitory factor promote primordial germ cell survival by suppressing programmed cell death (apoptosis). Development 118:1089–1094

Pesce M, Canipari R, Ferri GL, Siracusa G, de Felici M (1996) Pituitary adenylate cyclase-activating polypeptide (PACAP) stimulates adenylate cyclase and promotes proliferation of mouse primordial germ cells. Development 122:215–221

Rajpert-De Meyts E, Bartkova J, Samson M, Hoei-Hansen CE, Frydelund-Larsen L, Bartek J, Skakkebaek NE (2003) The emerging phenotype of the testicular carcinoma in situ germ cell. Apmis 111:267–278; discussion 278–269

Rapley EA, Crockford GP, Teare D, Biggs P, Seal S, Barfoot R, Edwards S, Hamoudi R, Heimdal K, Fossa SD, Tucker K, Donald J, Collins F, Friedlander M, Hogg D, Goss P, Heidenreich A, Ormiston W, Daly PA, Forman D, Oliver TD, Leahy M, Huddart R, Cooper CS, Bodmer JG, Easton DF, Stratton MR, Bishop DT (2000) Localization to Xq27 of a susceptibility gene for testicular germ-cell tumours. Nat Genet 24:197–200

Resnick JL, Bixler LS, Cheng L, Donovan PJ (1992) Long-term proliferation of mouse primordial germ cells in culture. Nature 359:550–551

Resnick JL, Ortiz M, Keller JR, Donovan PJ (1998) Role of fibroblast growth factors and their receptors in mouse primordial germ cell growth. Biol Reprod 59:1224-1229

Richards AJ, Enders GC, Resnick JL (1999) Activin and TGFbeta limit murine primordial germ cell proliferation. Dev Biol 207:470–475

Rosenberg C, Mostert MC, Schut TB, van de Pol M, van Echten J, de Jong B, Raap AK, Tanke H, Oosterhuis JW, Looijenga LH (1998) Chromosomal constitution of human spermatocytic seminomas: comparative genomic hybridization supported by conventional and interphase cytogenetics. Genes Chromosomes Cancer 23:286–291

Rugh R (1968) The mouse. Oxford Univ Press, Oxford

Sakurai T, Katoh H, Moriwaki K, Noguchi T, Noguchi M (1994) The ter primordial germ cell deficiency mutation maps near Grl-1 on mouse chromosome 18. Mamm Genome 5:333–336

Shamblott MJ, Axelman J, Wang S, Bugg EM, Littlefield JW, Donovan PJ, Blumenthal PD, Huggins GR, Gearhart JD (1998) Derivation of pluripotent stem cells from cultured human primordial germ cells. Proc Natl Acad Sci USA 95:13726–13731

Skakkebaek NE (1972) Possible carcinoma-in situ of the testis. Lancet 2:516–517

Sotillo R, Dubus P, Martin J, de la Cueva E, Ortega S, Malumbres M, Barbacid M (2001) Wide spectrum of tumors in knock-in mice carrying a Cdk4 protein insensitive to INK4 inhibitors. EMBO J 20:6637–6647

Stevens LC (1958) Studies on transplantable testicular teratomas of strain 129 mice. J Natl Cancer Inst 20:1257–1275

Stevens LC (1964) Experimental production of testicular teratomas in mice. Proc Natl Acad Sci USA 52:654–661

Stevens LC (1967a) Origin of testicular teratomas from primordial germ cells in mice. J Natl Cancer Inst 38:549–552

Stevens LC (1967b) The biology of teratomas. Adv Morphog 6:1–31

Stevens LC (1973) A new inbred subline of mice (129-terSv) with a high incidence of spontaneous congenital testicular teratomas. J Natl Cancer Inst 50:235–242

Stevens LC (1980) Teratocarcinogenesis and spontaneous parthenogenesis in mice. Res Prob Cell Differ 11:265–274

Stevens LC, Hummel KP (1957) A description of spontaneous congenital testicular teratomas in strain 129 mice. J Natl Cancer Inst 18:719–747

Stewart CL, Gadi I, Bhatt H (1994) Stem cells from primordial germ cells can reenter the germ line. Dev Biol 161:626–628

Stewart TA, Mintz B (1981) Successive generations of mice produced from an established culture line of euploid teratocarcinoma cells. Proc Natl Acad Sci USA 78:6314–6318

Stiles B, Groszer M, Wang S, Jiao J, Wu H (2004) PTENless means more. Dev Biol 273:175–184

Sulis ML, Parsons R (2003) PTEN: from pathology to biology. Trends Cell Biol 13:478–483

Surani MA (2001) Reprogramming of genome function through epigenetic inheritance. Nature 414:122–128

Szabo PE, Hubner K, Scholer H, Mann JR (2002) Allele-specific expression of imprinted genes in mouse migratory primordial germ cells. Mech Dev 115:157–160

Tada T, Tada M, Hilton K, Barton SC, Sado T, Takagi N, Surani MA (1998) Epigenotype switching of imprintable loci in embryonic germ cells. Dev Genes Evol 207:551–561

Takabayashi S, Nozaki M, Ishikawa K, Noguchi M (2001a) The ter/ter gonadal somatic cells cause apoptosis in ter/ter primordial germ cells (PGCs) with normal survivability and proliferation ability in the mouse: evidence from PGC-somatic cell »exchange-co-culture«. Zool Sci 18:695–704

Takabayashi S, Sasaoka Y, Yamashita M, Tokumoto T, Ishikawa K, Noguchi M (2001b) Novel growth factor supporting survival of murine primordial germ cells: evidence from conditioned medium of *ter* fetal gonadal somatic cells. Mol Reprod Dev 60:384–396

Thomson JA, Itskovitz-Eldor J, Shapiro SS, Waknitz MA, Swiergiel JJ, Marshall VS, Jones JM (1998) Embryonic stem cell lines derived from human blastocysts. Science 282:1145–1147

Toyoda-Ohno H, Obinata M, Matsui Y (1999) Members of the ErbB receptor tyrosine kinases are involved in germ cell development in fetal mouse gonads. Dev Biol 215:399–406

Turnpenny L, Brickwood S, Spalluto CM, Piper K, Cameron IT, Wilson DI, Hanley NA (2003) Derivation of human embryonic germ cells: an alternative source of pluripotent stem cells. Stem Cells 21:598–609

Von Eyben FE (2004) Chromosomes, genes, and development of testicular germ cell tumors. Cancer Genet Cytogenet 151:93–138

Wylie C (2000) Germ cells. Curr Opin Genet Dev 10:410–413

Yamazaki Y, Mann MR, Lee SS, Marh J, McCarrey JR, Yanagimachi R, Bartolomei MS (2003) Reprogramming of primordial germ cells begins before migration into the genital ridge, making these cells inadequate donors for reproductive cloning. Proc Natl Acad Sci USA 100:12207–12212

Youngren KK, Nadeau JH, Matin A (2003) Testicular cancer susceptibility in the 129.MOLF-Chr19 mouse strain: additive effects, gene interactions and epigenetic modifications. Hum Mol Genet 12:389–398

Tumor Suppressors APC and VHL: Gatekeepers of the Intestine and Kidney

Rachel H. Giles, Emile E. Voest

1 Introduction and Background

The isolation of the *APC* (Adenomatous Polyposis Coli) gene in 1991 and its establishment as the main regulator of the development of colorectal cancer (CRC) marked a milestone for the entire field of cancer molecular genetics. It was the first time that mutations in a single gene were found to result in most tumors arising in a particular anatomical venue. Germline mutations of the *APC* gene are responsible for the familial adenomatous polyposis (FAP), an inherited autosomal dominant condition leading not only to the development of multiple adenomas of the colorectum, but also of extra-colonic lesions of ectodermal, mesodermal, and endodermal origin. Of great interest is the fact that the *APC* gene is also mutated in the vast majority of sporadic colorectal tumors regardless of their degree of malignancy. Alterations of *APC* inappropriately activate the Wnt/β-catenin signal transduction pathway, in addition to affecting cellular events that could lead to tumorigenesis. Due to the extraordinary prevalence of *APC* mutations in CRC, APC has been dubbed the »gatekeeper« of proliferation of the colorectum.

Following in APC's footsteps, another »gatekeeper« gene was identified that regulates epithelial growth in the kidney: the von Hippel-Lindau (VHL) tumor suppressor. Like *APC*, heterozygous mutation of *VHL* results in an infrequent cancer syndrome, often associated with clear cell renal carcinoma (RCC). The majority of sporadic clear cell RCC samples also manifest biallelic *VHL* mutations, suggesting that this tumor suppressor holds the key to tumorigenesis in the majority of kidney cancers. Alterations of *VHL* inappropriately activate the Hypoxia Inducible factor (HIF) pathway, in addition to deregulating cytoskeletal processes that could influence tumorigenesis.

In this chapter we review the events leading from initial *APC* mutations to the development of CRC and attempt to draw parallels to the lesser-charac-

R.H. Giles (e-mail: r.giles@azu.nl), E.E. Voest
Department Medical Oncology. Rm. F02.126, University Medical Center Utrecht, Heidelberglaan 100, 3584 CX Utrecht, The Netherlands

Progress in Molecular and Subcellular Biology
Alvaro Macieira-Coelho (Ed.)
Development Biology of Neoplastic Growth

terized step-wise progression from *VHL* mutation to RCC (Fig. 1). We speculate on the similar roles these tumor suppressors play in signal transduction as weighted against the effects these proteins exercise on the cytoskeleton. On the basis of the numerous functions that APC and VHL carry out that are crucial for the maintenance of gut and kidney epithelium, respectively, it is perhaps not surprising that loss of fully functional APC or VHL produces a highly penetrant phenotype in these tissues.

2 Colorectal Cancer: Current Understanding of the Molecular Mechanism

The American Cancer Society estimates that colorectal cancer (CRC) will kill over 50,000 people in the United States this year alone. The remarkable thing about CRCs – the most common cause of non-smoking-related cancer deaths in the Western world – is that the molecular mechanisms underlying virtually all of these cases are uniform. Greater than 90% of all CRCs will have an activating mutation of the Wnt signaling pathway, ultimately leading to the stabilization and accumulation of β-catenin in the nucleus of a cell. Nuclear β-catenin is the hallmark of an active canonical Wnt pathway; the presence of

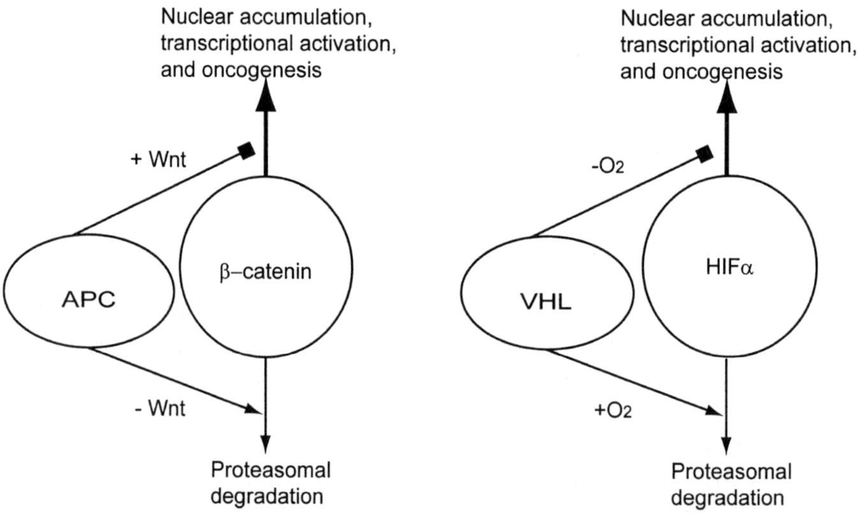

Fig. 1. Schematic illustration of the similar roles tumor suppressors APC and VHL play in the regulation of β-catenin and HIF-α, respectively. *Left* APC binds β-catenin in the absence (–) of Wnt signaling and targets it for proteasomal degradation. In the presence (+) of Wnt signaling, APC is no longer able to bind β-catenin, which then translocates to the nucleus and becomes transcriptionally active. *Right* In normoxia (+O_2) VHL binds HIFα and targets it for proteasomal degradation. However, under hypoxia (–O_2) VHL cannot bind HIFα, which then translocates to the nucleus and becomes transcriptionally active.

nuclear β-catenin is evident in even the smallest detectable lesions resulting from Wnt mutations (Kongkanuntn et al. 1999). The ubiquitous nature of Wnt signaling activation in CRC makes this cancer an attractive model for molecular intervention.

2.1 WNT/β-Catenin Signaling

This signaling pathway, named for its upstream Wnt ligands, ultimately results in the transcription of a suite of genes promoting cellular proliferation and repressing differentiation. Wnt signaling is essential for vertebrate embryonic development, but inappropriate activation of this pathway in post-embryonic cells results in carcinogenesis. In the absence of Wnt signal, a multiprotein complex phosphorylates β-catenin, marking it for subsequent ubiquitination and degradation (Orford et al. 1997). The genetic program initiated by β-catenin in concert with TCF/LEF transcription factors specifies the transcription of a subset of genes, mainly determining cell fate and regulating proliferation. Signaling through this pathway is present during embryogenesis, where it has been shown to regulate many developmental patterning events in organisms ranging from worm to man. In the developing vertebrate embryo, the formation of the dorsal-ventral axis depends on the activity of the Wnt signaling pathway. Misregulation of Wnt/β-catenin signaling causes developmental defects.

Under physiological circumstances, Wnt/β-catenin signaling is initiated following Wnt ligand binding to a member of the frizzled (Fz) family of seven-span transmembrane receptors (Bhanot et al. 1996; Wang et al. 1996; He et al. 1997; Dale 1998) together with the co-receptors LRP5 or 6, members of the low-density lipoprotein-receptor-related protein family (LRP) (Wehrli et al. 2000). Upon binding of Wnt, axin translocates to the membrane and interacts with the intracellular tail of LRP5 (Mao et al. 2001). Accordingly, LRP5 mutants lacking the extracellular domain function behave as constitutively active forms that bind and destabilize axin (Mao et al. 2001). As described below, axin – like APC – is a scaffold protein integral to the protein complex responsible for marking β-catenin for degradation. By destabilizing axin, β-catenin levels accumulate and TCF/LEF target genes are transactivated (Mao et al. 2001). In parallel but by an unknown mechanism, Wnt binding to Fz results in hyperphosphorylation of Disheveled (Dsh), which inhibits the activity of GSK3β (Yanagawa et al. 1995).

The cytoplasmic »β-catenin degradation complex« consists of GSK3β, axin, and the tumor suppressor APC. Axin and APC form a structural scaffold that allows GSK3β to specifically phosphorylate β-catenin. Although APC mutations are more frequently associated with the inappropriate activation of Wnt/β-catenin signaling, axin equally functions as a tumor suppressor in this system. Axin was originally identified as an inhibitor of Wnt signaling when mutations in murine axin were observed to cause axis duplication in homozygous mouse embryos (Zeng et al. 1997). Furthermore, injection of axin mRNA into frog embryos inhibits dorsal axis formation (Zeng et al. 1997). The relevance of axin's function is illustrated by studies showing that truncating mutations in *AXIN1* leading to nuclear accumulation of β-catenin are found in hepatocellular carcinomas (Satoh et al. 2000). Adenoviral transfer of wild-type *AXIN1* into these cells, as well as HepG2 cells that express mutant β-catenin, decreased nuclear accumulation of β-catenin and lower TCF/LEF-mediated transcriptional activity (Satoh et al. 2000).

2.2 Step-Wise CRC Progression

CRC develops through a series of genetic alterations involving the accumulation of mutations in a number of genes and progression through the adenoma-carcinoma sequence (Kinzler and Vogelstein 1996; Fig. 2). The underlying hypothesis is that the tumor microenvironment is an additional driving force of tumor progression. This tumor progression model was deduced from comparison of genetic alterations seen in normal colon epithelium, adenomas of progressively larger size, and malignancies (Vogelstein and Kinzler 1993).

APC Loss or β-Catenin Activation. Activating mutations of the Wnt pathway are the only known genetic alterations present in early premalignant lesions in the intestine, such as aberrant crypt foci and small polyps (Powell et al. 1992). Human *APC*, located on chromosome 5q, has monoallelic inactivating mutations in patients with familial adenomatous polyposis (FAP), an inherited autosomal dominant condition leading to the development of multiple adenomas in the colorectum (Groden et al. 1991; Nishisho et al. 1991). In these patients, polyps also develop in the upper gastrointestinal tract, and malig-

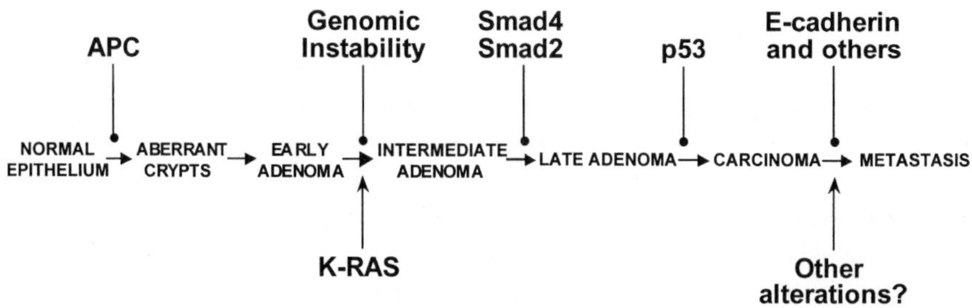

Fig. 2. The adenoma–carcinoma sequence for colorectal cancer. A mutation in APC results in the activation of the Wnt signaling pathway, triggering tumor formation. Subsequent progression towards malignancy is accompanied by sequential mutations in *KRAS*, deletion of chromosome 18q affecting genes encoding SMAD2 and SMAD4, *p53*, and genes involved in tumor invasiveness such as E-cadherin.

nancies may occur at other sites including the brain and thyroid. The remaining wild-type *APC* allele is mutated in most FAP patient polyps. Although FAP is a rare disease, colorectal cancer is not, and up to 85% of all sporadic colorectal cancers have mutations in *APC*. Over 300 different disease-associated mutations of the *APC* gene have been reported (Laurent-Puig et al. 1998). The vast majority of these changes are insertions, deletions, and nonsense mutations that lead to frameshifts and/or premature stop codons in the resulting transcript of the gene.

APC is generally considered a classical tumor suppressor: both alleles must be affected for loss of growth suppressing activity. However, CRCs with *APC* mutations do not completely lack APC protein. In the vast majority of cases, one allele acquires a truncating mutation while the second undergoes either a loss of heterozygosity or a second truncating mutation (Smits et al. 2000; Fearnhead et al. 2001). In particular, mutations close to codon 1300 are usually associated with allelic loss of the second allele, whereas tumors with mutations outside this region tend to harbor a second truncating mutation (Lamlum et al. 1999; Rowan et al. 2000). This interdependence of mutations suggests a strong selective process for tumor progression. Upon closer analysis of the mutation spectra in CRCs, it appears that the retention of polypeptides with residual activity in regulating β-catenin levels provide a selective advantage (Munemitsu et al. 1995; Rubinfeld et al. 1997; Dihlmann et al. 1999; Smits et al. 2000; Su et al. 2000; Albuquerque et al. 2002). A recent hypothesis dubbed the »just-right signaling model« suggests that APC function must be impaired sufficiently to allow the accumulation of β-catenin, but not above a certain limit (Albuquerque et al. 2002).

Activating mutations in the gene encoding β-catenin (*CTNNB1*) are present in approximately 5–10% of the remaining CRC tumors. *APC* and *CTNNB1* mutations are mutually exclusive, consistent with the notion that mutation of either gene has more or less the same effect on β-catenin stability and TCF transactivation. However, small adenomas with β-catenin mutations do not appear to be as likely to progress to larger adenomas and invasive carcinomas, nor are β-catenin mutations present in 5–10% of invasive cancers. Thus *APC* and β-catenin mutations do not appear to be functionally identical (Samowitz et al. 1999). Most β-catenin mutations occur in or around exon 3 of the *CTNNB1* gene, affecting the putative phosphorylation sites for GSK3β, making it refractory to degradation (Morin et al. 1997; Polakis 1999).

Target Genes. Several individual TCF/LEF target genes have been identified, some of which provide insight into the role of Wnt signaling in cancer. Among the most prevalent, Wnt signaling promotes the expression of c-MYC (He et al. 1998). A detailed list of target genes can be found at the Wnt home page hosted by R. Nusse (www.stanford.edu/~rnusse/wntwindow.html). Recently, two large-scale analyses using DNA microarray technology have begun to unravel the downstream genetic program activated by β-catenin/TCF in embryonic carcinoma cells and CRC cells, respectively (Willert et al. 2002; van de

Wetering et al. 2002;). The first of these studies, Wnt-3A-conditioned or control medium was applied to human teratocarcinoma cells for various lengths of time before RNA was isolated and the expression profiles compared by microarray hybridization. Several genes were identified in this and in earlier studies that interfere with the differentiation of cells, e.g., the *ID, MSX,* and *REST/NRSF* family of genes (Rockman et al. 2001; van de Wetering et al. 2002; Willert et al. 2002). Consistent with these results, van de Wetering et al. confirmed the generalization that β-catenin/TCF targets often work to repress differentiation. Microarray expression profiling was performed using CRC cell lines carrying doxycycline-inducible expression plasmids encoding dominant-negative TCF (dnTCF) proteins (van de Wetering et al. 2002). Upon induction with doxycycline these dnTCF proteins act as potent inhibitors of the endogenous β-catenin/TCF complexes present in CRC cells, imposing a robust cell cycle and proliferation arrest. Genes downregulated by the induction of dnTCF are thus putative targets of β-catenin/TCF and include the genes encoding c-MYC, c-MYB, ETS-2, and ephrin receptors EphB2 and EphB3 (van de Wetering et al. 2002). Reintroduction of c-MYC rescued the cell cycle arrest by blocking the expression of cyclin-dependent kinase inhibitor $p21^{CIP/WAF1}$. Furthermore, upon expression of dnTCF and subsequent G1 arrest, these cells recapitulate the physiological differentiation program of normal intestinal cells in a $p21^{CIP1/WAF1}$-dependent manner (van de Wetering et al. 2002). Many of the genes upregulated by dnTCF represent differentiation markers of mucososecretory and/or absorptive intestinal cells.

Genetic instability. A normal cell needs to comply with two essential requirements to develop into a cancer: it must acquire a selective advantage to allow for the initial clonal expansion, and genetic instability to allow for multiple hits at other genes aiding tumor progression and malignant transformation. CRCs resulting from activated Wnt signaling fulfill the first requirement by transactivating TCF target genes such as *c-MYC*.

In the intestine, fulfillment of the second requirement – genetic instability – often manifests as aneuploidy or hyperploidy, supernumerary centrosomes and abnormal mitotic spindle assembly (Lengauer et al. 1998; Pihan et al. 1998; Ghadimi et al. 2000; Kuo et al. 2000; Sato et al. 2001). These phenomena lead to multipolar cell divisions, incomplete or abnormal chromosome segregation, and occasionally to chromosomal breakage (Duensing and Munger 2001). There are at least two mechanisms inducing the molecular events that lead to colorectal cancer. About 85% of CRCs exhibit chromosomal instability (CIN), whereas the remaining 15% are due to events that result in microsatellite instability (MIN; Lengauer et al. 1997; Kinzler and Vogelstein 1998; Lindblom 2001). A strong correlation between *APC* mutation in CRC and the presence of CIN, coupled with the association of *CTNNB1* mutations occurring with MIN led to the hypothesis that mutations in *APC* not only confer a growth advantage but also promote CIN in CRCs. Mouse ES cells expressing truncated APC show CIN, aberrant spindle formation, and supernumerary centrosomes (Fodde et al. 2001; Kaplan et al. 2001). Moreover, during mitosis, APC localizes to the ends of microtubules embedded in kinetochores and forms a complex with the checkpoint proteins Bub1 and Bub3. In vitro, APC is a high-affinity substrate for Bub kinases (Kaplan et al. 2001). Based on these observations, the hypothesis has been put forward that a truncating mutation in *APC* initiates CIN thus accelerating the loss of the second allele, which ultimately leads to deregulated cellular proliferation.

Characteristic secondary mutations. Although genetic instability is global, cells with specific losses are selected for clonal growth. Generally, CRCs manifesting CIN acquire sequential chromosome losses at the loci for *APC* (chromosome 5q; Levy et al. 1994), *DCC/DPC/JV18* (chromosome 18q), and *p53* (chromosome 17p), and activating mutations of *K-ras* (Vogelstein and Kinzler 1993). Accordingly, Min/+ mice deficient for p53 enhanced the number and size of intestinal adenomas (Halberg et al. 2000). The paradigm of sequential genetic alterations driving tumor progression is not totally inflexible; not every CRC tumor needs to acquire every mutation, nor do the mutations always need to occur in a specific order. However, the type of mutation may influence the rate or type of pathologic change in the tumors. As opposed to CIN, tumors manifesting MIN are characterized by a largely intact chromosome complement, but they have acquired defects in DNA repair. Thus, mutations that may occur in important cancer-associated genes are allowed to persist. These types of cancers are detectable at the molecular level by alterations in repeating units of DNA that occur normally throughout the genome, known as a DNA microsatellite. Many of the same genes are affected in MIN tumors as in CIN tumors.

2.3 The APC Protein

APC mutations are most commonly associated with the development of CRC. APC is a large (312-kDa) protein that is known to interact with at least ten protein partners including β-catenin, axin, end-binding protein EB-1, and tumor-suppressor protein disks-large (DLG; Fearnhead et al. 2001). It can also form homodimers, the functional relevance of which has not yet been determined (Joslyn et al. 1993; Su et al. 1993; Pyles et al. 1998). APC has multiple, diverse functions in cell migration and adhesion, in cell cycle regulation, and in chromosome stability (Peifer and Polakis 2000). However, its critical role in the control of cellular levels of β-catenin, thus acting as a negative regulator of the Wnt-signaling pathway, appears to be critical in tumorigenesis (Fodde 2002). Although *APC* is mutated in 85% of all CRCs, familial and sporadic, it is not absolutely necessary for the proper functioning of the β-catenin degradation complex. Overexpression of axin can compensate for the absence of functional APC (Hart et al. 1998; Nakamura et al. 1998). In addition, APC has been shown to capture and escort nuclear β-catenin to the cytoplasmic destruction machinery (Henderson 2000; Neufeld et al. 2000; Rosin-Arbesfeld et al. 2000). Inclusion of APC in the β-catenin degradation complex likely results in an improved presentation of β-catenin to GSK3β, leading to more efficient phosphorylation and subsequent destruction.

A paralogue of APC termed APC2 or APCL was identified in humans and mouse (Nakagawa et al. 1998; van Es et al. 1999). *APC2* localizes to human chromosome band 19p13.3 and encodes a 2303 amino acid protein that appears to be ubiquitously expressed in normal tissues although clear differences in expression levels are present (van Es et al. 1999). APC2 bears significant homology to APC at the N-terminus and central region but not at the C-terminus. The dimerization domain is conserved suggesting that APC2 can also form homo- and/or hetero-dimers with APC itself. APC2 can bind β-catenin and regulate its concentrations as efficiently as APC (Nakagawa et al. 1998; van Es et al. 1999). The C-terminus of APC2, but not of APC, has been shown to bind to the p53-binding protein 2, suggesting that distinct functions of these proteins may be mediated by the C-terminus (Nakagawa et al. 2000a). All other protein partners appear to be conserved, including members of the EB family of microtubule binding proteins (Nakagawa et al. 2000b; van Es et al. 2001). Given the similarities between APC and APC2 it seems likely that APC2 participates in Wnt signaling and consequently tumorigenesis. No specific mutations in *APC2* in tumors have yet been reported (Nakagawa et al. 1999), although one recent report does document allelic imbalance of the *APC2* locus in 19/20 ovarian carcinomas (Jarrett et al. 2001).

Apc Mouse Models. The generation of mouse models has provided the best tools for dissecting the individual functions of Wnt pathway components in cancer. Chemical mutagenesis produced a mouse model for FAP stably expressing truncated Apc due to a nonsense mutation at codon 850. Mice heterozygous for this mutation develop multiple intestinal neoplasia (Min) and seldom live longer than 3 months (Moser et al. 1990; Su et al. 1992). Of note is the fact that while the Min adenomas resemble those found in FAP patients, they arise in the small intestine, not the colon. Adenomas harvested from Min mice harbor somatic mutations of the wild-type Apc allele (Luongo et al. 1994). Female Min mice that live long enough may also develop mammary tumors. Transplantation of mammary cells from Min/+ or +/+ donors into +/+ hosts demonstrated that the propensity to develop mammary tumors is intrinsic to the Min/+ mammary cells (Moser et al. 1993). Mice homozygous for the Min mutation die in utero (Moser et al. 1995). Alternatively, transgenic mouse models inducibly expressing constitutively active β-catenin suffer intestinal tumors indistinguishable from Min mice (Harada et al. 1999; Romagnolo et al. 1999). Thus, constitutive β-catenin signaling is sufficient for intestinal tumor induction.

Although perhaps most widely used, Min mice are not the only model generated to study Apc. Conventional gene targeting has also generated mice with truncated alleles at codons 474 and 1309, respectively (Quesada et al. 1998; Sasai et al. 2000). Both mice share phenotypes with the Min mouse. Oshima et al. used homologous recombination to generate mice expressing Apc truncated at residue 716 (Oshima et al. 1995). Like Min mice, the $Apc^{\Delta 716}$ heterozygotes developed numerous adenomas throughout the intestinal tract, mostly in the small intestine. Genotyping even the smallest tumors revealed that all had lost their wild-type *Apc* allele. This study was pivotal in two respects: it determined that loss of *Apc* is an early, if not initiating event in tumorigenesis, and it established that these microadenomas stem from single crypts by forming abnormal outpockets protruding into the neighboring villus (Oshima et al. 1995). Shibata et al. created a conditional Apc model in which *Apc* exon 14 is deleted upon Cre recombinase expression in the colon, resulting in Apc truncated at codon 580. The mice developed adenomas within 4 weeks, implying that inactivation of *Apc* is sufficient to drive polyp formation (Shibata et al. 1997).

Two Apc mouse models make use of an introduced termination mutation at residue 1638. The first, called $Apc^{1638\,N}$, introduces the PGK–neomycin gene at residue 1638 in the opposite transcriptional orientation of *Apc*, resulting in greatly reduced levels of the truncated polypeptide (Fodde et al. 1994). Although undetectable by western blot, the allele is leaky and approximately 2% of the 182-kDa $Apc^{1638\,N}$ is present and functional (Fodde et al. 1994; Kielman et al. 2002). In the same genetic background, $Apc^{1638\,N}$/+ mice develop considerably fewer intestinal tumors than Min mice and live considerably longer, but their tumors tend to develop higher up in the gastrointestinal tract, at the transition from stomach to small intestine (periampullary region; van der Houven van Oordt et al. 1997; Smits et al. 1998). Interestingly, $Apc^{1638\,N}$/+ mice also develop high numbers of extracolonic lesions, particularly cutaneous follicular cysts and benign desmoid growths. Human FAP patients with APC mutations located between codons 1445 and 1578 are also often associated with severe desmoids as well as osteomas, epidermoid cysts, and polyps of the upper gastrointestinal tract (Caspari et al. 1995), making this mouse model indispensable for studying these particular extraintestinal manifestations. Homozygosity for the $Apc^{1638\,N}$ mutation is not compatible with adult life (Fodde et al. 1994; Kielman et al. 2002).

When the same the PGK–neomycin gene at residue 1638 is introduced in alignment with the transcriptional orientation of *Apc*, it results in a truncated allele, called 1638T. The truncated protein lacks the C-terminal domains binding to tubulin, EB/RP proteins, and DLG, but can regulate β-catenin levels efficiently. Heterozygous as well as homozygous Apc^{1638T} mice are not only viable and fertile, but they remain tumor free (Smits et al. 1999). Because the C-terminus of Apc binds microtubules, EB-family proteins, and DLG, Apc^{1638T} mice provide an excellent model for delineating the critical domains of Apc involved in tumorigenesis and embryonic development. Embryonic stem cells homozygous for APC^{1638T} exhibit chromosome instability underscoring the role of the C-terminus in chromosome segregation (Fodde et al.

2001). Evidently, the predisposition to chromosomal instability is not sufficient to drive tumor formation in these mice.

2.4 APC in Cytoskeletal Dynamics

APC has been recently implicated in cell morphology and migration through modifications of the actin cytoskeleton, through interaction with the Rac-specific guanine nucleotide exchange factor Asef. This interaction stimulated cell flattening, membrane ruffling, and lamellipodia formation in MDCK cells induced by exogenous expression of Asef (Kawasaki et al. 2000). Microtubules are essential for cellular processes including migration, organelle transport, and chromosome segregation during cell division by formation of the mitotic spindle. APC has been localized to clusters near the distal ends of microtubules at the edges of migrating epithelial cells (Nathke et al. 1996), and other studies have shown APC to bind microtubules both directly and indirectly (Munemitsu et al. 1994; Smith et al. 1994; Su et al. 1995; Deka et al. 1998). The C-terminus of Apc is important for the stabilization of microtubule structures; mice heterozygous for *Apc* demonstrate a significant decrease in apico–basal arrays of microtubule bundles (Mogensen et al. 2002). Because APC associates with microtubules and accumulates at their growing tips even in the absence of its C-terminal sequences (including both EB- and microtubule-binding domains; Mimori-Kiyosue et al. 2000), it has been suggested that another domain at the N-terminus targets APC to the microtubular cytoskeleton (Mimori-Kiyosue and Tsukita 2001; Zumbrunn et al. 2001). Jimbo et al. demonstrated an interaction between the arm repeats of APC and the kinesin superfamily-associated protein 3 (KAP3), an adaptor for the kinesin superfamily KIF3A/3B plus-end directed microtubule motor proteins (Jimbo et al. 2002).

2.5 Architectural Changes in the Intestine

Wnt signaling appears to regulate the complex balance of proliferation, migration, and differentiation essential to normal functioning of the rapidly proliferating intestinal epithelium. Any perturbation of this balance appears to disrupt normal intestinal homeostasis leading to tumor development. The colorectal mucosa contains large numbers of invaginations termed the crypts of Lieberkühn. Epithelial cells are constantly being renewed in these crypts in a coordinated series of events involving proliferation, differentiation, and cell migration toward the intestinal lumen. Pluripotent stem cells are believed to reside at the bottom positions of the crypt. From these stem cells, progenitors are generated that occupy the lower third of the crypt, the amplification compartment. Cells in this compartment divide approximately every 12 h until their migration brings them to the mid-crypt region. Here, they cease proliferating and differentiate into one of the functional cell types of the colon. At the surface epithelium, cells undergo apoptosis and/or extrusion into the lumen. Apc mutations result in an extended proliferative compartment and reduced cellular turnover of the normal intestinal epithelium of the Min mouse (Mahmoud et al. 1997, 1999) by providing an early mechanism for disease progression: an increased number of cells in the crypt–villus compartment would allow for opportunities for a second hit.

In the gut, TCF and β-catenin inversely control the expression of the EphB2/EphB3 tyrosine kinase receptors and their ligand, ephrin-B1 along the crypt–villus axis (Batlle et al. 2002; van de Wetering et al. 2002). The interactions between Eph receptors and ephrin ligands involve direct cell-to-cell interactions and frequently result in repulsion. Eph-ephrin signaling provides repulsive cues in a wide range of developmental phenomena, including axon pathfinding, the migration of neural crest cells to their target tissues, and boundary formation between adjacent cell populations in segmented structures such as rhombomeres (Wilkinson 2001). It has been strongly suggested that Eph receptors and ephrin ligands transduce repulsive signals and that this bi-directional signaling is required to maintain boundaries at the interface of adjacent cell compartments (Xu et al. 1999). Polyps in the small intestine of Min mice develop at the crypt–villus junction and form pockets that migrate inside the normal epithelium of the villus (Oshima et al. 1997). These cells proliferate inside the mucosa as a disorganized mass that will eventually give rise to a tumor. This abnormal migratory behavior is likely to be the outcome of the β-catenin/Tcf target gene program autonomously activated in the *Apc* mutant cells. Intriguingly, this initial outpocketing arises in the crypt–villus junction where *APC* mutant cells that overexpress EphB receptors encounter the maximum threshold of ephrin-mediated repulsion. These observations suggest that the initial founding polyp cells expressing high levels of EphB receptors likely initiate an abnormal migration inside the villus to avoid the cells expressing high levels of ephrin-B. Future analysis of the Min phenotype in an EphB deficient background will determine the role of the ephrin/EPH system in architectural changes associated with neoplasm.

In short. A large body of literature is beginning to sketch out the mechanisms by which active Wnt signaling causes cancer. Nuclear β-catenin transactivates TCF/LEF target genes, simultaneously promoting cellular growth and repressing differentiation programs. Once a growth advantage has been provided, architectural changes provide a harbor for tumor progression. By controlling the EFN/ephrin system, cells with an activated Wnt signal might form outpockets, allowing pools of undifferentiated cells to foster conditions beneficial for the acquisition of additional mutations (Batlle et al. 2002). Genetic instability, such as that incurred in the presence of APC mutations, destabilize the mitotic spindle and promote loss of tumor-suppressor genes (Fodde et al. 2001; Kaplan et al. 2001). Changes in cell adhesion and migration, brought on by APC or CTNNB1 mutation, contribute to the invasiveness of a particular tumor (Hajra and Fearon 2002).

Mutations in several components of the Wnt signaling cascade cause cancer in a large number of anatomical venues, yet –despite all the variables – they all manifest a single molecular defect: the presence of β-catenin in the cell nucleus. The inventory of tumors testifies to the robustness of clonal selection. Furthermore, identifying nuclear β-catenin clinically will aid molecular profiling of tumor types. For example, the presence of nuclear β-catenin is correlated with poor prognosis in patients with hepatoblastoma (Park et al. 2001). The most widely used method to detect nuclear β-catenin, immunohistochemistry, is inexpensive and rapid. Although not as widely available as immunohistochemistry, real-time PCR allows β-catenin levels to be quantified and followed during treatment.

3 Clear Cell Renal Carcinoma: Current Understanding of the Molecular Mechanism

Renal cell carcinoma (RCC) is the 7th most common form of malignant cancer, and its incidence seems to be rising. RCCs are morphologically and genetically heterogeneous; the most prevalent class of RCC tumors affecting adults is that of clear-cell carcinomas (ccRCCs), accounting for approximately 85% of all RCCs.

The ccRCCs are frequently characterized by the loss of chromosome arm 3p; these deletions encompass the locus of the von Hippel-Lindau gene (*VHL*), cloned in 1993 (Latif et al. 1993). Even in the absence of gross chromosomal rearrangements, the vast majority of ccRCCs carry inactivating mutations of *VHL*. This near-perfect correlation implicates a critical role of VHL in this disease as well as in the maintenance of normal kidney physiology (Foster et al. 1994; Gnarra et al. 1994; Kaelin 2002).

3.1 The von Hippel-Lindau (VHL) Disease

VHL disease is an autosomal dominant cancer syndrome, characterized by extensively vascularized tumors and cysts in different organs. The most frequent manifestations of the disease are hemangioblastomas of the central nervous system and retina and the development of clear cell RCC. VHL disease is characterized by a complex genotype–phenotype correlation and can be classified into two types, based on predisposition to neoplasms in the adrenal glands, called pheochromocytomas. Type 1 mutations predispose families to RCC and hemangioblastoma but not to pheochromocytoma, whereas type 2 mutations predispose mainly to pheochromocytoma. Type 1 disease correlates with *VHL* whole-gene deletions or truncating mutations, whereas type 2 patients generally harbor relatively subtle missense mutations (Chen et al. 1995). Type 2 VHL can be further subdivided into types 2A, 2B, and 2C. Type 2A families rarely display RCC but develop hemangioblastomas in addition to pheochromocytoma, while type 2B families are prone to RCC, hemangioblastoma, and pheochromocytoma. Type 2C mutations predispose families to pheochromocytoma, without evidence of other tumors (Friedrich 2001). Studies of the type 2C associated mutations suggest that in contrast with the other disease associated *VHL* mutations, these alleles are competent for HIF capture and ubiquitination (Hoffman et al. 2001), implying a distinct mode of tumor promotion for at least the pheochromocytoma. For example, it could be speculated that either the development of pheochromocytoma is incompatible with complete loss of VHL function, or that some gain of function is acquired by VHL tumors that promote this type of tumor. VHL might thus carry out distinct, perhaps tissue-specific, tumor-suppressor mechanisms.

In support, close examination of the *VHL* mutation spectrum in familial and sporadic tumors suggest that mutations specifically occurring in the p30 isoform of VHL do not necessarily affect HIF regulation and are sufficient for tumorigenesis. Literature and database searches for *VHL* mutations in familial and sporadic tumors (VHL mutation database, www.umd.be:2020/) unveiled 32 familial and sporadic mutations reported in the first 59 residues of p30. These mutations have been reported almost exclusively in pheochromocytomas and renal cell carcinomas, suggesting a previously unappreciated tumor-suppressor role of the acidic domain in these polarized cell types (Giles and Voest, unpubl.). Strikingly, 12 mutations are predicted to only affect the p30 isoform, i.e., translation of p19 should not be affected as a result of the mutation. It is thus worth noting that mutations in the first 159 nucleotides of the open reading frame of VHL should not be considered null alleles, as p19 will not be affected. Moreover, because these patient tumors will not harbor biallelic inactivating mutations of p19, the p19 isoform of VHL fails to conform to the classic definition of a tumor-suppressor protein (Sherr 2004). Moreover, the most common and early neoplasm accounting for the first manifestation of the disease in 83% of VHL patients (Melmon and Rosen 1964; Maher et al. 1990) – retinal or CNS angiomas – appear to be absent or occur with late-onset in patients with p30-specific mutations. However, the small number of patients with available clinical data in this group holds no statistical weight and is merely suggestive of a genotype–phenotype correlation. The under-representation of angiomas implies that upon somatic inactivation of the second *VHL* allele, the p19 allele produced from an allele with a 5′ mutation is sufficient to suppress angioma development. Our analysis supports the notion put forth by a number of groups that VHL-mediated tumor suppression is HIF independent, mediated by an unknown function of the N-terminal acidic domain (Fig. 3).

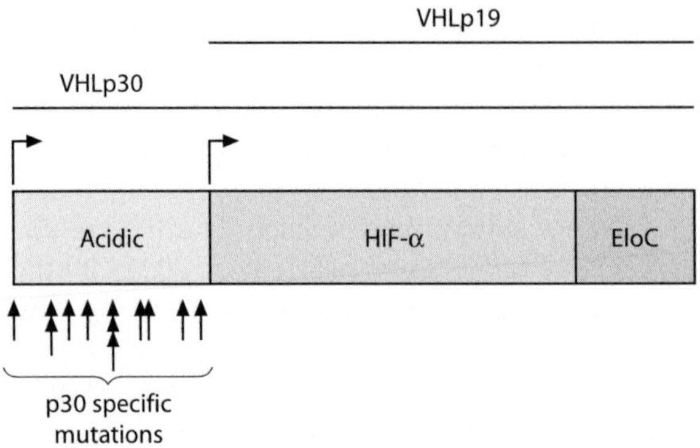

Fig. 3. N-terminal mutations in VHL suggest that only the p30 isoform functions as a tumor suppressor. *VHL* contains two functional translational initiation sites resulting in two VHL isoforms; a full-length protein (VHLp30) with an N-terminal acidic domain of unknown function, and an downstream reading frame (VHL p19) starting at methionine 54. Both isoforms contain sequences responsible for binding to HIF-α and elongin C (EloC). Although the vast majority of *VHL* mutations are downstream of the second initiation site, thereby altering both isoforms, 12 mutations (*vertical arrows*) have been described that should have no effect on the structure or function of the smaller pi 9 isoform.

3.2 HIF Signaling

Insights into the molecular basis of tumor formation by the inactivation of *VHL* have come mainly from the identification of proteins that associate with VHL (Duan et al. 1995; Kibel et al. 1995; Lonergan et al. 1998; Ohh et al. 1998; Kamura et al. 1999; Pause et al. 1999; Hergovich et al. 2003). Immunoprecipitation studies revealed that VHL binds to elongin B and elongin C (Duan et al. 1995; Kibel et al. 1995). Elongins B and C are transcriptional elongation factors that form a tripartite complex with elongin A, collectively known in the literature as elongin (SIII; Aso et al. 1995). Elongin (SIII) activates transcription elongation of RNA polymerase II, the major polymerase for messenger RNA synthesis. VHL has a small region that is homologous with elongin A and important for the interaction with elongin B and C. The interaction site of elongin C and VHL is frequently altered by *VHL*-germline mutations, indicating the functional importance of this complex for the tumor-suppressor function of VHL (Kibel et al. 1995; Kishida et al. 1995; Zbar et al. 1996). Recently, it has been demonstrated that VHL is stabilized by association with both elongin B and C. VHL also interacts with Cullin-2 (Cul-2), orthologous to the yeast protein Cdc53. Cdc53 participates in the SCF complex (Skp1-Cdc 53/Cul-1-F-box protein), which functions as an E3-like ubiquitin protein ligase complex targeting proteins for degradation by the ubiquitin proteolytic pathway (Kipreos et al. 1996). Accordingly, the VHL/elongins B, C/Cul-2 (VBC-Cul2) complex structurally resembles the SCF complex, and elongin B appears to be a ubiquitin-like protein. Further studies demonstrated that the endogenous VHL complex in rat liver also includes Rbx1 (Kamura et al. 1999). Rbx1 is an evolutionarily conserved protein containing a RING-H2 fingerlike motif that interacts with cullins (Zachariae et al. 1998). All these observations suggested that VHL might function in an E3-ubiquitin ligase complex. Iwai et al. showed that the VBC-Cul2 complex indeed exhibits ubiquitin ligase activity (Iwai et al. 1999). Rbx1 is crucial for the ligase activity. VHL acts as the substrate recognition unit (F-box-like) of the E3 ligase complex.

For normal cell growth and viability oxygen is important. Physiological tissue and cellular responses to hypoxia are the induction of angiogenesis, erythropoiesis, and glycolysis (Carmeliet et al. 1998; Semenza 2001). Many genes involved in these processes such as VEGF, erythropoietin, glycolytic enzymes, and inducible nitric-oxide synthase are upregulated during hypoxia. The promoter regions of these so called hypoxia-inducible genes contain hypoxia-responsive elements (HRE) reviewed by Semenza (2001). The hypoxia-inducible factor transcription factor HIF-1 binds HREs of hypoxia-inducible genes and promotes transcription. HIF-1 is a heterodimeric member of the basic helix-loop-helix PAS family composed of α and β subunits (Semenza 2001). The activity of HIF-1 is mainly determined by the stability of the α subunit. Under normoxic conditions the α subunits are constitutively degraded by the ubiquitination/proteasome pathway. Under hypoxic conditions HIF-1α is stabilized and translocated to the nucleus. In the nucleus it forms a complex with the HIF-1β subunit also called ARNT (arylhydrocarbon nuclear translocator) (Semenza 2001). HIF-1β is abundantly expressed and its expression is not influenced by oxygen tension. In *VHL*-defective cells it has been observed that HIFα-subunits are constitutively stabilized. Reintroduction of VHL resulted in downregulation of HIF and restored the oxygen-dependent instability of these cells (Maxwell et al. 1999). It has been shown that VHL binds to HIF-1α, HIF-2α, and HIF-3α subunits in normoxic cells.

How does the understanding of VHL function in HIF signaling elucidate the etiology of RCC and other cancers in VHL patients? The obvious connection is through the extensive vascularization characteristic of VHL-related tumors and RCC. Vascular endothelial growth factor (VEGF) and it receptors have been found to be highly upregulated in VHL-related and sporadic hemangioblastomas and RCCs (Wizigmann-Voos et al. 1995; Los et al. 1997). VEGF is a key angiogenic factor, whose expression is induced by hypoxia, steroid hormones, protein kinase C agonists, reactive oxygen intermediates and iron deficiency (Shweiki et al. 1992; Beerepoot et al. 1996). In vitro studies have shown that human renal cell carcinoma cells that lack the endogenous wild-type (wt) *VHL* gene express high levels of VEGF mRNA and protein under normoxic and hypoxic conditions (Iliopoulos et al. 1996; Siemeister et al. 1996). Reintroduction of wt VHL, but not mutant VHL, resulted in a significant downregulation of VEGF mRNA under normoxic conditions. There is convincing evidence that regulation of VEGF by VHL occurs at the level of transcription. The production of hypoxia-inducible mRNAs such as VEGF is regulated by the HIF (Maxwell et al. 1999).

Although these findings explain the high vascularization of *VHL*-related tumors, upregulation of HIF-α alone does not explain the development of tumors in VHL disease. For example, a subset of VHL cancer patients (type 2C, presenting exclusively with pheochromocytoma) have been shown to carry germline mutations of *VHL*, which do not compromise the protein's ability to regulate HIF-α levels, yet these patients develop tumors (Hoffman et al. 2001). Furthermore, naturally occurring stabilizing mutations of any of the three HIF-α gene parologs (*HIF1A, HIF2A, HIF3A*) have never been reported in tumor or other tissues. Most convincingly, however, is the existence of a distinct VHL congenital syndrome, namely, the autosomal recessive Chuvash polycythemia, caused by constitutively »mild« homozygous *VHL* mutations (Ang et al. 2002). These patients manifest thrombosis and vascular abnormalities, due to elevated normoxic levels of HIF-α commensurate with their elevated serum erythropoietin and VEGF concentrations. However, despite upregulated HIF-α levels, Chuvash polycythemia patients are not predisposed to cancer (Gordeuk et al. 2004). Thus, upregulation of HIF-α alone does not appear to be oncogenic in humans, suggesting an alternate – and as yet undiscovered – tumor-suppressor function for VHL.

3.3 Step-wise RCC Progression

As described in the previous section, a normal cell needs to comply with two essential requirements to develop into a cancer: it must acquire selective advantage to allow for the initial clonal expansion, and genetic instability to allow for multiple hits at other genes aiding tumor progression and malignant transformation. It is not known how tumors resulting from inactivated VHL fulfill the first requirement, although it is believed that transactivation of HIF target genes plays a role.

Tumor development in VHL patients is linked to somatic inactivation of the remaining wild-type allele. Biallelic inactivation of the *VHL* gene by somatic mutations has been found in 57% and up to 98% of sporadic RCCs, respectively (Gnarra et al. 1994), whereas *VHL* is transcriptionally silenced by methylation of the promoter in about 20% of RCCs (Herman et al. 1994). It can thus be concluded that VHL loss is a virtual prerequisite for ccRCC development. VHL is conserved in mouse, *Drosophila*, zebrafish, and *C. elegans*. The VHL protein consists of 213 amino acid residues and is 30 kDa in size. A second protein is generated by translation initiation of an internal methionine at position 54 of the open reading frame. This protein (p19) has a molecular weight of approximately 19 kDa (Iliopoulos et al. 1998; Blankenship et al. 1999). That VHL acts as a tumor suppressor became clear from nude mice xenograft assays. In this model subcutaneous injection of RCC cells lacking a functional *VHL* gene results in tumor growth. Reintroduction of wild-type VHL in these cells significantly suppresses tumor growth (Iliopoulos et al. 1995).

Fulfillment of the second requirement for RCC occurs through a series of genetic alterations involving the accumulation of mutations in a number of genes and progression. This tumor progression model was deduced from comparison of genetic alterations seen in normal kidney epithelium, tumors of progressively larger size, and malignancies (Phillips et al. 2001). Although genetic instability is global, cells with specific losses are selected for clonal growth (Lengauer et al. 1998). Generally, RCCs manifesting chromosome instability (CIN) acquire sequential chromosome losses at the loci for *VHL* (chromosome 3p; Latif et al. 1993; Prowse et al. 1997), followed by loss of chromosome arm 6q, and then losses of chromosomes 9, 18q, 8p, and 14q, or gain of chromosome 19 (Thrash-Bingham et al. 1995; Gijtenbeek et al. 2002). These chromosome losses are thought to perpetrate further instability at the molecular and chromosomal level (Lengauer et al. 1998).

What is the relationship between biallelic loss of VHL and the accumulation of secondary mutations? Like CRC tumors with *APC* mutations, RCCs with *VHL* mutations are generally aneuploid but chromosomally stable, an indication of clonal selection after chromosomal instability (www.atcc.org). Several reports have included comments on the frequency of CIN in RCC samples and cell lines (reviewed in Meloni-Ehrig 2002), where 87.5% cases showed an abnormal DNA content (Pepe et al. 2000). RCCs have been hypothesized to undergo DNA aneuploidization during tumor expansion based on the fact that the tumors of multiple aneuploid clones were mainly found in large-sized tumors (Li et al. 2002). Supporting the notion that *VHL* mutation may cause CIN, we have determined that while the *VHL*-defective 786–0 cell line is hypertriploid, re-introduction of wild-type *VHL* returns the DNA content to a near-diploid value (unpubl. observ.).

3.4
VHL Regulates Microtubule Stability

Microtubules are essential for cellular processes including migration, organelle transport, and chromosome segregation during cell division by formation of the mitotic spindle. Hergovich and colleagues (2003) found that cytoplasmic VHL is largely associated with the microtubule (MT) apparatus. In in vivo assays, VHL can protect MTs from nocodazole-induced depolymerization in several cell lines (Hergovich et al. 2003). Intriguingly, they found that this function requires amino acids 95–123 and is independent of E3 complex formation, which suggests that MT-binding and stabilization may be an independent function of VHL. In an attempt to place these data within the context of VHL disease, Hergovich et al. tested VHL mutations associated with different patterns of tumor development for MT stabilization properties. They found that although many missense mutations retained the ability to stabilize MTs, the disease-associated mutations Y98H and Y112H disrupted this function. Both of these mutations are associated with type 2A disease, predisposing the families to pheochromocytoma and hemangioblastoma (Hergovich et al. 2003).

Pivotal evidence for another VHL function contributing to neoplastic growth was provided in one seminal paper in which re-expression of the wild-type *VHL* in VHL-deficient RCC lines had no effect on monolayer growth, but reduced growth of such cells as tumors in nude mice (Iliopoulos et al. 1995). These data indicate that the tumor-suppressor action of VHL is in some way dependent on conditions manifested during solid tumor growth, but not during standard monolayer cell culture. It was thus hypothesized that VHL may have a function in extracellular matrix and cell–cell interactions (Ohh et al. 1998; Koochekpour et al. 1999; Davidowitz et al. 2001). A puzzling phenomenon reported by Lee et al. (1996) gave weight to this hypothesis; they reported that transport of the VHL protein into and/or out of the nucleus is tightly regulated and cell-density-dependent. In densely grown cells, it is predominantly in the cytoplasm, whereas in sparse cultures, most of the protein can be detected in the nucleus. They identified a putative nuclear localization signal in the first 60 and first 28 amino acids of the human and rat VHL protein, respectively. Sequences in the C-terminal region of VHL protein may also be required for localization to the cytosol. The findings indicated a novel cell-density-dependent pathway responsible for the regulation of VHL cellular localization.

Fibronectin matrix assembly has been shown to inhibit tumorigenesis and metastasis in vivo (Shaub 1999). Ohh et al. demonstrated that fibronectin coimmunoprecipitates with normal VHL protein but not tumor-derived VHL mutants (Ohh et al. 1998). Immunofluorescence and biochemical fractionation experiments showed that fibronectin colocalized with a fraction of VHL associated with the endoplasmic reticulum, and cold competition experiments suggested that complexes between fibronectin and VHL protein exist in intact cells. Assembly of an extracellular fibronectin matrix by *VHL* –/– RCC cells, as determined by immunofluorescence and ELISA assays, was grossly defective compared with *VHL* +/+ renal carcinoma cells. Reintroduction of wild-type, but not mutant, VHL protein into *VHL* –/– renal carcinoma cells partially corrected this defect. Extracellular fibronectin matrix assembly by *Vhl* –/– mouse embryos and mouse embryonic fibroblasts, unlike their *Vhl* +/+ counterparts, was grossly impaired. They concluded that altered matrix organization contributes to the aggressive behavior of *VHL*-defective tumors. Hoffman et al. reported that the products of four different type 2C *VHL* alleles, such as L188 V or K159E, retain the ability to downregulate HIF but are defective for promotion of fibronectin matrix assembly (Hoffman et al. 2001). We have recently reported that VHL positively regulates the transcription of *FN1* in an HIF-independent manner (Bluyssen et al. 2004). Recently, modification of three lysines in VHL by the ubiquitin-like molecule NEDD8 has been shown to regulate fibronectin binding, where the inhibition of VHL neddylation prohibited fibronectin binding (Stickle et al. 2004). A possible consequence of faulty fibronectin matrix assembly could be profound; VHL mutations expedite cell migration across transwell filters, and show defects in differentiation and morphogenesis (Lieubeau-Teillet et al. 1998; Davidowitz et al. 2001). Furthermore, we have demonstrated that phosphorylation of three serines in the N-terminus of VHL by casein kinase 2 (CK2) regulates FN deposition to the ECM and mediates tumor suppression in mice (unpubl.).

3.5 Renal Cysts: Precursors to RCC?

VHL patients frequently develop multiple cysts in the kidney, pancreas, and liver. The development of renal cysts in VHL patients is almost invariably correlated with the »second hit« inactivating the wild-type *VHL* gene (Lubensky et al. 1996). However, the development of renal cysts is not sufficient to cause cancer; polycystic kidney disease does not increase the risk of developing renal tumors (Truong et al. 2003). Similarly, patients with type 2A VHL disease can develop cystic disease in their kidneys, pancreas, and/or liver, in addition to other VHL-associated neoplasms, but they are not predisposed to renal tumors. *VHL* mutations associated with this patient subgroup demonstrate aberrations in MT stability (Hergovich et al. 2003). VHL thus appears to embody two essential and independent tumor-suppressor functions in the kidney: preventing the advent of cystic disease through localizing MT stability, and regulating HIF levels. Interestingly, the type 2A mutants have residual capacity for HIF – particularly HIF2α regulation (Clifford et al. 2001). These mutant VHL alleles might prevent the advent of tumor formation through maintaining low HIF levels, but are deficient in regulating proper microtubule cytoskeleton. In contrast, type 2B patient alleles are unable to regulate HIF levels and are associated with renal tumor formation. This analysis takes an important feature of the VHL disease – cysts – into account.

3.6 VHL Mouse Models

Gene targeted *Vhl –/–* embryos die between E10.5 and E12.5 with lack of placental vasculogenesis (Gnarra et al. 1997). In this particular study, the *Vhl +/–* littermate mice were normal and did not show increased incidence of tumors. However, *Vhl +/–* mice of a different background and bearing a different targeted Vhl mutation showed increased susceptibility to form vascular lesions (Haase et al. 2001). After 12 months, the majority of these animals developed vascular lesions and cavernous hemangiomas in the liver. Mice with a liver-specific mutation of *Vhl* appear sick and die at 6–12 weeks. Vascular histopathologies, such as hepatocellular steatosis and foci of increased vascularization within the hepatic parenchyma, were observed, but not tumors. HIF-2α and VEGF expression was increased in *Vhl –/–* hepatocytes.

Mice lacking *Vhl* in the cartilaginous elements as result of the Cre-loxP strategy displayed a unique cartilage phenotype and indicate a crucial role in endochondral bone development (Pfander 2004). This phenotype was characterized by hypocellularity, a dramatic decrease of cell proliferation, an augmented amount of matrix, and an increased cell size in the resting and proliferating zones. After birth, *Vhl* null mice showed a delayed appearance of the secondary ossification center and severe dwarfism (Pfander 2004)). It thus seems likely that murine Vhl, while exerting similar functions to its human counterpart in angiogenesis and cellular proliferation, does not function as a classical tumor-suppressor protein. New models for the analysis of VHL in RCC development are currently being sought by many groups.

4
Discussion

4.1
APC and VHL: Overlap in Function

Tumor-suppressor genes form an arbitrary group of proteins whose biallelic mutation ultimately results in neoplastic transformation. Both VHL and APC have been dubbed »gatekeepers« thought to negatively control signals promoting unbridled growth. APC, however, has recently confounded this categorization by studies showing that mutated APC actively plays a role in producing secondary mutations, by initiating CIN (Fodde et al. 2001; Kaplan et al. 2001). CIN is also associated with VHL loss in RCCs. Like APC, VHL also binds and stabilizes MTs, binds with the KIF3A/KAP3 motor complex, controls the regulation and nuclear localization of a transcriptional activator, and displays a complex genotype–phenotype correlation, suggestive of at least two functions involved in tumorigenesis. Table 1 summarizes some of the remarkable similarities in VHL vs. APC genetics and gene function.

The most remarkable difference between these two tumor suppressors is size. APC is ten times the size of VHL. The compactness of VHL has made it tempting to speculate that its tumor-suppressor properties are encapsulated in a single function. However, strong evidence argues for the presence of an HIF-independent, cytoskeletal-dependent mechanism for VHL-mediated tumor suppression.

Table 1. Comparison of APC and VHL tumor suppressor function

APC	VHL
Mutated in a rare genetic disease (FAP) with predisposition to colon cancer	Mutated in a rare genetic disease (VHL) with predisposition to renal cancer
Mutated in >80% sporadic colorectal cancers	Mutated in >80% sporadic clear cell renal cancers
Loss of APC is the first detectable genetic defect in colorectal neoplasia	Loss of VHL is the first detectable genetic defect in renal neoplasia
Tumorigenesis is a slow process requiring secondary mutations	Tumorigenesis is a slow process requiring secondary mutations
Genotype–phenotype correlations	Genotype–phenotype correlations
Mutation affects polarized cell types	Mutation affects polarized cell types
Localized predominantly in the cytoplasm, but also in the nucleus and Golgi apparatus	Localized predominantly in the cytoplasm, but also in the nucleus and Golgi apparatus
Regulates stability of β-catenin through targeting for ubiquitination and subsequent degradation.	Regulates stability of HIFα through ubiquitination and subsequent degradation.
Functions in a scaffold-like manner to assemble components essential for the ubiquitination of β-catenin	Functions in a scaffold-like manner to assemble components essential for the ubiquitination of HIFα
β-catenin nuclear export is APC-dependent	HIFα nuclear export is VHL-dependent
APC stabilizes microtubules	VHL stabilizes microtubules
APC binds microtubule motor proteins KIF-3A and KAP3	VHL binds microtubule motor proteins KIF-3A and KAP3
Mutations in APC in colorectal cancer cause chromosome instability	Mutations in VHL in renal cell cancer are associated with chromosome instability
Restoration of wild-type APC in mutant cells induces differentiation	Restoration of wild-type VHL in mutant cells induces differentiation
Apc –/– ES cells suffer a profound differentiation block and fail to produce teratocarcinomas	Vhl –/– ES cells suffer a profound differentiation block and produce meager teratocarcinomas
APC inhibits cell motility, APC mutations activate motility	VHL inhibits cell motility, VHL mutations activate motility?

4.2 Balancing Acts: Signaling vs. Cytoskeleton Regulation

In both APC and VHL, the domains regulating binding to the KIF3A/KAP3 complex lie at the N-terminus, while the domains regulating β-catenin and HIF-α, respectively, are situated toward the C-terminus. Given the non-random nature of neoplastic mutations in these genes, this distribution might reflect an intrinsic balance between C- and N-terminal regions of APC or VHL. Most CRC and RCC tumors select for the retention of one truncated or missense allele of *APC* or *VHL*, respectively, while promoting total loss of the other allele. When C-terminal regions are missing as a result of tumor-associated truncations, this balance might be tipped toward interaction with KIF3A/KAP3, resulting in a dominant negative effect of the remaining N-terminal domain.

Inappropriate activation of Wnt or hypoxia signaling can lead to inappropriate proliferation and lack of differentiation. APC and VHL affect cytoskeletal regulation through interactions with microtubules. Loss of APC or VHL may also result in an increased incidence in mitotic errors due to microtubule defects in mitotic spindles. This could subsequently contribute to problems with genetic stability and lead to aneuploidy, an important component in tumor progression.

It is interesting to speculate that these two critical pathways interact at some level, or perhaps even compete for kinesin II components KIF3A/KAP3. A few research groups are looking at the cooperative effects of VHL and APC on microtubule stability. Future understanding of the function of these pathways in normal physiology will provide leads into how related processes go awry during neoplasm.

References

American Cancer Society (2002) www.cancer.org

Albuquerque C, Breukel C, van der Luijt R, Fidalgo P, Lage P, Slors FJ, Leitao CN, Fodde R, Smits R (2002) The 'just-right' signaling model: APC somatic mutations are selected based on a specific level of activation of the beta-catenin signaling cascade. Hum Mol Genet 11:1549–1560

Ang, SO, Chen H, Hirota K, Gordeuk VR, Jelinek J, Guan Y, Liu E, Sergueeva AI, Miasnikova GY, Mole D, Maxwell PH, Stockton DW, Semenza GL, Prchal JT (2002) Disruption of oxygen homeostasis underlies congenital Chuvash polycythemia. Nat Genet 32:614-621

Aso T, Lane WS, Conaway JW, Conaway RC (1995) Elongin (SIII): a multisubunit regulator of elongation by RNA polymerase II. Science 269:1439–1443

Batlle E, Henderson JT, Beghtel H, van den Born MM. W, Sancho E, Huls G, Meeldijk J, Robertson J, van de Wetering M, Pawson T, Clevers H (2002) Beta-catenin and TCF mediate cell positioning in the intestinal epithelium by controlling the expression of the EPHB/EphrinB system. Cell 111:251–263

Beerepoot LV, Shima DT, Kuroki M, Yeo KT, Voest EE (1996) Up-regulation of vascular endothelial growth factor production by iron chelators. Cancer Res 56:3747–3751

Bhanot P, Brink M, Samos CH, Hsieh JC, Wang Y, Macke JP, Andrew D, Nathans J, Nusse R (1996) A new member of the frizzled family from Drosophila functions as a Wingless receptor. Nature 382:225–230

Blankenship C, Naglich JG, Whaley JM, Seizinger B, Kley N (1999) Alternate choice of initiation codon produces a biologically active product of the von Hippel Lindau gene with tumor-suppressor activity. Oncogene 18:1529–1535

Bluyssen HA, Lolkema MP, van Beest M, Boone M, Snijckers CM, Los M, Gebbink MF, Braam B, Holstege FC, Giles RH, Voest EE (2004) Fibronectin is a hypoxia-independent target of the tumor suppressor VHL. FEBS Lett 556:137–142

Carmeliet P, Dor Y, Herbert JM, Fukumura D, Brusselmans K, Dewerchin M, Neeman M, Bono F, Abramovitch R, Maxwell P, Koch CJ, Ratcliffe P, Moons L, Jain RK, Collen D, Keshert E, Keshet E (1998) Role of HIF-1alpha in hypoxia-mediated apoptosis, cell proliferation and tumour angiogenesis. Nature 394:485–490

Caspari R, Olschwang S, Friedl W, Mandl M, Boisson C, Boker T, Augustin A, Kadmon M, Moslein G, Thomas G et al (1995) Familial adenomatous polyposis: desmoid tumours and lack of ophthalmic lesions (CHRPE) associated with APC mutations beyond codon 1444. Hum Mol Genet 4:337–340

Chen F, Kishida T, Yao M, Hustad T, Glavac D, Dean M, Gnarra JR, Orcutt ML, Duh FM, Glenn G et al (1995) Germline mutations in the von Hippel-Lindau disease tumor-suppressor gene: correlations with phenotype. Hum Mutat 5:66–75

Clifford SC, Cockman ME, Smallwood AC, Mole DR, Woodward ER, Maxwell PH, Ratcliffe PJ, Maher ER (2001) Contrasting effects on HIF-1alpha regulation by disease-causing pVHL mutations correlate with patterns of tumourigenesis in von Hippel-Lindau disease. Hum Mol Genet 10:1029–1038

Dale TC (1998) Signal transduction by the Wnt family of ligands. Biochem J 329:209–223

Davidowitz EJ, Schoenfeld AR, Burk RD (2001) VHL induces renal cell differentiation and growth arrest through integration of cell-cell and cell-extracellular matrix signaling. Mol Cell Biol 21:865–874

Deka J, Kuhlmann J, Muller O (1998) A domain within the tumor-suppressor protein APC shows very similar biochemical properties as the microtubule-associated protein tau. Eur J Biochem 253:591–597

Dihlmann S, Gebert J, Siermann A, Herfarth C, von Knebel Doeberitz M (1999) Dominant negative effect of the APC1309 mutation: a possible explanation for genotype-phenotype correlations in familial adenomatous polyposis. Cancer Res 59:1857–1860

Duan DR, Humphrey JS, Chen DY, Weng Y, Sukegawa J, Lee S, Gnarra JR, Linehan WM, Klausner RD (1995) Characterization of the VHL tumor-suppressor gene product: localization, complex formation, and the effect of natural inactivating mutations. Proc Natl Acad Sci USA 92:6459–6463

Duensing S, Munger K (2001) Centrosome abnormalities, genomic instability and carcinogenic progression. Biochim Biophys Acta 2:M81–M88

Fearnhead NS, Britton MP, Bodmer WF (2001) The ABC of APC. Hum Mol Genet 10:721–733

Fodde R (2002) The APC gene in colorectal cancer. Eur J Cancer 38:867–871

Fodde R, Edelmann W, Yang K, van Leeuwen C, Carlson C, Renault B, Breukel C, Alt E, Lipkin M, Khan PM et al. (1994) A targeted chain-termination mutation in the mouse Apc gene results in multiple intestinal tumors. Proc Natl Acad Sci USA 91:8969–8973

Fodde R, Kuipers J, Rosenberg C, Smits R, Kielman M, Gaspar C, van Es JH, Breukel C, Wiegant J, Giles RH, Clevers H (2001) Mutations in the APC tumour suppressor gene cause chromosomal instability. Nat Cell Biol 3:433–438

Foster K, Prowse A, van den Berg A, Fleming S, Hulsbeek MM, Crossey PA, Richards FM, Cairns P, Affara NA, Ferguson-Smith MA et al (1994) Somatic mutations of the von Hippel-Lindau disease tumour suppressor gene in non-familial clear cell renal carcinoma. Hum Mol Genet 3:2169–2173

Friedrich CA (2001) Genotype-phenotype correlation in von Hippel-Lindau syndrome. Hum Mol Genet 10:763–767

Ghadimi BM, Sackett DL, Difilippantonio MJ, Schrock E, Neumann T, Jauho A, Auer G, Ried T (2000) Centrosome amplification and instability occurs exclusively in aneuploid, but not in diploid colorectal cancer cell lines, and correlates with numerical chromosomal aberrations. Genes Chromosomes Cancer 27:183–190

Gijtenbeek JM, Jacobs B, Sprenger SH, Eleveld MJ, van Kessel AG, Kros JM, Sciot R, van Calenbergh F, Wesseling P, Jeuken JW (2002) Analysis of von Hippel-Lindau mutations with comparative genomic hybridization in sporadic and hereditary hemangioblastomas: possible genetic heterogeneity. J Neurosurg 97:977–982

Gnarra JR, Tory K, Weng Y, Schmidt L, Wei MH, Li H, Latif F, Liu S, Chen F, Duh FM et al (1994) Mutations of the VHL tumour suppressor gene in renal carcinoma. Nat Genet 7:85–90

Gnarra JR, Ward JM, Porter FD, Wagner JR, Devor DE, Grinberg A, Emmert-Buck MR, Westphal H, Klausner RD, Linehan WM (1997) Defective placental vasculogenesis causes embryonic lethality in VHL-deficient mice. Proc Natl Acad Sci USA 94:9102–9107

Gordeuk VR, Sergueeva AI, Miasnikova GY, Okhotin D, Voloshin Y, Choyke PL, Butman JA, Jedlickova K, Prchal JT, Polyakova LA (2004) Congenital disorder of oxygen sensing: association of the homozygous Chuvash polycythemia VHL mutation with thrombosis and vascular abnormalities but not tumors. Blood 103:3924–3932

Groden J, Thliveris A, Samowitz W, Carlson M, Gelbert L, Albertsen H, Joslyn G, Stevens J, Spirio L, Robertson M et al. (1991) Identification and characterization of the familial adenomatous polyposis coli gene. Cell 66:589–600

Haase VH, Glickman JN, Socolovsky M, Jaenisch R (2001) Vascular tumors in livers with targeted inactivation of the von Hippel-Lindau tumor suppressor. Proc Natl Acad Sci USA 98:1583–1588

Hajra KM, Fearon ER (2002) Cadherin and catenin alterations in human cancer. Genes Chromosomes Cancer 34:255–268

Halberg RB, Katzung DS, Hoff PD, Moser AR, Cole CE, Lubet RA, Donehower LA, Jacoby RF, Dove WF (2000) Tumorigenesis in the multiple intestinal neoplasia mouse: redundancy of negative regulators and specificity of modifiers. Proc Natl Acad Sci USA 97:3461–3466

Harada N, Tamai Y, Ishikawa T, Sauer B, Takaku K, Oshima M, Taketo MM (1999) Intestinal polyposis in mice with a dominant stable mutation of the beta-catenin gene. EMBO J 18:5931–5942

Hart MJ, de los Santos R, Albert IN, Rubinfeld B, Polakis P (1998) Downregulation of beta-catenin by human axin and its association with the APC tumor suppressor, beta-catenin and GSK3 beta. Curr Biol 8:573–581

He TC, Sparks AB, Rago C, Hermeking H, Zawel L, da Costa LT, Morin PJ, Vogelstein B, Kinzler KW (1998) Identification of c-MYC as a target of the APC pathway. Science 281:1509–1512

He X, Saint-Jeannet JP, Wang Y, Nathans J, Dawid I, Varmus H (1997) A member of the Frizzled protein family mediating axis induction by Wnt- 5A. Science 275:1652–1654

Henderson BR (2000) Nuclear-cytoplasmic shuttling of APC regulates beta-catenin subcellular localization and turnover. Nat Cell Biol 2:653–660

Hergovich A, Lisztwan J, Barry R, Ballschmieter P, Krek W (2003) Regulation of microtubule stability by the von Hippel-Lindau tumour suppressor protein pVHL. Nat Cell Biol 5:64–70

Herman JG, Latif F, Weng Y, Lerman MI, Zbar B, Liu S, Samid D, Duan DS, Gnarra JR, Linehan WM et al (1994) Silencing of the VHL tumor-suppressor gene by DNA methylation in renal carcinoma. Proc Natl Acad Sci USA 91:9700–9704

Hoffman MA, Ohh M, Yang H, Klco JM, Ivan M, Kaelin WG Jr (2001) von Hippel-Lindau protein mutants linked to type 2C VHL disease preserve the ability to downregulate HIF. Hum Mol Genet 10:1019–1027

Iliopoulos O, Kibel A, Gray S, Kaelin WG Jr (1995) Tumour suppression by the human von Hippel-Lindau gene product. Nat Med 1:822–826

Iliopoulos O, Levy AP, Jiang C, Kaelin WG Jr, Goldberg MA (1996) Negative regulation of hypoxia-inducible genes by the von Hippel-Lindau protein. Proc Natl Acad Sci USA 93:10595–10599

Iliopoulos O, Ohh M, Kaelin WG Jr (1998) pVHL19 is a biologically active product of the von Hippel-Lindau gene arising from internal translation initiation. Proc Natl Acad Sci USA 95:11661–11666

Iwai K, Yamanaka K, Kamura T, Minato N, Conaway RC, Conaway JW, Klausner RD, Pause A (1999) Identification of the von Hippel-Lindau tumor-suppressor protein as part of an active E3 ubiquitin ligase complex. Proc Natl Acad Sci USA 96:12436–12441

Jarrett CR, Blancato J, Cao T, Bressette DS, Cepeda M, Young PE, King CR, Byers SW (2001) Human APC2 localization and allelic imbalance. Cancer Res 61:7978–7984

Jimbo T, Kawasaki Y, Koyama R, Sato R, Takada S, Haraguchi K, Akiyama T (2002) Identification of a link between the tumour suppressor APC and the kinesin superfamily. Nat Cell Biol 4:323–327

Joslyn G, Richardson DS, White R, Alber T (1993) Dimer formation by an N-terminal coiled coil in the APC protein. Proc Natl Acad Sci USA 90:11109–11113

Kaelin WG Jr (2002) Molecular basis of the VHL hereditary cancer syndrome. Nat Rev Cancer 2:673–682

Kamura T, Koepp DM, Conrad MN, Skowyra D, Moreland RJ, Iliopoulos O, Lane WS, Kaelin WG Jr, Elledge SJ, Conaway RC, Harper JW, Conaway JW (1999) Rbx1, a component of the VHL tumor suppressor complex and SCF ubiquitin ligase. Science 284:657–661

Kaplan KB, Burds AA, Swedlow JR, Bekir SS, Sorger PK, Nathke IS (2001) A role for the Adenomatous Polyposis Coli protein in chromosome segregation. Nat Cell Biol 3:429–432

Kawasaki Y, Senda T, Ishidate T, Koyama R, Morishita T, Iwayama Y, Higuchi O, Akiyama T (2000) Asef, a link between the tumor suppressor APC and G-protein signaling. Science 289:1194–1197

Kibel A, Iliopoulos O, DeCaprio JA, Kaelin WG Jr (1995) Binding of the von Hippel-Lindau tumor suppressor protein to Elongin B and C. Science 269:1444–1446

Kielman MF, Rindapaa M, Gaspar C, van Poppel N, Breukel C, van Leeuwen S, Taketo M, Roberts S, Smits R, Fodde R (2002) Apc modulates embryonic stem-cell differentiation by controlling the dosage of beta-catenin signaling. Nat Genet 32:594–605

Kinzler KW, Vogelstein B (1996) Lessons from hereditary colorectal cancer. Cell 87:159–170

Kinzler KW, Vogelstein B (1998) Landscaping the cancer terrain. Science 280:1036–1037

Kipreos ET, Lander LE, Wing JP, He WW, Hedgecock EM (1996) cul-1 is required for cell cycle exit in C. elegans and identifies a novel gene family. Cell 85:829–839

Kishida T, Stackhouse TM, Chen F, Lerman MI, Zbar B (1995) Cellular proteins that bind the von Hippel-Lindau disease gene product: mapping of binding domains and the effect of missense mutations. Cancer Res 55:4544–4548

Kongkanuntn R, Bubb VJ, Sansom OJ, Wyllie AH, Harrison DJ, Clarke AR (1999) Dysregulated expression of beta-catenin marks early neoplastic change in Apc mutant mice, but not all lesions arising in Msh2 deficient mice. Oncogene 18:7219–7225

Koochekpour S, Jeffers M, Wang PH, Gong C, Taylor GA, Roessler LM, Stearman R, Vasselli JR, Stetler-Stevenson WG, Kaelin WG Jr, Linehan WM, Klausner RD, Gnarra JR, Vande Woude GF (1999) The von Hippel-Lindau tumor suppressor gene inhibits hepatocyte growth factor/scatter factor-induced invasion and branching morphogenesis in renal carcinoma cells. Mol Cell Biol 19:5902–5912

Kuo KK, Sato N, Mizumoto K, Maehara N, Yonemasu H, Ker CG, Sheen PC, Tanaka M (2000) Centrosome abnormalities in human carcinomas of the gallbladder and intrahepatic and extrahepatic bile ducts. Hepatology 31:59–64

Lamlum H, Ilyas M, Rowan A, Clark S, Johnson V, Bell J, Frayling I, Efstathiou J, Pack K, Payne S, Roylance R, Gorman P, Sheer D, Neale K, Phillips R, Talbot I, Bodmer W, Tomlinson I (1999) The type of somatic mutation at APC in familial adenomatous polyposis is determined by the site of the germline mutation: a new facet to Knudson's 'two-hit' hypothesis. Nat Med 5:1071–1075

Latif F, Tory K, Gnarra J, Yao M, Duh FM, Orcutt ML, Stackhouse T, Kuzmin I, Modi W, Geil L et al (1993) Identification of the von Hippel-Lindau disease tumor suppressor gene. Science 260:1317–1320

Laurent-Puig P, Beroud C, Soussi T (1998) APC gene: database of germline and somatic mutations in human tumors and cell lines. Nucleic Acids Res 26:269–270

Lee S, Chen DY, Humphrey JS, Gnarra JR, Linehan WM, Klausner RD (1996) Nuclear/cytoplasmic localization of the von Hippel-Lindau tumor suppressor gene product is determined by cell density. Proc Natl Acad Sci USA 93:1770–1775

Lengauer C, Kinzler KW, Vogelstein B (1997) Genetic instability in colorectal cancers. Nature 386:623–627

Lengauer C, Kinzler KW, Vogelstein B (1998) Genetic instabilities in human cancers. Nature 396:643–649

Levy DB, Smith KJ, Beazer-Barclay Y, Hamilton SR, Vogelstein B, Kinzler KW (1994) Inactivation of both APC alleles in human and mouse tumors. Cancer Res 54:5953–5958

Li G, Cottier M, Sabido O, Gentil-Perret A, Lambert C, Passebosc-Faure K, Genin C, Tostain J (2002) The in vivo DNA aneuploidization during expansion of conventional renal cell carcinoma. In Vivo 16:341–344

Lieubeau-Teillet B, Rak J, Jothy S, Iliopoulos O, Kaelin W, Kerbel RS (1998) von Hippel-Lindau gene-mediated growth suppression and induction of differentiation in renal cell carcinoma cells grown as multicellular tumor spheroids. Cancer Res 58:4957–4962

Lindblom A (2001) Different mechanisms in the tumorigenesis of proximal and distal colon cancers. Curr Opin Oncol 13:63–69

Lonergan KM, Iliopoulos O, Ohh M, Kamura T, Conaway RC, Conaway JW, Kaelin WG Jr (1998) Regulation of hypoxia-inducible mRNAs by the von Hippel-Lindau tumor suppressor protein requires binding to complexes containing elongins B/C and Cul2. Mol Cell Biol 18:732–741

Los M, Aarsman CJ, Terpstra L, Wittebol-Post D, Lips CJ, Blijham GH, Voest EE (1997) Elevated ocular levels of vascular endothelial growth factor in patients with von Hippel-Lindau disease. Ann Oncol 8:1015–1022

Lubensky IA, Gnarra JR, Bertheau P, Walther MM, Linehan WM, Zhuang Z (1996) Allelic deletions of the VHL gene detected in multiple microscopic clear cell renal lesions in von Hippel-Lindau disease patients. Am J Pathol 149:2089–2094

Luongo C, Moser AR, Gledhill S, Dove WF (1994) Loss of Apc+ in intestinal adenomas from Min mice. Cancer Res 54:5947–5952

Maher, ER, Yates JR, Harries R, Benjamin C, Harris R, Moore AT, Ferguson-Smith MA 1990) Clinical features and natural history of von Hippel-Lindau disease. Q J Med 77: 1151–1163

Mahmoud NN, Boolbol SK, Bilinski RT, Martucci C, Chadburn A, Bertagnolli MM (1997) Apc gene mutation is associated with a dominant-negative effect upon intestinal cell migration. Cancer Res 57:5045–5050

Mahmoud NN, Bilinski RT, Churchill MR, Edelmann W, Kucherlapati R, Bertagnolli MM (1999) Genotype-phenotype correlation in murine Apc mutation: differences in enterocyte migration and response to sulindac. Cancer Res 59:353–359

Mao J, Wang J, Liu B, Pan W, Farr GH 3rd, Flynn C, Yuan H, Takada S, Kimelman D, Li L, Wu D (2001) Low-density lipoprotein receptor-related protein-5 binds to axin and regulates the canonical Wnt signaling pathway. Mol Cell 7:801–809

Maxwell PH, Wiesener MS, Chang GW, Clifford SC, Vaux EC, Cockman ME, Wykoff CC, Pugh CW, Maher ER, Ratcliffe PJ (1999) The tumour suppressor protein VHL targets hypoxia-inducible factors for oxygen-dependent proteolysis. Nature 399:271–275

Melmon KL, Rosen SW (1964) Lindau's disease. Review of the literature and study of a large kindred. Am J Med 36:595–617

Meloni-Ehrig AM (2002) Renal cancer: cytogenetic and molecular genetic aspects. Am J Med Genet 115:164–172

Mimori-Kiyosue Y, Tsukita S (2001) Where is APC going? J Cell Biol 154:1105–1109

Mimori-Kiyosue Y, Shiina N, Tsukita S (2000) Adenomatous polyposis coli (APC) protein moves along microtubules and concentrates at their growing ends in epithelial cells. J Cell Biol 148:505–518

Mogensen MM, Tucker JB, Mackie JB, Prescott AR, Nathke IS (2002) The adenomatous polyposis coli protein unambiguously localizes to microtubule plus ends and is involved in establishing parallel arrays of microtubule bundles in highly polarized epithelial cells. J Cell Biol 157:1041–1048

Morin PJ, Sparks AB, Korinek V, Barker N, Clevers H, Vogelstein B, Kinzler KW (1997) Activation of beta-catenin-Tcf signaling in colon cancer by mutations in beta-catenin or APC. Science 275:1787–1790

Moser AR, Pitot HC, Dove WF (1990) A dominant mutation that predisposes to multiple intestinal neoplasia in the mouse. Science 247:322–324

Moser AR, Mattes EM, Dove WF, Lindstrom MJ, Haag JD, Gould MN (1993) ApcMin, a mutation in the murine Apc gene, predisposes to mammary carcinomas and focal alveolar hyperplasias. Proc Natl Acad Sci USA 90:8977–8981

Moser AR, Shoemaker AR, Connelly CS, Clipson L, Gould KA, Luongo C, Dove WF, Siggers PH, Gardner RL (1995) Homozygosity for the Min allele of Apc results in disruption of mouse development prior to gastrulation. Dev Dyn 203:422–433

Munemitsu S, Souza B, Muller O, Albert I, Rubinfeld B, Polakis P (1994) The APC gene product associates with microtubules in vivo and promotes their assembly in vitro. Cancer Res 54:3676–3681

Munemitsu S, Albert I, Souza B, Rubinfeld B, Polakis P (1995) Regulation of intracellular beta-catenin levels by the adenomatous polyposis coli (APC) tumor-suppressor protein. Proc Natl Acad Sci USA 92:3046–3050

Nakagawa H, Murata Y, Koyama K, Fujiyama A, Miyoshi Y, Monden M, Akiyama T, Nakamura Y (1998) Identification of a brain-specific APC homologue, APCL, and its interaction with beta-catenin. Cancer Res 58:5176–5181

Nakagawa H, Koyama K, Monden M, Nakamura Y (1999) Analysis of APCL, a brain-specific adenomatous polyposis coli homologue, for mutations and expression in brain tumors. Jpn J Cancer Res 90:982–986

Nakagawa H, Koyama K, Murata Y, Morito M, Akiyama T, Nakamura Y (2000a) APCL, a central nervous system-specific homologue of adenomatous polyposis coli tumor suppressor, binds to p53-binding protein 2 and translocates it to the perinucleus. Cancer Res 60:101–105

Nakagawa H, Koyama K, Murata Y, Morito M, Akiyama T, Nakamura Y (2000b) EB3, a novel member of the EB1 family preferentially expressed in the central nervous system, binds to a CNS-specific APC homologue. Oncogene 19:210–216

Nakamura T, Hamada F, Ishidate T, Anai K, Kawahara K, Toyoshima K, Akiyama T (1998) Axin, an inhibitor of the Wnt signalling pathway, interacts with beta- catenin, GSK-3beta and APC and reduces the beta-catenin level. Genes Cells 3:395–403

Nathke IS, Adams CL, Polakis P, Sellin JH, Nelson WJ (1996) The adenomatous polyposis coli tumor suppressor protein localizes to plasma membrane sites involved in active cell migration. J Cell Biol 134:165–179

Neufeld KL, Zhang F, Cullen BR, White RL (2000) APC-mediated downregulation of beta-catenin activity involves nuclear sequestration and nuclear export. EMBO Rep 1:519–523

Nishisho I, Nakamura Y, Miyoshi Y, Miki Y, Ando H, Horii A, Koyama K, Utsunomiya J, Baba S, Hedge P (1991) Mutations of chromosome 5q21 genes in FAP and colorectal cancer patients. Science 253:665–669

Ohh M, Yauch RL, Lonergan KM, Whaley JM, Stemmer-Rachamimov AO, Louis DN, Gavin BJ, Kley N, Kaelin WG Jr, Iliopoulos O (1998) The von Hippel-Lindau tumor suppressor protein is required for proper assembly of an extracellular fibronectin matrix. Mol Cell 1:959–968

Orford K, Crockett C, Jensen JP, Weissman AM, Byers SW (1997) Serine phosphorylation-regulated ubiquitination and degradation of beta-catenin. J Biol Chem 272:24735–24738

Oshima H, Oshima M, Kobayashi M, Tsutsumi M, Taketo MM (1997) Morphological and molecular processes of polyp formation in Apc(delta716) knockout mice. Cancer Res 57:1644–1649

Oshima M, Oshima H, Kitagawa K, Kobayashi M, Itakura C, Taketo M (1995) Loss of Apc heterozygosity and abnormal tissue building in nascent intestinal polyps in mice carrying a truncated Apc gene. Proc Natl Acad Sci USA 92:4482–4486

Park WS, Oh RR, Park JY, Kim PJ, Shin MS, Lee JH, Kim HS, Lee SH, Kim SY, Park YG, An WG, Jang JJ, Yoo NJ, Lee JY (2001) Nuclear localization of beta-catenin is an important prognostic factor in hepatoblastoma. J Pathol 193:483–490

Pause A, Peterson B, Schaffar G, Stearman R, Klausner RD (1999) Studying interactions of four proteins in the yeast two-hybrid system: structural resemblance of the pVHL/elongin BC/hCUL-2 complex with the ubiquitin ligase complex SKP1/cullin/F-box protein. Proc Natl Acad Sci USA 96:9533–9538

Peifer M, Polakis P (2000) Wnt signaling in oncogenesis and embryogenesis – a look outside the nucleus. Science 287:1606–1609

Pepe S, Ruggiero A, d'Acquisto M, de Laurentiis M, de Placido S, Sandomenico C, Staibano S, de Rosa G, Lucariello A, D'Armiento M, Bianco AR (2000) Nuclear DNA content-derived parameters correlated with heterogeneous expression of p53 and bcl-2 proteins in clear cell renal carcinomas. Cancer 89:1065–1075

Pfander D, Kobayashi T, Knight MC, Zelzer E, Chan DA, Olsen BR, Giaccia AJ, Johnson RS, Haase VH, Schipani E (2004) Deletion of Vhlh in chondrocytes reduces cell proliferation and increases matrix deposition during growth plate development. Development 131: 2497–2508

Phillips JL, Ghadimi BM, Wangsa D, Padilla-Nash H, Worrell R, Hewitt S, Walther M, Linehan WM, Klausner RD, Ried T (2001) Molecular cytogenetic characterization of early and late renal cell carcinomas in von Hippel-Lindau disease. Genes Chromosomes Cancer 31:1–9

Pihan GA, Purohit A, Wallace J, Knecht H, Woda B, Quesenberry P, Doxsey SJ (1998) Centrosome defects and genetic instability in malignant tumors. Cancer Res 58:3974–3985

Polakis P (1999) The oncogenic activation of beta-catenin. Curr Opin Genet Dev 9:15–21

Powell SM, Zilz N, Beazer-Barclay Y, Bryan TM, Hamilton SR, Thibodeau SN, Vogelstein B, Kinzler KW (1992) APC mutations occur early during colorectal tumorigenesis. Nature 359:235–237

Prowse AH, Webster AR, Richards FM, Richard S, Olschwang S, Resche F, Affara NA, Maher ER (1997) Somatic inactivation of the VHL gene in Von Hippel-Lindau disease tumors. Am J Hum Genet 60:765–771

Pyles RB, Santoro IM, Groden J, Parysek LM (1998) Novel protein isoforms of the APC tumor suppressor in neural tissue. Oncogene 16:77–82

Quesada CF, Kimata H, Mori M, Nishimura M, Tsuneyoshi T, Baba S (1998) Piroxicam and acarbose as chemopreventive agents for spontaneous intestinal adenomas in APC gene 1309 knockout mice. Jpn J Cancer Res 89:392–396

Rockman SP, Currie SA, Ciavarella M, Vincan E, Dow C, Thomas RJ, Phillips WA (2001) Id2 is a target of the beta-catenin/T cell factor pathway in colon carcinoma. J Biol Chem 276:45113–45119

Romagnolo B, Berrebi D, Saadi-Keddoucci S, Porteu A, Pichard AL, Peuchmaur M, Vandewalle A, Kahn A, Perret C (1999) Intestinal dysplasia and adenoma in transgenic mice after overexpression of an activated beta-catenin. Cancer Res 59:3875–3879

Rosin-Arbesfeld R, Townsley F, Bienz M (2000) The APC tumour suppressor has a nuclear export function. Nature 406:1009–1012

Rowan AJ, Lamlum H, Ilyas M, Wheeler J, Straub J, Papadopoulou A, Bicknell D, Bodmer WF, Tomlinson IP (2000) APC mutations in sporadic colorectal tumors: a mutational »hotspot« and interdependence of the »two hits". Proc Natl Acad Sci USA 97:3352–3357

Rubinfeld B, Albert I, Porfiri E, Munemitsu S, Polakis P (1997) Loss of beta-catenin regulation by the APC tumor suppressor protein correlates with loss of structure due to common somatic mutations of the gene. Cancer Res 57:4624–4630

Samowitz WS, Powers MD, Spirio LN, Nollet F, van Roy F, Slattery ML (1999) Beta-catenin mutations are more frequent in small colorectal adenomas than in larger adenomas and invasive carcinomas. Cancer Res 59:1442–1444

Sasai H, Masaki M, Wakitani K (2000) Suppression of polypogenesis in a new mouse strain with a truncated Apc(Delta474) by a novel COX-2 inhibitor, JTE-522. Carcinogenesis 21:953–958

Sato N, Mizumoto K, Nakamura M, Maehara N, Minamishima YA, Nishio S, Nagai E, Tanaka M (2001) Correlation between centrosome abnormalities and chromosomal instability in human pancreatic cancer cells. Cancer Genet Cytogenet 126:13–19

Satoh S, Daigo Y, Furukawa Y, Kato T, Miwa N, Nishiwaki T, Kawasoe T, Ishiguro H, Fujita M, Tokino T, Sasaki Y, Imaoka S, Murata M, Shimano T, Yamaoka Y, Nakamura Y (2000) AXIN1 mutations in hepatocellular carcinomas, and growth suppression in cancer cells by virus-mediated transfer of AXIN1. Nat Genet 24:245–250

Semenza GL (2001) HIF-1 and mechanisms of hypoxia sensing. Curr Opin Cell Biol 13:167–171

Shaub A (1999) Unravelling the extracellular matrix. Nat Cell Biol 1:E173–E175

Sherr, CJ (2004) Principles of tumor suppression. Cell 116:235–246

Shibata H, Toyama K, Shioya H, Ito M, Hirota M, Hasegawa S, Matsumoto H, Takano H, Akiyama T, Toyoshima K, Kanamaru R, Kanegae Y, Saito I, Nakamura Y, Shiba K, Noda T (1997) Rapid colorectal adenoma formation initiated by conditional targeting of the Apc gene. Science 278:120–123

Shweiki D, Itin A, Soffer D, Keshet E (1992) Vascular endothelial growth factor induced by hypoxia may mediate hypoxia-initiated angiogenesis. Nature 359:843–845

Siemeister G, Weindel K, Mohrs K, Barleon B, Martiny-Baron G, Marme D (1996) Reversion of deregulated expression of vascular endothelial growth factor in human renal carcinoma cells by von Hippel-Lindau tumor suppressor protein. Cancer Res 56:2299–2301

Smith KJ, Levy DB, Maupin P, Pollard TD, Vogelstein B, Kinzler KW (1994) Wild-type but not mutant APC associates with the microtubule cytoskeleton. Cancer Res 54:3672–3675

Smits R, van der Houven van Oordt W, Luz A, Zurcher C, Jagmohan-Changur S, Breukel C, Khan PM, Fodde R (1998) Apc1638 N: a mouse model for familial adenomatous polyposis-associated desmoid tumors and cutaneous cysts. Gastroenterology 114:275–283

Smits R, Kielman MF, Breukel C, Zurcher C, Neufeld K, Jagmohan-Changur S, Hofland N, van Dijk J, White R, Edelmann W, Kucherlapati R, Khan PM, Fodde R (1999) Apc1638T:

a mouse model delineating critical domains of the adenomatous polyposis coli protein involved in tumorigenesis and development. Genes Dev 13:1309–1321
Smits R, Hofland N, Edelmann W, Geugien M, Jagmohan-Changur S, Albuquerque C, Breukel C, Kucherlapati R, Kielman MF, Fodde R (2000) Somatic Apc mutations are selected upon their capacity to inactivate the beta-catenin downregulating activity. Genes Chromosomes Cancer 29:229–239
Stickle NH, Chung J, Klco JM, Hill RP, Kaelin, WG Jr. Ohh M (2004) pVHL modification by NEDD8 is required for fibronectin matrix assembly and suppression of tumor development. Mol Cell Biol 24:3251-3261
Su LK, Kinzler KW, Vogelstein B, Preisinger AC, Moser AR, Luongo C, Gould KA, Dove WF (1992) Multiple intestinal neoplasia caused by a mutation in the murine homolog of the APC gene. Science 256:668–670
Su LK, Johnson KA, Smith KJ, Hill DE, Vogelstein B, Kinzler KW (1993) Association between wild type and mutant APC gene products. Cancer Res 53:2728–2731
Su LK, Burrell M, Hill DE, Gyuris J, Brent R, Wiltshire R, Trent J, Vogelstein B, Kinzler KW (1995) APC binds to the novel protein EB1. Cancer Res 55:2972–2977
Su LK, Barnes CJ, Yao W, Qi Y, Lynch PM, Steinbach G (2000) Inactivation of germline mutant APC alleles by attenuated somatic mutations: a molecular genetic mechanism for attenuated familial adenomatous polyposis. Am J Hum Genet 67:582–590
Thrash-Bingham CA, Salazar H, Freed JJ, Greenberg RE, Tartof KD (1995) Genomic alterations and instabilities in renal cell carcinomas and their relationship to tumor pathology. Cancer Res 55:6189–6195
Truong, LD, Choi YJ, Shen SS, Ayala G, Amato R, Krishnan B (2003) Renal cystic neoplasms and renal neoplasms associated with cystic renal diseases: pathogenetic and molecular links. Adv Anat Pathol 10:135–159.
Van de Wetering M, Sancho E, Verweij C, de Lau W, Oving IM, Hurlstone A, van der Horn K, Batlle E, Coudreuse DY, Haramis AP, Tjon-Pon-Fong M, Moerer P, van den Born MM. W, Soete G, Pals ST, Eilers M, Medema RH, Clevers H (2002) The beta-catenin/TCF4 complex controls the proliferation/differentiation switch in colon epithelium through c-MYC-mediated repression of p21CIP1/WAF1. Cell 111:241–250
Van der Houven van Oordt CW, Smits R, Williamson SL, Luz A, Khan PM, Fodde R, van der Eb AJ, Breuer ML (1997) Intestinal and extra-intestinal tumor multiplicities in the Apc1638 N mouse model after exposure to X-rays. Carcinogenesis 18:2197–203
Van Es JH, Kirkpatrick C, van de Wetering M, Molenaar M, Miles A, Kuipers J, Destree O, Peifer M, Clevers H (1999) Identification of APC2, a homologue of the adenomatous polyposis coli tumour suppressor. Curr Biol 9:105–108
Van Es JH, Giles RH, Clevers HC (2001) The many faces of the tumor suppressor gene APC. Exp Cell Res 264:126–134
Vogelstein B, Kinzler KW (1993) The multistep nature of cancer. Trends Genet 9:138–141
Wang Y, Macke JP, Abella BS, Andreasson K, Worley P, Gilbert DJ, Copeland NG, Jenkins NA, Nathans J (1996) A large family of putative transmembrane receptors homologous to the product of the Drosophila tissue polarity gene frizzled. J Biol Chem 271:4468–4476
Wehrli M, Dougan ST, Caldwell K, O'Keefe L, Schwartz S, Vaizel-Ohayon D, Schejter E, Tomlinson A, DiNardo S (2000) arrow encodes an LDL-receptor-related protein essential for Wingless signalling. Nature 407:527–530
Wilkinson DG (2001) Multiple roles of EPH receptors and ephrins in neural development. Nat Rev Neurosci 2:155–164
Willert J, Epping M, Pollack JR, Brown PO, Nusse R (2002) A transcriptional response to Wnt protein in human embryonic carcinoma cells. BMC Dev Biol 2:8
Wizigmann-Voos S, Breier G, Risau W, Plate KH (1995) Up-regulation of vascular endothelial growth factor and its receptors in von Hippel-Lindau disease-associated and sporadic hemangioblastomas. Cancer Res 55:1358–1364

Xu Q, Mellitzer G, Robinson V, Wilkinson DG (1999) In vivo cell sorting in complementary segmental domains mediated by Eph receptors and ephrins. Nature 399:267–271

Yanagawa S, van Leeuwen F, Wodarz A, Klingensmith J, Nusse R (1995) The dishevelled protein is modified by wingless signaling in Drosophila. Genes Dev 9:1087–1097

Zachariae W, Shevchenko A, Andrews PD, Ciosk R, Galova M, Stark MJ, Mann M, Nasmyth K (1998) Mass spectrometric analysis of the anaphase-promoting complex from yeast: identification of a subunit related to cullins. Science 279:1216–1219

Zbar B, Kishida T, Chen F, Schmidt L, Maher ER, Richards FM, Crossey PA, Webster AR, Affara NA, Ferguson-Smith MA, Brauch H, Glavac D, Neumann HP, Tisherman S, Mulvihill JJ, Gross DJ, Shuin T, Whaley J, Seizinger B, Kley N, Olschwang S, Boisson C, Richard S, Lips CH, Lerman M et al (1996) Germline mutations in the Von Hippel-Lindau disease (VHL) gene in families from North America, Europe, and Japan. Hum Mutat 8:348–357

Zeng L, Fagotto F, Zhang T, Hsu W, Vasicek TJ, Perry WL, 3rd, Lee JJ, Tilghman SM, Gumbiner BM, Costantini F (1997) The mouse Fused locus encodes axin, an inhibitor of the Wnt signaling pathway that regulates embryonic axis formation. Cell 90:181–192

Zumbrunn J, Kinoshita K, Hyman AA, Nathke IS (2001) Binding of the adenomatous polyposis coli protein to microtubules increases microtubule stability and is regulated by GSK3 beta phosphorylation. Curr Biol 11:44–49

Hormonal and Stromal Regulation of Normal and Neoplastic Prostatic Growth

W.A. Ricke, Y. Wang, T. Kurita, S.W. Hayward, G.R. Cunha

1 Introduction

The World Health Organization estimates that prostate cancer (PRCA) will be detected in 250,000 men worldwide in 2004. Many of these patients will undergo treatment to halt or slow PRCA progression. Many more will be diagnosed with advanced PRCA for which there is no cure. Moreover, this year more than 30,000 men will die of this disease in the United States alone. Goals of early detection, surgery, hormone therapy, chemotherapy, and chemoprevention are to decrease morbidity and mortality and increase life span. To achieve these goals, useful models of human PRCA progression must be developed and tested. To develop and evaluate useful animal models it is necessary to understand both normal growth and development as well as processes associated with carcinogenesis.

One approach to cancer research focuses on the abnormal traits of the cancer cells themselves. While this approach has yielded a wealth of information, especially on the genetic alterations associated with cancer development, the process of carcinogenesis can also be examined in the broader context of tissue formation and loss of homeostasis. In this way, a more complete insight is gained into how tissues undergo malignant conversion, and hence how this process may be controlled or possibly reversed. Cancer and tumor progression is not just a disease of uncontrolled cell growth. The hallmarks of tumorigenesis also include loss of normal tissue architecture, inflammation, nuclear

W.A. Ricke (rickew@itsa.ucsf.edu), G.R. Cunha
Department of Anatomy, University of California, San Francisco, California 94143–0452, USA

Y. Wang, T. Kurita
Department of Cancer Endocrinology, BC Cancer Agency, 601 West 10th Avenue, Vancouver, British Columbia, V5Z 1L3 Canada

S.W. Hayward
Department of Urologic Surgery, Vanderbilt University Medical Center, A1302, Medical Center North, 1161 21st Ave South, Nashville, Tennessee 37232–2765, USA

Progress in Molecular and Subcellular Biology
Alvaro Macieira-Coelho (Ed.)
Development Biology of Neoplastic Growth

atypia, genetic alterations, destruction of tissue boundaries and invasion to surrounding tissues, stromal changes, angiogenesis, and compromise of distant organs through metastatic spread. Cancer results from dysregulation of the finely orchestrated processes that determine how cells are integrated into normal tissues and tissues into organs (Ingber and Jamieson 1982).

This chapter focuses on the role of the stromal microenvironment in normal prostate development and prostatic carcinogenesis, and in a broader context will discuss mechanisms of microenvironmental influences that occur during carcinogenesis. The process of carcinogenesis can be defined as the process whereby benign epithelial cells acquire characteristics of malignancy. All epithelial cells are associated with connective tissue, which plays a critical role in epithelial development and differentiation. Accordingly, we first will review the role of epithelial–mesenchymal interactions in the developing prostate and then discuss carcinogenesis of the prostate.

2 Prostatic Growth and Development

2.1 Mesenchymal–Epithelial Interactions in Prostatic Development

In all species the prostate develops from the urogenital sinus, which is composed of endodermal urogenital sinus epithelia and urogenital sinus mesenchyme. The urogenital sinus is derived from the caudal terminus of the hindgut called the cloaca. The urorectal septum subdivides the cloaca into the urogenital sinus ventrally and the rectum and anal canal dorsally. The urogenital sinus is an ambisexual embryonic rudiment, which in males develops into the bladder, prostate, prostatic urethra, and bulbourethral glands; in females the urogenital sinus develops into the bladder, lower vagina, and urethra (Staack et al. 2003). The earliest event in prostatic development is specification of prostatic epithelial identity. Before sexual differentiation of the urogenital sinus, urogenital sinus mesenchyme (but not urogenital sinus epithelium) expresses androgen receptors in both sexes and thus acquires the capacity to undergo masculine (prostatic) development (Cooke et al. 1991a; Takeda and Chang 1991).

2.2 Bud Formation

In response to fetal testicular androgens, epithelial buds emerge from the wall of the endodermal urogenital sinus and grow into the surrounding stroma. The mechanisms involved in prostatic bud formation are poorly understood. When first recognizable, prostatic buds are small spherical protrusions about

45 μm in diameter extending from the urogenital sinus epithelium into the surrounding urogenital sinus mesenchyme. These small epithelial buds contain the progenitor cells for generation of the extensively branched ductal trees that subsequently develop into the secretory prostate gland. While only a small number of epithelial cells are present within a newly emerged prostatic bud, evidence from analysis of chimeric mice reveals that at least some of the buds are polyclonal in origin (Lipschutz et al. 1999). Thus, each prostatic bud may contain multiple progenitor lineages. Localized proliferation in the embryonic epithelium does not precede initial bud formation, and there is no local thickening or epithelial placode formation preceding bud formation. The only morphological indication of where buds will form are small indentations of the basal lamina (Donjacour et al. 2003). In early elongating buds within embryonic and neonatal developing prostate, the distal tips have a higher Ki67 (proliferation marker) labeling index than do the proximal bud segments, and therefore allow for outward growth of ducts (Donjacour et al. 2003; Sugimura et al. 1986a).

In the perinatal period of rodents, the solid prostatic buds elongate and then undergo a process of branching morphogenesis, which is completed by the end of puberty (Sugimura et al. 1986b; Kinbara and Cunha 1995). Initially, prostatic buds and prostatic ducts are solid before becoming a secretory epithelium. Beginning shortly after birth the solid epithelial cords begin the canalization process. Canalization begins at the urethra and progresses distally towards the ductal tips (Marker et al. 2003). During ductal canalization, epithelial cells differentiate into luminal and basal cells. Secretory epithelial cytodifferentiation occurs postnatally in rodents, and prostate-specific secretory proteins are initially detected in rodents at 12 to 20 days postnatal (Lopes et al. 1996). Prostatic epithelial differentiation is accompanied by differentiation of the predominantly fibroblastic mesenchyme into smooth muscle cells (Hayward et al. 1996b, 1997).

2.3 Prostate Lobes and Regions

Androgen-regulated prostatic bud development in rodents occurs in a precise spatial pattern, and thus establishes the lobar subdivisions of the prostate into dorsal-lateral, ventral, and anterior prostates based on their anatomical location. Each lobe has a characteristic ductal branching pattern (Price 1963; Sugimura et al. 1986b; Kinbara and Cunha 1995; Marker et al. 2003). Due to lobe-specific differences in the patterns of branching morphogenesis, the overall gross shape of each lobe is unique. In addition, the lobes have distinct histological features, with extensive epithelial infolding in the anterior prostate, a lesser amount of epithelial infolding in the dorsal-lateral prostate, and minimal infolding in the ventral prostate. The prostatic lobes also express distinct secretory proteins. Within each lobe, regional differences in cellular morphology, DNA synthesis, and secretory activity are also observed along the proximal-distal (urethra to ductal tip) axis within prostatic ducts (Lee et al. 1990; Sugimura et al. 1986a). The anterior prostate (also known as the coagulating gland) grows in close association with the seminal vesicle. Indeed, the progenitor epithelium of the seminal vesicle and the anterior prostate develop within a common mass of mesenchyme, historically designated seminal vesicle mesenchyme (Cunha 1972). In addition to inducing the seminal vesicle, seminal vesicle mesenchyme is a prostatic inducer. In heterotypic tissue recombinants constructed with either Wolffian-duct urogenital-sinus-derived epithelium, and either urogenital sinus or seminal vesicle mesenchyme, the specific tissue response is determined by germ layer derivation of the epithelium (Donjacour and Cunha 1995; Higgins et al. 1989; Risbridger et al. 2001a,b).

In contrast to the rodent prostate, the adult human prostate is a round gland without distinct lobes. It is roughly the size of a small egg (20 g and 4×2.5 cm). The adult human prostate clearly exhibits – though not lobes – distinct anatomical regions, now described as three zones: the central, transition, and peripheral zones, reflecting three distinct sets of ducts (McNeal 1983). Comparative observation of prostatic development in rodents and humans demonstrates that prostatic morphogenesis occurs in an analogous manner in the two species, with several distinct sets of epithelial buds growing out of the urethra/urogenital sinus into the surrounding stroma (Price 1963; Timms et al. 1994). Compelling molecular evidence for homology between specific rodent prostatic lobes and human prostatic zones has yet to be identified, and little is known about the molecular basis for the lobe- and region-specific features observed within the prostate. Nevertheless, this aspect of prostatic biology is important because prostatic diseases occur in a highly zone-specific manner. Human prostatic adenocarcinoma is predominantly a disease of the peripheral zone, and benign prostatic hyperplasia is predominantly a disease of the transition zone.

One gene that is involved in lobe specificity is *homeobox* a10 (*Hoxa10*). *Hoxa10* encodes a transcription factor that is expressed within both epithelial and mesenchymal compartments of the developing prostate. Mice lacking functional *Hoxa10* genes have reduced branching in the anterior prostate and a partial anterior prostate to dorsal-lateral prostate transformation based on ductal morphology and branching pattern (Podlasek et al. 1999). These morphological changes implicate *Hoxa10* in the establishment of the anterior-prostate-specific pattern of branching morphogenesis. A second gene, *fucosyltransferase 1* (*Fut1*) is also involved in region-specific differences in epithelial proliferation during prostatic development. *Fut1* is a transmembrane-carbohydrate-modifying enzyme present within the secretory pathway and at the cell surface. *Fut1* is expressed in the developing epithelium of all prostatic lobes in rodents. However, within each lobe, *Fut1* has been localized only to a subset of epithelial ducts. Inhibitory antibodies directed against the cell-surface portion of the *Fut1* protein reduces epithelial cell proliferation during prostatic development (Marker et al. 2001). These observations implicate *Fut1* as part of the molecular mechanism involved in region-specific heterogeneity within the prostate.

2.4 Androgens and Prostatic Development

Prostatic development is dependent on fetal androgens and their effects on epithelial-mesenchymal interactions. Interactions between urogenital sinus mesenchyme and urogenital sinus epithelium are necessary for prostatic development and maturation. When urogenital sinus epithelium and urogenital sinus mesenchyme are isolated and grown by themselves, prostatic development does not occur, even in the presence of circulating androgens. However, if urogenital sinus mesenchyme and urogenital sinus epithelium are reassociated in an androgenic milieu, prostatic development proceeds. During androgen-dependent prostatic development, urogenital sinus mesenchyme specifies the identity of prostatic epithelial cells, induces epithelial bud formation, elicits epithelial bud growth and ductal branching, promotes epithelial cytodifferentiation into secretory epithelial cells, and determines the types of secretory products expressed (Cunha 1984; Cunha et al. 1980, 1983a,b, 1987; Cunha and Young 1991). These conclusions have been drawn from the analysis of heterotypic tissue recombinants composed of urogenital sinus mesenchyme plus epithelium of the urinary bladder. The fact that adult bladder epithelium can be induced to undergo this remarkable change in cytodifferentiation (bladder to prostate) implies a continued importance of epithelial–stromal interactions in the maintenance of adult epithelial cells. The induced bladder epithelium appears to be prostatic at both light and electron microscopic levels. Androgen receptors appear in the normally androgen receptor-negative bladder epithelium of urogenital sinus mesenchyme plus bladder epithelium tissue recombinants, and prostate-specific secretory proteins are expressed (Cunha 1984; Cunha et al. 1980, 1983a). Mechanistically, urogenital sinus-mesenchyme-induced conversion of bladder into prostate appears to represent a change of differentiated adult bladder epithelial cells into a prostatic epithelium. Another interpretation may be that undifferentiated embryonic stem cells reside in adult bladder epithelium, which under the influence of urogenital sinus mesenchyme are induced to differentiate into prostatic epithelium.

The urogenital sinus-mesenchyme-derived factors that induce prostatic epithelial differentiation are currently unknown. Induction of prostatic epithelial identity is thought to be an early event, occurring prior to the formation of prostatic buds. In this regard, the homeobox gene, Nkx3.1, appears in urogenital sinus epithelium of male mouse fetuses two days prior to the formation of prostatic buds. Nkx3.1 is an androgen-inducible transcription factor expressed in the male, but not female urogenital sinus epithelium (Sciavolino et al. 1997). Thus, Nkx3.1 is currently the earliest prostatic identifier, its expression occurring when prostatic epithelial identity is being acquired.

While Nkx3.1 is the earliest known prostatic marker, studies of Nkx3.1 mutant mice reveal that prostatic development can occur in the absence of Nkx3.1, although subsequent growth and differentiation of the prostate are adversely affected (Bhatia-Gaur et al. 1999).

During prostate development, androgens acting via mesenchymal and epithelial androgen receptors induce prostate development. In this sense, the field of prostatic development lies at the crossroads of developmental biology and reproductive endocrinology. To merge these two fields of study as they pertain to prostatic development, the analysis of the ontogeny of androgen receptors has been investigated. Several laboratories have demonstrated that prior to and during prostatic development, the urogenital sinus mesenchyme expresses high levels of androgen receptors (Cooke et al. 1991a; Takeda and Chang 1991). Initially, when prostatic buds grow from the urogenital sinus epithelium into the surrounding urogenital sinus mesenchyme, androgen receptors are undetectable in the induced epithelial buds. This observation suggests that only mesenchymal androgen receptors are involved in the early phases of prostatic development. Androgen receptors are initially detected in developing prostatic epithelium after birth in rodents as the ducts begin to canalize during the first week postnatally (Moeller et al. 1987; Prins and Birch 1995; Prins et al. 1996). Thus, the appearance of androgen receptors within the epithelium coincides with later aspects of prostatic development. To help identify the respective roles of epithelial and mesenchymal androgen receptors in prostatic development, tissue recombinants were analyzed that were constructed with epithelium and mesenchyme from wild-type and androgen receptor-negative testicular feminization mice (see Fig. 1). Tissue recombinants composed of wild-type urogenital sinus mesenchyme plus wild-type epithelium, and wild-type urogenital sinus mesenchyme plus testicular-feminization epithelium, both underwent prostatic development. The common feature of these two tissue recombinants is the expression of stromal androgen receptors. The other two possible tissue recombinants (testicular-feminization urogenital sinus mesenchyme plus testicular-feminization epithelium, and testicular-feminization urogenital sinus mesenchyme plus wild-type epithelium) did not develop into prostatic epithelium in the presence of testicular androgens.

The absence of prostatic development in these two tissue recombinants prepared with testicular-feminization urogenital sinus mesenchyme suggested a critical role for mesenchymal androgen receptors in prostatic development. This idea was confirmed through analysis of tissue recombinants composed of wild-type urogenital sinus mesenchyme plus testicular-feminization epithelium, which demonstrated that when combined with wild-type urogenital sinus mesenchyme, androgen-receptor-deficient testicular feminization epithelium can develop into normal prostate at the morphological and biochemical levels (Cunha et al. 1992). This strongly suggests that certain »androgenic effects« on epithelium are independent of epithelial androgen receptors. Instead, many androgenic effects on epithelium are elicited by paracrine factors or "*andromedins*" produced by androgen receptor-positive mesenchyme/ stroma. The role of epithelial androgen receptors has been revealed by comparison of tissue recombinants composed of androgen receptor-positive wild-type urogenital sinus mesenchyme plus androgen receptor-positive wild-type epithelium versus androgen receptor-positive wild-type urogenital sinus mesenchyme plus androgen receptor-negative testicular-feminization epithelium. Such experiments have demonstrated that while the androgen receptor-defi-

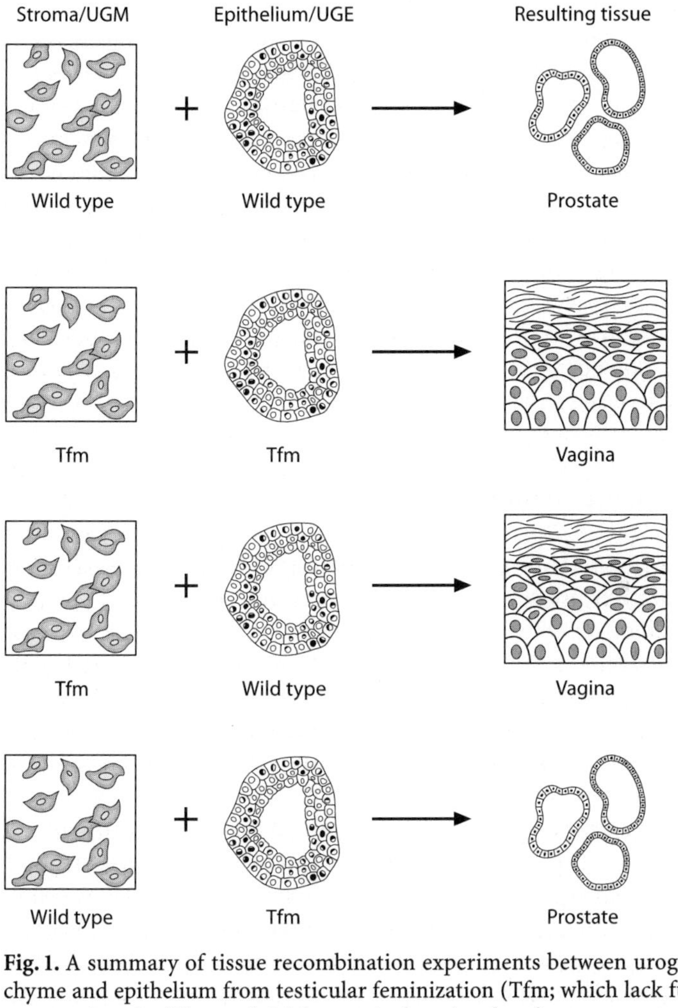

Fig. 1. A summary of tissue recombination experiments between urogenital sinus mesenchyme and epithelium from testicular feminization (Tfm; which lack functional androgen receptors) and wild-type (which contain functional androgen receptors) embryos. An androgenic response (i.e., prostate differentiation) occurs when wild-type mesenchyme is grown in association with either wild-type or Tfm epithelium. Conversely, vagina-like morphogenesis occurs when either wild-type or Tfm epithelium is grown in association with Tfm mesenchyme. These data demonstrate the necessity of stromal androgen receptors for prostate development. (Redrawn with permission from Cunha et al. 2002)

cient epithelium was induced to undergo prostatic differentiation, it failed to express specific prostatic secretory proteins (Cunha and Young 1991; Donjacour and Cunha 1993). Thus, epithelial androgen receptors are necessary for production of androgen receptor-dependent prostatic secretory proteins. In a broader context, it is known that the sex steroids such as estradiol and progesterone also regulate epithelial proliferation in their respective target organs of the male and female reproductive tracts. Comparable tissue recombinant studies have been carried out using tissues from wild-type, estrogen receptor-α-null-, or progesterone receptor-null mice. Such experiments demonstrated that for all three classes of steroids epithelial proliferation in vivo is regulated via the appropriate hormone receptors within the stromal cells (Sugimura et al. 1986a; Cooke et al. 1997; Kurita et al. 1998). Thus, regulation of epithelial proliferation in normal and possibly cancerous tissues may occur via paracrine mechanisms. Conversely, hormonal regulation of epithelial differentiation and function requires direct hormone action mediated by hormone receptors residing within the epithelium (Buchanan et al. 1998a,b). It should be emphasized that these conclusions are applicable to normal epithelia. Similar studies are currently being evaluated in carcinogenic models. In this regard, androgenic regulation of malignant human prostatic epithelial cells appears to involve a change from a paracrine to an autocrine mechanism of androgen-stimulated growth (Gao et al. 2001).

2.5 Stromal–Epithelial Interactions in Mature Prostate

The adult prostate is divided into two compartments: a glandular epithelial compartment and a fibromuscular stroma. The stroma of the adult prostate contains smooth muscle cells as well as fibroblasts, both of which are derived from embryonic urogenital sinus mesenchyme. In the mature prostate, cells of both the epithelial and the stromal compartments are highly differentiated. Under steady-state androgenic conditions of the intact adult male, the prostate contains fully differentiated smooth muscle cells in intimate contact with differentiated and functional epithelial cells. Both cell types are essentially growth-quiescent with minimal levels of cellular proliferation and of cell death. It should be noted that this adult growth-quiescent, homeostatic state exists in the presence of high circulating levels of androgens, which at developmental stages would be profoundly growth stimulatory. This quiescence of growth in the adult prostate is likely to be dependent upon maintenance of the normal architectural associations between epithelium and stroma. This interpretation is based upon the observations that epithelial and stromal cultures proliferate when established from growth-quiescent adult prostatic tissue. Proliferative activity of prostatic epithelial and stromal cells in vitro is associated with marked changes in the differentiation profile of both cell types. Thus, alteration of the normal stromal–epithelial interactions during carcinogenesis causes dedifferentiation and elicits proliferation of both adult prostatic stromal and epithelial cells.

Another means of altering the cell–cell communication in the adult prostate is to associate adult growth-quiescent epithelium with embryonic or neonatal prostatic inductive mesenchyme. This was demonstrated in experiments in which tissue recombinants were prepared utilizing rat urogenital sinus mesenchyme plus a 300 μm segment of an adult mouse prostatic duct (urogenital sinus mesenchyme plus mouse prostatic epithelium). These fragments of adult prostatic ducts contain approximately 5,000 epithelial cells. When grown for 1 month under the renal capsule of a intact male hosts, urogenital sinus mesenchyme plus mouse prostatic epithelium recombinants form about 30–50 mg wet weight of prostatic tissue, an estimated 30.0×10^6 mouse prostatic epithelial cells (Hayashi et al. 1993; Norman et al. 1986). Another important observation occurred in tissue recombinants consisting of adult ventral prostatic ducts recombined with urogenital sinus mesenchyme, or seminal vesicle mesenchyme. These tissue recombinants developed prostatic epithelium expressing secretory proteins specific to the dorsal-lateral and anterior prostate zones (Norman et al. 1986; Hayashi et al. 1993). This observation of mesenchyme-induced re-specification of lobar identity in adult prostatic epithelium, in conjunction with the ability of embryonic mesenchyme to induce prostatic differentiation of adult bladder epithelium (Donjacour and Cunha 1988), provides yet another example of the plasticity of adult epithelia to respond to hormonal and stromal stimuli. The adult stromal microenvironment is thus thought to play a major role in the regulation of adult epithelial proliferation and differentiation. The corollary to this idea is that perturbation of the normal homeostatic stromal–epithelial interaction may play a central role in pathogenetic activity in the prostate and other glands.

Maintenance of adult prostatic epithelium via stromal–epithelial interactions is dependent upon the presence of testosterone. Androgen deprivation elicited by physical castration triggers apoptosis of prostatic epithelium (Lee 1981; Kyprianou et al. 2000). Thus, failure to occupy androgen receptors initiates prostatic epithelial apoptosis (Isaacs 1984). The mature prostate contains epithelial and stromal cells, both of which are positive for androgen receptors (Prins et al. 1991). This raises the question as to whether occupation of either epithelial or stromal androgen receptors prevents prostatic epithelial apoptosis. To address this question, heterotypic tissue recombinants were constructed with epithelium from wild-type and testicular-feminization mice combined with rat urogenital sinus mesenchyme (Kurita et al. 2001). The resultant tissue recombinants were grafted into intact male immuno-compromised mouse hosts. At one month post grafting, both rat urogenital sinus mesenchyme plus testicular-feminization epithelium and rat urogenital sinus mesenchyme plus wild-type epithelium tissue recombinants formed prostate tissue. The hosts were then surgically castrated, which increased prostatic epithelial cell apoptosis comparably in both the rat urogenital sinus mesenchyme plus testicular- feminization epithelium and rat urogenital sinus mesenchyme plus wild-type epithelium tissue recombinants. Moreover, re-administration of androgens inhibited post-castration-induced epithelial apoptosis in both groups of tissue recombinants. These results imply that epithelial androgen receptors are not necessary to regulate apoptosis of the prostatic epithelium. At the time it was not possible to test whether stromal androgen receptors were essential to inhibit castration-induced prostatic epithelial apoptosis by androgen, because stromal androgen receptors are essential to induce prostate formation. Therefore, it is conceivable that androgens may have prevented apoptosis of prostatic epithelium through a systemic effect on the host mouse. This issue can be settled in the future by tissue recombination experiments utilizing an androgen receptor knockout mouse under an inducible system, with which stromal and/or epithelial androgen receptor genes can be inactivated after full growth of the prostate (Yeh et al. 2002).

2.6 Smooth Muscle of the Prostate

During prostatic growth and development, interactions between urogenital sinus mesenchyme and urogenital sinus epithelium are reciprocal in that urogenital sinus mesenchyme induces prostatic epithelial differentiation, and the developing prostatic epithelium induces smooth muscle development and differentiation in the urogenital sinus mesenchyme. Data supporting a role of prostatic epithelium as the inducer of smooth muscle differentiation is based on the observation that minimal amounts of smooth muscle developed when urogenital sinus mesenchyme was grafted alone without any epithelium. In contrast, when rodent urogenital sinus mesenchyme was recombined with epithelium of either adult prostate, bladder, or embryonic urogenital sinus, prostatic ducts developed which were surrounded by α-actin-positive smooth muscle cells (Cunha 1994). Studies of tissue recombinants composed of rat urogenital sinus mesenchyme plus human prostatic epithelium have demonstrated that the human prostatic epithelium can also induce the rat stroma to undergo smooth muscle differentiation. The human prostatic epithelium also determined the spatial patterning of the smooth muscle (Hayward et al. 1996c, 1998). Thus, androgen-dependent prostatic development results from reciprocal cell–cell interactions in which the stroma induces prostatic epithelial differentiation, and developing prostatic epithelium induces and patterns of smooth muscle differentiation in the stroma. The final result of these reciprocal epithelial–mesenchymal interactions is the differentiation of mature prostatic tissue in which the growth-quiescent epithelium develops into differentiated secretory columnar cells, and the mesenchyme develops into a mature prostatic stroma composed predominantly of smooth muscle cells.

2.7 Estrogen Action in the Prostate

While the prostate is primarily considered to be an androgen target organ, it is also sensitive to estrogens and phyto-estrogens. One well-recognized effect of chronic long-term exposure to exogenous estrogens on the prostate is squamous metaplasia (Risbridger et al. 2001a). Estrogen's effects are mediated via ligand binding to estrogen receptors α and β, which are expressed in the developing and adult prostate (Prins et al. 1991; Kuiper et al. 1996; Couse et al. 1997; Lau et al. 1998). The paracrine actions known to occur with androgens in the prostate are mediated via stromal androgen receptors, this raises the possibility that estrogenic effects on the prostate may be similarly elicited via stromal estrogen receptors. Estrogenic effects on the prostate are complicated and may involve both indirect and direct actions. In intact males, pharmacologic doses of estrogen elicit androgen deprivation by suppressing pituitary luteinizing hormone secretion, thus reducing the production of testosterone by the testes (Walsh 1975). This indirect effect of estrogen has profound implications on the prostate because in response to high levels of exogenous estrogen two processes occur: (1) the prostate is first deprived of androgens as a result of indirect effects on the pituitary; (2) estrogens then act directly on the gland unopposed by androgens to elicit squamous metaplasia of the prostate. Experiments using estrogen receptor-α and estrogen receptor-β mutant mice demonstrated that estrogen receptor-α, but not estrogen receptor-β, is crucial for estrogen induction of prostatic squamous metaplasia (Risbridger et al. 2001a,b). To determine the respective roles of epithelial versus stromal estrogen receptor-α in the development of squamous metaplasia of the prostate, the following tissue recombinants were constructed with prostatic epithelium and stroma from wild-type and α-estrogen receptor knockout mice: wild-type stroma plus wild-type prostatic epithelium, α-estrogen receptor knockout stroma plus α-estrogen receptor knockout prostatic epithelium, wild-type stroma plus α-estrogen receptor knockout prostatic epithelium and α-estrogen receptor knockout stroma plus wild-type prostatic epithelium. Diethylstilbestrol, a synthetic estrogen, elicited squamous metaplasia only in wild-type stroma plus wild-type prostatic epithelium tissue recombinants. Tissue recombinants containing α-estrogen receptor knockout prostatic epithelium and/or α-estrogen receptor knockout stroma (i.e., α-estrogen receptor knockout stroma plus α-estrogen receptor knockout prostatic epithelium, wild-type stroma plus α-estrogen receptor knockout prostatic epithelium and α-estrogen receptor knockout stroma plus wild-type prostatic epithelium) failed to develop squamous metaplasia of the prostate. Therefore, estrogen induction of prostatic squamous metaplasia requires the action of estrogen via stromal estrogen receptor-α (paracrine mechanism), as well as direct action of estrogen via epithelial estrogen receptor-α. The importance of paracrine interaction via estrogen signaling in the prostate has also been suggested in studies on neonatal imprinting of the prostate (Chang et al. 1999; Prins et al.

2001). Immunohistochemical and steroid autoradiographic studies have demonstrated the presence of estrogen receptor-α and estrogen receptor-β in both prostatic stromal and epithelial cells during various stages of development (Prins and Birch 1997; Prins et al. 1998, 2001; Weihua et al. 2001). Nonetheless, the tissue recombinant studies formally infer the importance of both stromal and epithelial estrogen receptors in estrogenic response in the prostate.

2.8 Prostatic Epithelial Cytodifferentiation

All of the above aspects of prostatic development (specification of prostatic epithelial identity, induction of epithelial bud formation, and prostatic bud growth and ductal branching) are induced by the urogenital sinus mesenchyme and mediated by androgens through the androgen receptors. This in turn promotes prostatic epithelial differentiation into secretory epithelial cells and specifies the types of secretory proteins expressed. Prostatic development involves the emergence of solid epithelial buds from the stratified epithelium of the urogenital sinus, followed in turn by their canalization to form ducts lined by a simple columnar secretory epithelium. Concurrent with ductal elongation and branching morphogenesis, epithelial cytodifferentiation begins shortly after birth in rodents. Urogenital sinus epithelial cells and the developing solid prostatic buds express a wide array of cytokeratins (including cytokeratins 5, 8, 14, 18, and 19) and p63 (Wang et al. 2001b). As the solid epithelial cords develop and expand into the surrounding mesenchyme, ductal canalization is initiated beginning at the urethra and proceeding distally toward the ductal tips. In rodents, as the solid epithelial cords canalize, the epithelium reorganizes into tall columnar luminal cells and a discontinuous layer of basal cells in rodents, whereas in humans, basal cells line the basement membrane and support the overlying luminal cells. The luminal cells express cytokeratins 8, 18, and 19 and differentiate into columnar secretory cells (Hayward et al. 1996a). Another epithelial cell type that develops in the prostate is the neuroendocrine cell. Neuroendocrine cells make up only a small proportion of the total number of epithelial cells and are characterized by the expression of functional markers such as chromogranin A and synaptophysin (Cohen et al. 1993; Xue et al. 2000). The third epithelial cell type is the basal cell; these cells express cytokeratins 5 and 14, and p63 and are located along the basement membrane (Hayward et al. 1996a; Wang et al. 2001a). During the differentiation process, the solid epithelial buds (co-expressing both luminal and basal cell markers) develop into the distinct luminal and basal cell lineages with characteristic phenotypes and functional roles, each expressing their characteristic subset of cytokeratins and other distinct markers (Hayward et al. 1996a; Wang et al. 2001a). Thus the basal cell (or a population within) has been implicated as the putative prostatic stem cell (Signoretti et al. 2000; Wang et al. 2001a). The fourth epithelial cell type is the prostatic stem cell. The stem cell is important in two contexts. First, after androgen ablation the prostate regresses and massive prostatic epithelial cell apoptosis occurs. Readministration of testosterone leads to repopulation of the prostatic epithelium, presumably by proliferation of the residing stem cell population. Secondly, stem cells are long lived and are thought to collect a number of genetic lesions; thus they may be the progenitor cell of prostate cancer. To help elucidate the role of prostatic basal cells, prostates of p63 null mice were analyzed (Signoretti et al. 2000; Kurita et al. 2004). A homologue of the p53 tumor suppressor gene, p63, is expressed in basal cells of the adult prostate. It has been reported that p63 null mice, dying shortly after birth, have no prostate (Signoretti et al. 2000). However, prostatic rudiments (i.e., urogenital sinus) can be transplanted and grown in immuno-compromised mice (Kurita et al. 2004). Urogenital sinuses isolated from wild-type and p63 null mouse embryos both developed into prostate when grafted under the renal capsule of adult male nude mice, thus suggesting that p63 is not necessary for prostate development. Prostatic tissue that grew from p63 null UGS grafts contained both neuroendocrine and luminal cells, but basal cells were not observed. These data strongly suggest that p63 is essential for differentiation of prostatic basal cells, but that p63, and thus basal cells, are not essential for differentiation of prostatic neuroendocrine and secretory luminal epithelial cells. To evaluate the role of basal cells in prostatic biology, wild-type and p63 null urogenital sinuses were grown for 1 month under the kidney capsule of intact athymic male mice that were then were castrated. Regression of p63 null prostatic tissue was far more severe than that of wild-type prostates, with almost complete loss of ducts, resulting in the formation of residual cystic structures devoid of epithelium. However, subsequent testosterone replacement stimulated ductal growth and development in the residual p63 null prostatic tissue. These experiments suggest that basal cells play critical roles in maintaining ductal integrity and survival of luminal cells. Clearly, p63-positive basal stem cells are not required for the development and differentiation of luminal and neuroendocrine cells.

3
Prostate Cancer

3.1
The Role of Stroma in Carcinogenesis: a Historical Perspective

The idea that stroma may play a role in initiation, promotion, and maintenance of cancer has been considered for many years, and is based on pathology literature demonstrating that »*tumor stroma*« is commonly different from normal stroma (Bosman et al. 1993; Seljelid et al. 1999). It has been demonstrated that »*peritumoral dermis*« can be mitogenic on embryonic epidermis (Redler and Lustig 1968). Studies demonstrated neoplastic transformation of embryonic mouse submandibular gland epithelium by polyoma virus was observed when salivary gland epithelium was recombined with embryonic salivary gland mesenchyme, but not when isolated epithelium was grown by itself (Dawe 1972). This early work suggested a role of the stromal microenvironment in carcinogenesis. A pillar in the field of microenvironmental influences in cancer was the book, *Tissue Interactions in Carcinogenesis*, which described the contributions of the microenvironment to neoplastic growth (Tarin 1972). This collection of reviews emphasized that a perturbation of stromal–epithelial interactions may contribute to multiple stages of carcinogenesis and that the prolonged interaction of the carcinoma cell with its abnormal stromal microenvironment played an important role in the carcinogenic process.

Once cancer develops, the neoplastic epithelial cells may continue to be responsive to stromal influences that exacerbate the malignancy and thus may have possible therapeutic significance. In this regard, de Cosse et al. (1973, 1975) reported that mammary carcinoma cells exhibit a more orderly histodifferentiation and a lower proliferative rate when grown in association with normal mammary stroma. When basal cell carcinomas are recombined with normal stroma, the malignant cells differentiated with an apparent loss of their former malignant properties (Cooper and Pinkus 1977). Transitional cell carcinomas of the bladder recombined with urogenital sinus mesenchyme differentiated into prostate adenocarcinomatous structures (Fujii et al. 1982). In a similar fashion, colon carcinomas differentiated in response to rodent intestinal mesenchyme (Fukamachi et al. 1986, 1987). Perhaps most striking of all of these studies is the mesenchymal alteration of Dunning tumor epithelium.

The idea that paracrine interactions between normal mesenchyme and malignant carcinomas could possibly be used as a therapeutic strategy to inhibit tumorigenic growth was examined by combining rat Dunning prostatic adenocarcinoma R3327 epithelium with urogenital sinus mesenchyme or seminal vesicle mesenchyme. It should be noted that the Dunning tumor consists of carcinoma cells associated with an immortal stroma of uncertain origin that has been passaged along with the malignant epithelial cells through countless generations of serial transplantation. This Dunning tumor stroma is abnormal, being composed primarily of fibroblastic cells and containing little or no smooth muscle. Additionally, the basement membrane between the malignant epithelium and the stroma is discontinuous and/or excessively reduplicated (Wong et al. 1992). Thus, the stromal–epithelial interactions within the transplanted Dunning tumors are clearly abnormal. To determine whether malignant Dunning tumor epithelial cells might be modified by a »*normal*« stromal environment, Dunning tumor epithelial cells were grown for 1 month either alone or in combination with normal prostatic mesenchyme (Cunha 1972; Risbridger et al. 2001b). Grafts of Dunning tumor demonstrated the characteristic histopathology for prostate carcinomas, forming tumors containing small ducts lined by one or more layers of undifferentiated epithelial cells. In contrast, when Dunning tumor epithelium alone was recombined with urogenital sinus mesenchyme the Dunning tumor epithelium differentiated into tall columnar epithelial cells arranged in large cystic ducts (Hayashi et al. 1990, 1996; Hayashi and Cunha 1991; Tam et al. 1997). These changes in histodifferentiation of the Dunning tumor epithelium induced by stroma were associated with a marked decrease in tumorigenesis and a significantly lower proliferation rate in comparison to the parental Dunning tumor (Hayashi and Cunha 1991). Upon harvest of mesenchyme plus Dunning tumor epithelium tissue recombinants, the highly differentiated ducts within the tissue recombinants were grafted directly into new male hosts or were combined with fresh mesenchyme to form a second generation mesenchyme plus Dunning tumor epithelial tissue recombinant. Ducts previously induced by seminal vesicle mesenchyme and secondary seminal vesicle mesenchyme plus Dunning tumor epithelium tissue recombinants exhibited relatively little growth during 12 weeks in vivo, and maintained a highly differentiated state. Conversely, when grafts of Dunning tumor alone (Dunning tumor stroma plus Dunning tumor epithelium) were grown, they formed large (5 to 7 g) invasive tumors during the same time period. The highly differentiated state of the mesenchyme-induced Dunning tumor epithelium cells and diminishment of tumorigenesis were associated with a dramatic decrease in epithelial ^{3}H-thymidine labeling index (Hayashi and Cunha 1991). Significantly, smooth muscle cells, apparently derived from the added mesenchyme, were found in close association to the highly differentiated, relatively growth-quiescent Dunning tumor epithelium in these secondary mesenchyme plus Dunning tumor epithelium tissue recombinants (Wong et al. 1992).

Taken together, the above examples serve to emphasize the importance of stromal–epithelial interactions in carcinogenesis. Continued responsiveness of carcinoma cells, such as the Dunning tumor epithelial cells, to their stromal counterpart may provide a means of regulating both differentiation and proliferation of a carcinoma, perhaps to therapeutic benefit.

The paracrine effects of both normal and »*tumor stroma*« on differentiation of carcinoma cells implies an alteration in the cellular and extracellular matrix of the stroma immediately adjacent to carcinoma cells. These stromal changes have been observed for several other tumors (Yee et al. 1989; Chiquet-Ehrismann et al. 1986; Basset et al. 1993; Wright et al. 1994; Singer et al. 1995). For example, in breast tumors, the histopathology and growth kinetics of carcinoma-associated fibroblasts (fibroblasts found within a tumor) differ to that of fibroblasts associated with normal or benign breast epithelial cells (Ronnov-Jessen et al. 1996). Other observations attributed to carcinoma-associated fibroblasts include altered migratory behavior in vitro (Schor et al. 1988), alterations in cell surface molecules (Chaudhuri et al. 1975; Oishi et al. 1981), aberrant expression of a variety of growth factors (Yee et al. 1989; Yan et al. 1993; Ellis et al. 1994; Ponten et al. 1994; Frazier and Grotendorst 1997; Nakamura et al. 1997), expression of prostaglandin-synthesizing enzymes (Shattuck-Brandt et al. 1999; Pupa et al. 2002), and alterations in extracellular matrix (Werb et al. 1996; Pupa et al. 2002). While these phenotypic changes have been observed in carcinoma-associated fibroblasts from a variety of tumors, their role in tumor growth, progression, and differentiation are poorly understood.

3.2 Stromal–Epithelial Interactions During Prostate Cancer Progression

Many studies have analyzed epithelial or carcinoma cells during the various stages of carcinogenesis. This approach has yielded a plethora of information, especially at the genetic level. However, carcinogenesis is a complex process that can also be examined more broadly in the context of loss of homeostatic control over normal tissue architecture, nuclear atypia, genetic alterations, destruction of tissue boundaries, stromal changes, angiogenesis, and destruction of distant organs by metastatic cells. Indeed, from a more global perspective it is apparent that carcinogenesis is a disease of tissue organization resulting from deregulation of the finely orchestrated processes that determine how cells are integrated into normal tissues and tissues into organs (Ingber and Jamieson 1982). From this view several new models of prostatic carcinogenesis have emerged (Wang et al. 2000, 2001b; Hayward et al. 2001), which may be instrumental in deciphering the mechanisms of progression from normal cellular behavior to tumorigenesis and thence to metastasis.

Prostatic growth and development culminate in a mature gland composed of differentiated secretory luminal epithelial cells and highly differentiated contractile smooth muscle cells (see Fig. 2). In the absence of pathological perturbations, both the epithelium and stroma are essentially growth-quiescent in the adult prostate. It has been postulated that the highly differentiated growth-quiescent state of the adult prostate is maintained by reciprocal homeostatic interactions between epithelium and the smooth muscle stroma (Cunha et al. 2002). Thus, in adulthood, homeostasis is maintained through specific and reciprocal smooth muscle–epithelial interactions whose outcome is the maintenance of functional differentiation and growth quiescence of the mature gland. Thus it has been proposed that carcinogenesis occurs, in part, as a result of perturbation of these stromal–epithelial interactions. It is likely that prostatic carcinogenesis is initiated by genetic lesions present within the epithelial or stem cell population. Therefore, it is conceivable that perturbation of reciprocal homeostatic smooth muscle–epithelial interactions may induce de-differentiation of both the emerging prostatic carcinoma cells and the smooth muscle cells. Thus, following genetic insult to prostatic epithelium, the epithelial cells fail to signal appropriately to the adjacent stroma, which in turn begins to de-differentiate toward a more fibroblastic phenotype that stimulates carcinogenesis. As smooth muscle begins to de-differentiate into a fibroblastic cell type, signaling from this abnormal prostatic stroma to the epithelium becomes anomalous, resulting in progressive loss of control over epithelial differentiation and proliferation. Accordingly, a vicious cycle is established during prostate cancer progression in which both epithelium and

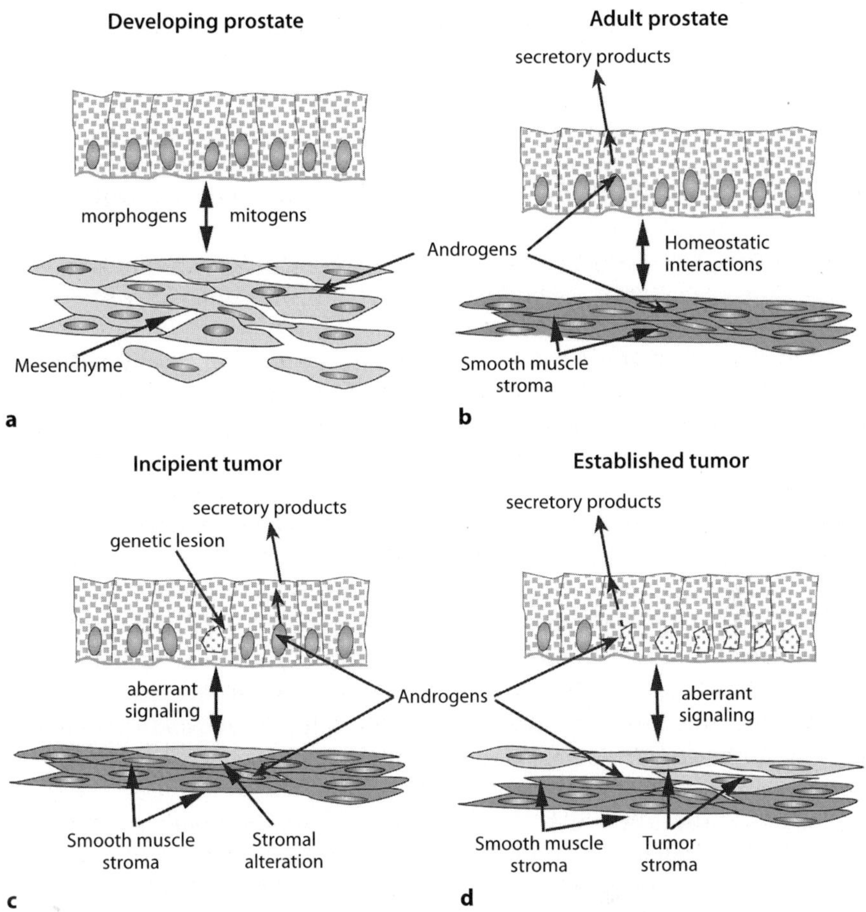

Fig. 2. Interactions between the epithelial and the stromal compartments of the prostate during growth and development and adulthood. In neonates prostatic development occurs with **a** low levels of androgens, acting through the mesenchymal androgen receptor, to stimulate proliferation and differentiation of the prostatic epithelium. Concurrently and reciprocally, epithelium induces differentiation of the fibroblastic mesenchyme to smooth muscle. In the normal adult (**b**), high levels of circulating testosterone, acting through stromally located androgen receptors, maintain the fully differentiated and growth-quiescent adult epithelium. Secretory function of luminal cells is elicited by direct epithelial androgenic stimulation of the differentiated columnar epithelium. A mature epithelial and smooth muscle stroma differentiated state is maintained by reciprocal paracrine-acting homeostatic factors. In early tumors it has been hypothesized that: following genetic insult to the epithelium or stem cell, **c** signaling between the epithelium and the underlying smooth muscle becomes abnormal, leading **d** to the formation of a fibroblastic »tumor stroma«, which responds to androgenic stimulation by producing paracrine-acting mitogens, stimulating a cycle of tumor proliferation and stromal dedifferentiation. (Redrawn with permission from Cunha et al. 2003).

stroma de-differentiate and proliferate in an uncontrolled fashion. In vivo the mature prostatic epithelium and smooth muscle are maintained under normal reciprocal interactions requiring intimate association of these two cell types. If adult prostatic epithelial and stromal cells are cultured separately in vitro, both the epithelial and smooth muscle cells rapidly de-differentiate and proliferate. This de-differentiation can be remedied in part by growing epithelial or smooth muscle cells on various extracellular matrices (Bissell and Hall 1987; Thyberg 1996). Thus, the intimate cell–cell and cell–extracellular matrix interactions seen in vivo are likely to be necessary for maintenance of the differentiated state found in the adult prostate.

The hypothesis that prostatic epithelium undergoes neoplastic transformation and loses its ability to maintain (and induce) smooth muscle differentiation was tested in a number of different experiments. This possibility was examined in experimental tissue recombinants in which various normal or neoplastic prostatic epithelia were grown in combination with embryonic prostatic stroma. It was found that only normal (non-neoplastic) epithelia were capable of inducing differentiation of prostatic smooth muscle in urogenital sinus mesenchyme (Hayward et al. 1996c). This observation was consistent with the notion that one aspect of carcinogenesis is the loss of the ability of the epithelium to induce and maintain smooth muscle differentiation. It is now apparent that smooth muscle–epithelial interactions are imperative for homeostatic growth and maintenance of the adult prostate, and that smooth muscle–epithelial interactions play key roles in regulating epithelial differentiation, proliferation, and carcinogenesis.

The mechanisms by which stromal cells promote tumorigenesis are poorly understood. Differential gene expression in normal stroma versus carcinoma-associated fibroblasts by factors affecting the local microenvironment have been shown to play key roles in carcinogenesis. In this regard, Tuxhorn et al. (2002) demonstrated that reactive stromal cells supported growth of tumors with enhanced angiogenesis when grafted subcutaneously. Stromal cells associated within tumors are known to produce a variety of enzymes, which may affect tumor initiation, growth, migration, angiogenesis, apoptosis, invasion, and metastasis (Lynch and Matrisian 2002). Future work on elucidating the cellular and molecular mechanisms of stromal–carcinoma cell interactions will undoubtedly provide the basis for new therapeutic strategies for regulating carcinoma growth and/or apoptosis.

3.3 The Role of Carcinoma-Associated Fibroblasts in the Progression of Human Prostate Cancer

As reviewed above, stromal–epithelial interactions play a critical role in normal development and adult function (Cunha et al. 1992). Experiments reviewed above demonstrated that normal stroma can slow the growth of epithelium from carcinomas and stimulate differentiation of cancer cells. The reciprocal experiment would be to evaluate non-tumorigenic epithelium recombined with fibroblasts from tumors. This experiment was performed by recombining human prostatic carcinoma-associated fibroblasts with the non-tumorigenic human prostatic epithelial cells (BPH-1; Olumi et al. 1999). In these experiments carcinoma-associated fibroblasts cells isolated from human prostate tumors and recombined with the non-tumorigenic SV40T human prostatic epithelial cell line BPH-1 (Hayward et al. 1995) promoted carcinogenesis, while normal prostatic fibroblasts did not. Human carcinoma-associated fibroblasts plus BPH-1 tissue recombinants formed large poorly differentiated invasive tumors (Olumi et al. 1999), while growth was meager and non-invasive in tissue recombinants composed of normal human prostatic fibroblasts plus BPH-1 cells. Thus, unlike normal adult prostatic stroma, carcinoma-associated fibroblast cells do not respond to androgens by inducing growth quiescence and differentiation, but instead stimulate epithelial proliferation and carcinogenesis. In experimental models in which BPH-1 cells were combined with normal neonatal stroma, epithelial growth was also stimulated, but tumorigenesis did not ensue. This observation suggests that simple stimulation of epithelial proliferation is not the single determinant in carcinoma-associated fibroblast-induced promotion of tumorigenesis (Wang et al. 2001b). The inference is that stromal cells surrounding a human prostate carcinoma stimulate epithelial proliferation and dedifferentiation, while normal stroma promotes differentiation and inhibits epithelial growth.

The neoplastic process promoted by carcinoma-associated fibroblasts in the parental non-tumorigenic BPH-1 cells involved changes in gene expression and further genetic lesions, suggesting that carcinoma-associated fibroblast cells play an active, dynamic role in tumor progression. Specifically, human prostatic carcinoma-associated fibroblasts induced carcinogenesis and elicited a characteristic pattern of genetic change associated with malignant progression (Hayward et al. 2001; Phillips et al. 2001). A comparison of BPH-1 epithelial cells, induced to become tumorigenic by exposure to carcinoma-associated fibroblasts cells (BPH-1-CAFTD) versus BPH-1 cells, induced to become cancerous as a result of hormonal carcinogenesis (BPH-1-TETD; Wang et al. 2001b), revealed multiple common differences as well as changes unique to the mode of cancer induction. BPH-1-CAFTD and the BPH-1-TETD cells share chromosomal amplifications at 11q and 20q and chromosomal losses at 3q, 8p, and 10p. Conversely, the cells induced by carcinoma-associated fibroblasts contain multiple non-reciprocal translocations and a number of complex harlequin chromosomes based on chromosome 7, while BPH-1-TETD cells that underwent cancer progression were much more likely to contain reciprocal translocations and show harlequin chromosomes based on chromosome 5 (Hayward et al. 2001; Phillips et al. 2001). These data support the contention that interactions between the epithelium and the surrounding stroma within a given tumor influence genetic changes across tissue layer boundaries.

Others laboratories have examined stromal–epithelial interactions using immortalized fibroblasts or tumorigenic fibroblastic cell lines (Picard et al. 1986; Camps et al. 1990; Yan et al. 1993) rather than the mortal fibroblasts used in the aforementioned study (Olumi et al. 1999). Studies using immortalized fibroblasts or tumorigenic fibroblastic cell lines have indicated that such fibroblasts promote tumorigenesis when co-inoculated with carcinoma cells of proven tumorigenicity. Additionally, studies evaluating the effects of sarcomal cell lines plus immortalized prostatic epithelial cells showed the development of adenosarcomas (Chung et al. 1989). Lastly, experiments performed evaluating the induction of bladder carcinogenesis utilizing tissue recombinants composed of carcinogen-treated epithelium plus carcinogen-treated bladder stroma (Momose et al. 1990; Uchida et al. 1990) are relevant to the hypothesis that stromal factors stimulate carcinogenesis. In those studies carcinogen-treated stromal cells stimulated carcinoma formation. Taken together these models firmly establish the idea that carcinoma-associated fibroblasts can play a central role in progression to full malignancy of initiated but non-tumorigenic epithelial cells.

The mechanisms by which »*tumor stroma*« promote epithelial tumorigenesis are not well defined, even though it is clear that neoplastic changes in the epithelium are frequently associated with concomitant alterations in the stroma. Thus, altered stromal–epithelial interactions, specifically by altering the stroma, fosters progression to malignancy. Future work on the cellular and molecular mechanisms of interaction between stromal microenvironment and carcinoma cells may provide new therapeutic strategies for regulating carcinoma growth and/or apoptosis to the benefit of patients suffering from cancer.

3.4 The Role of Steroid Hormones and Stroma in Prostate Cancer Progression

Epidemiologic studies have demonstrated a pivotal role for steroid hormones in cancer of the prostate. Because the prostate is a target for sex steroids, both testosterone and estrogen have been implicated in prostatic carcinogenesis by certain observations. First, prostate cancer does not occur in eunuchs, castrated early in life. Second, in species susceptible to prostate cancer (men and dogs), androgen secretion and plasma testosterone levels decline with age, while plasma-free estradiol levels are unchanged or elevated late in life when prostate cancer develops. Thus, elevation in the estrogen:testosterone ratio is temporally related to the development of benign prostatic hyperplasia and prostate cancer (Brendler et al. 1983; Hayes et al. 1992). Third, African Americans, who have the highest incidence of prostatic cancer, have high levels of both plasma-free testosterone and estrogen (Ross et al. 1986). Fourth, in combination with estrogens, testosterone induces prostatic hyperplasia and dysplasia in dogs and prostate cancer in rodents. Thus, androgens alone and in combination with estrogens play a key role in cancer of the prostate.

3.5 Androgens and Prostate Cancer

As discussed above, androgens are necessary for normal prostate development. Many androgenic responses in normal prostatic epithelial cells are mediated through stromal androgen receptors. To elucidate the role of stromal androgen receptors in prostate cancer progression, tissue recombinants were prepared with rodent urogenital sinus mesenchyme plus BPH-1 cells (Wang et al. 2001b). Androgen receptors are not detected in BPH-1 prostatic epithelial cells by immunohistochemical and RT-PCR techniques (Hayward et al. 1995), but are found within urogenital sinus mesenchyme (Cooke et al. 1991a; Takeda and Chang 1991). Estrogen receptor-α is also not found within BPH-1 cells but are expressed within urogenital sinus mesenchyme (Cooke et al. 1991b; Prins and Birch 1997; Cunha et al. 2003). BPH-1 cells grown in untreated or testosterone- plus estrogen-treated male hosts survive, but do not become neoplastic, presumably because they are not tumorigenic and possibly due to the absence of androgen receptors. When BPH-1 cells are combined with urogenital sinus mesenchyme and grown in untreated male mice, these tissue recombinants undergo prostatic development and form branched solid epithelial cords that canalize into ducts. However, when urogenital sinus mesenchyme plus BPH-1 tissue recombinants are grown in testosterone plus estrogen-treated hosts, BPH-1 cells undergo malignant transformation even without epithelial androgen receptors or estrogen receptors (Hayward et al. 1995; Wang et al. 2001b). Because androgen receptors and estrogen receptor-α are only expressed within the stroma of these tissue recombinants, these findings suggest an essential role of stromal hormone receptors in hormonal carcinogenesis of the prostate. Hence, hormonal carcinogenesis of the prostate appears to involve paracrine mechanisms, as is the case for many steroidal effects during normal prostatic development (see above).

3.6 Estrogens and Prostate Cancer

As discussed above, estrogens play an important role in prostate biology and histopathology. Estrogens act through estrogen receptor-α and estrogen receptor-β, which are both found in the prostate (Kuiper et al. 1996; Couse and Korach 1999). While epithelial estrogen receptor-α in the prostate is difficult to detect by immunohistochemical methods, estrogen receptor-α has been consistently detected in prostatic stroma (Prins and Birch 1997; Lau et al. 1998) and is also expressed within squamous metaplastic epithelium (Adams et al. 2002). Estrogen receptor-β is localized within both prostatic epithelium and stroma in the developing prostate, but in adulthood this receptor is found primarily in prostatic epithelial cells (Adams et al. 2002). The role of estrogens in the normal and diseased prostate is an important but poorly understood area. In wild-type mice treated with testosterone plus estrogen the prostate undergoes hyperplasia, dysplasia, and carcinoma in situ (Wang et al. 2001c). Induction of hormonal carcinogenesis by testosterone plus estrogen in wild-type mice implies that the underlying molecular mechanisms may be elucidated using α- and β-estrogen-receptor-knockout mice. Testosterone plus estrogen does not elicit prostatic hyperplasia or dysplasia in α-estrogen receptor knockout mice, suggesting a key role of estrogen receptor α in hormonal carcinogenesis (Wang et al. 2001c). Estrogen receptor-β knockout mice (which express estrogen receptor-α) developed prostatic hyperplasia, dysplasia, and carcinoma in situ comparable to that seen in wild-type mice when treated with testosterone plus estrogen. Additionally, mice that lack estrogen receptor-β have been reported to develop prostate cancer (Weihua et al. 2001), suggesting that estrogen receptor-β is a putative tumor-suppressor gene. Taken together, these data emphasize the importance of signaling studies using α-estrogen receptor knockout or β-estrogen receptor knockout mice for the elucidation of hormone action in prostate cancer progression.

3.7 Mechanisms of Microenvironmental Influences During Carcinogenesis

Research into mechanisms by which tumor microenvironment promotes progression from benign to tumorigenic is being actively pursued on many fronts. A frequent event that occurs during carcinogenesis is a major change in the cellular and extracellular composition of the stroma. In the rat (Wong and Tam 2002) and human prostate, this involves replacement of smooth muscle cells by a fibroblast or myofibroblast population, either through changes in the differentiation of smooth muscle cells or by recruitment of fibroblasts to the tumor site. In invasive breast cancer, stromal fibroblasts are replaced by, or differentiate into, abnormal myofibroblasts (Ronnov-Jessen et al. 1996). In other models, change in the cellular composition of the stroma involves an infiltration of inflammatory cells that replace the normal resident stromal cells. This is also the case in epidermal carcinogenesis in transgenic mice expressing the E6 and E7 oncogenes of human papilloma virus driven by the keratin-14 promoter. This transgenic mouse model results in a reproducible multistage carcinogenesis with a histopathology of invasive squamous cell carcinomas of the epidermis (Arbeit et al. 1994; Coussens et al. 1996). In this model, morphometric analysis has revealed replacement of resident dermal cells by mast cells and neutrophils, particularly in areas of hyperplasia, dysplasia, and in invasive fronts of carcinomas. Infiltrating neutrophils and mast cells produce an array of cytokines and growth factors capable of stimulating epithelial cells. Release of enzymes such as matrix metalloproteinases (MMPs), including MMP-9 (gelatinase B), by these inflammatory cells promoted carcinogenesis, presumably by stimulating growth factor pathways and/or other MMP pathways (van Kempen et al. 2002). Additionally, in the RIP-TAg transgenic mouse model of pancreatic carcinogenesis, based on targeted expression of SV40 T antigen via the rat insulin promoter, the resultant carcinomas depend on stromal MMP-9 to trigger the angiogenic switch that increases the bioavailability of vascular endothelial growth factor and propels tumor growth (Bergers et al. 2000). These experiments emphasize the importance of stromal factors, especially proteases, in cancer progression.

The stromal microenvironment surrounding tumor cells is 3-dimensional and consists of fibroblasts, myofibroblasts, vascular cells, and/or inflammatory cells. Moreover, the extracellular matrix contains both bound and soluble factors and serves as a reservoir for numerous growth factors. Cell–extracellular matrix interactions play key roles in both homeostasis and pathogenesis. It has been well established that these contacts influence tissue differentiation, proliferation, homeostasis, migration, function, and tissue specificity (Bissell et al. 2002). MMPs constitute a large family of enzymes consisting of over 25 members, which collectively can degrade all components of the extracellular matrix. Proper maintenance of extracellular matrix turnover requires strict control of MMP expression, activity, and enzyme degradation such that an equilibrium is established. Disease states such as cancer and arthritis are often associated with high levels of MMP activity at the epithelial–stromal interface, which is important in the rearrangement of the extracellular matrix required for tumor cell dissemination. While MMPs may be produced by the tumor cells themselves, it has become increasingly clear that MMPs are also produced by the stromal cells associated with carcinoma cells (Lynch and Matrisian 2002). The latter increase in MMP activity may be due to increased EMMPRIN production by tumor cells. EMMPRIN also known as CD147 or basigin, is an extracellular MMP inducer that is thought to play a key role in carcinoma expansion into the surrounding stroma. MMPs produced by tumor cells or the surrounding stromal fibroblasts and inflammatory cells have been implicated in a wide range of processes such as tumor initiation, growth, migration, angiogenesis, the selection of apoptosis resistant subpopulations, and in invasion and metastasis (Lynch and Matrisian 2002). While MMP activity may exacerbate carcinogenesis, the MMP status within a tumor may also restrict tumor aggressiveness – and thus inhibitors of MMPs are potential therapeutics.

In xenograft studies, MMPs of stromal origin have been shown to restrict tumor aggressiveness in that MMP-11 null fibroblasts do not support in vivo growth of breast cancer cells, whereas wild-type fibroblasts do (Masson et al. 1998). In a similar fashion, intestinal carcinogenesis is suppressed in mice lacking the metalloproteinase matrilysin (MMP-7; Wilson et al. 1997). Moreover, injection of cancer cells into the circulation of MMP-2 or MMP-9 null mice were less capable of establishing metastatic foci in their lungs than in those of wild-type mice (Itoh et al. 1998, 1999). In contrast, mice that overexpress MMP-1 or MMP-7 develop hyperproliferative disease and have an increased cancer susceptibility (Egeblad and Werb 2002). Targeted expression of MMP-3 (stromelysin) to mammary epithelium leads to dysregulation of the E-cadherin–β-catenin complex, epithelial–mesenchymal transition, and mammary carcinogenesis (Sternlicht et al. 1999). Thus, modification of the extracellular matrix via epithelial or stromal MMPs can profoundly affect the carcinogenic process. For decades, it has been appreciated that activated tumor stromal cells can promote growth of normal and neoplastic epithelium (Billingham et al. 1951; Redler and Lustig 1968; Gabbert 1985; Cornil et al. 1991; Gregoire and Lieubeau 1995), which suggests genetic or functional alterations within tumor stroma (Turner et al. 1997; Olumi et al. 1999). More recently, studies evaluating epidermal carcinogenesis using immortalized but non-tumorigenic human epidermal cells (HaCaT) revealed a complex interplay of growth-factor-signaling pathways that operate during the carcinogenic process. In these studies it has been demonstrated that platelet-derived growth factor produced by the epidermal cells induced the fibroblasts to produce keratinocyte growth factor (FGF-7), which in turn promoted carcinogenesis (Brauchle et al. 1994). Thus, an emerging picture is that tumor and stromal cells interact via a complex arrangement of diffusible paracrine factors resulting in reciprocal modulation of cell proliferation and differentiation that facilitates tumorigenesis in a permissive, tumor-promoting stromal microenvironment.

The interplay of growth-promoting signals is no less complicated in normal prostate and during conversion from benign to invasive growth. In normal prostate, the testicular feminization/wild-type experiments described above revealed that androgenic induction of prostatic epithelial growth does not require epithelial androgen receptors and is instead stimulated by paracrine growth factors (*»andromedins«*) produced by androgen-receptor-positive mesenchyme/stroma. Two members of the FGF family, FGF-7 and FGF-10, have been extensively studied for their function in the paracrine regulation of epithelial growth (Thomson 2001). In the prostate, FGF-7 and FGF-10 are stromally derived. The FGF receptor (FGFR2iiib) is found exclusively within the prostate epithelium (McKeehan et al. 1991, 1998; Thomson 2001). Thus, the spatial pattern of expression of FGFs and FGF receptors in the prostate is consistent with their role as an *»andromedin«* (Yan et al. 1992). Evidence supporting an important paracrine role of FGF-7 and FGF-10 in epithelial growth is as follows. Ribonucleic acid transcripts for both FGF-7 and FGF-10 are expressed in normal prostatic stroma, whereas localization of FGFR2iiib is found exclusively within the epithelium. Antibody neutralization experiments of endogenous prostatic FGF-7 inhibits androgen-stimulated prostatic epithelial growth and ductal branching morphogenesis. Alternatively, addition of exogenous FGF-7 and FGF-10 largely overcomes the requirement for testosterone during prostatic epithelial growth and ductal-branching morphogenesis. Although these data are circumstantial in light of data showing that FGF-

7 is not androgen regulated, it is important to note that while FGF-7 and FGF-10 are important paracrine mediators of androgen signaling on normal prostatic epithelium, androgenic stimulation of prostatic epithelial cells following malignant transformation appears to involve conversion from a paracrine to an autocrine mechanism of androgen-regulated growth (Gao et al. 2001). The molecular mechanism for this switch from paracrine to autocrine androgen regulation within prostatic epithelial cells that occurs during tumor progression is conceivably due to changes in the FGF axis. This idea is reinforced in the rat Dunning prostatic tumor system (see above), where FGFR2 in the epithelium switches from the FGFR2iiib isoform (that preferentially binds FGF-7 and FGF-10) to FGFR2iiic (that preferentially binds FGF-2). In parallel to this, FGF-2, FGF-3, and FGF-5 mRNA are expressed by the epithelium, creating an autocrine loop, which is associated with progression to a more malignant phenotype (Yan et al. 1993). Likewise, in clinical prostate cancer specimens, FGF-7 expression can switch from the stroma to epithelium (McGarvey and Stearns 1995). In the transgenic adenocarcinoma of the mouse prostate model, FGF-7 and FGF-10 are inappropriately expressed by epithelial cells concomitant with a change in the expression of FGF receptor from the FGFR2iiib isoform to the FGFR1 isoform (Foster et al. 1997, 2002). Finally, senescence of the stroma has recently been implicated as playing an important role in carcinogenesis. It has been postulated that the ability of stroma to instruct and maintain epithelial function is altered with age. Aged stromal cells secrete a number of different factors that can disrupt epithelial tissue architecture and/or stimulate epithelial proliferation. Thus, senescent fibroblasts are able to create a stromal microenvironment that can promote epithelial neoplasia (Krtolica and Campisi 2002).

4 Conclusion

The recurring theme emphasized in this chapter is that reciprocal interactions between the epithelium and the connective tissue stroma play a key role in both normal development and carcinogenesis. Through these cell–cell interactions epithelial and stromal morphogenesis, growth, differentiation, and function are elicited during normal development and progression to malignancy. In adulthood normal epithelial–stromal interactions are required to maintain the highly differentiated, growth-quiescent and functional state of both the epithelium and stroma. Carcinogenesis and tumor progression are misregulated in part when perturbation of stromal–epithelial interactions occur, this is turn fosters progression to malignancy. Prostatic epithelial–mesenchymal interactions during development and epithelial–stromal interactions in the mature gland are mediated by differential regulation of growth/differentiation factors and proteases that modulate the local microenvironment. As a result of altered cell–cell and cell–extracellular matrix communications, progression to malignancy is favored. Stromal cells associated with cancerous cells profoundly affect a plethora of processes that include tumor initiation, growth, migration, angiogenesis, apoptosis, invasion, malignant transformation, and metastasis (Lynch and Matrisian 2002). Future studies on the underlying mechanisms of interactions between stromal microenvironment and carcinoma cells may provide new therapeutic strategies for halting or diminishing carcinoma growth and accelerating apoptosis to the benefit of patients suffering from cancer.

References

Adams JY, Leav I, Lau KM, Ho SM, Pflueger SM (2002) Expression of estrogen receptor beta in the fetal, neonatal, and prepubertal human prostate. Prostate 52:69–81

Arbeit JM, Munger K, Howley PM, Hanahan D (1994) Progressive squamous epithelial neoplasia in K14-human papillomavirus type 16 transgenic mice. J Virol 68:4358–4368

Basset P, Wolf C, Chambon P (1993) Expression of the stromelysin-3 gene in fibroblastic cells of invasive carcinomas of the breast and other human tissues: a review. Breast Cancer Res Treat 24:185–193

Bergers G, Brekken R, McMahon G, Vu TH, Itoh T, Tamaki K, Tanzawa K, Thorpe P, Itohara S, Werb Z, Hanahan D (2000) Matrix metalloproteinase-9 triggers the angiogenic switch during carcinogenesis. Nat Cell Biol 2:737–744

Bhatia-Gaur R, Donjacour AA, Sciavolino PJ, Kim M, Desai N, Young P, Norton CR, Gridley T, Cardiff RD, Cunha GR, Abate-Shen C, Shen MM (1999) Roles for Nkx3.1 in prostate development and cancer. Genes Dev 13:966–977

Billingham R, Orr J, Woodhouse D (1951) Transplantation of skin components during chemical carcinogenesis with 20-methylcholanthrene. Br J Cancer 5:417–432

Bissell MJ, Hall HG (1987) Form and function in the mammary gland: The role of extracellular matrix. In: Neville MC, Daniel CW (eds) The mammary gland: development, regulation, and function. Plenum Press, New York, pp 97–146

Bissell MJ, Radisky DC, Rizki A, Weaver VM, Petersen OM (2002) The organizing principal: microenvironmental influences in the normal and malignant breast. Differentiation 60:537–546

Bosman FT, de Bruine A, Flohil C, van der Wurff A, ten Kate J, Dinjens WW (1993) Epithelial-stromal interactions in colon cancer. Int J Dev Biol 37:203–211

Brauchle M, Angermeyer K, Hubner G, Werner S (1994) Large induction of keratinocyte growth factor expression by serum growth factors and pro-inflammatory cytokines in cultured fibroblasts. Oncogene 9:3199–3204

Brendler CB, Berry SJ, Ewing LL, McCullough AR, Cochran RC, Strandberg JD, Zirkin BR, Coffey DS, Wheaton LG, Hiler ML, Bordy MJ, Niswender GD, Scott WW, Walsh PC (1983) Spontaneous benign prostatic hyperplasia in the beagle. Age-associated changes in serum hormone levels, and the morphology and secretory function of the canine prostate. J Clin Invest 71:1114–1123

Buchanan DL, Kurita T, Taylor JA, Lubahn DL, Cunha GR, Cooke PS (1998a) Role of stromal and epithelial estrogen receptors in vaginal epithelial proliferation, stratification and cornification. Endocrinology 139:4345–4352

Buchanan DL, Setiawan T, Lubahn DL, Taylor JA, Kurita T, Cunha GR, Cooke PS (1998b) Tissue compartment-specific estrogen receptor participation in the mouse uterine epithelial secretory response. Endocrinology 140:484–491

Camps JL, Chang SM, Hsu TC, Freeman MR, Hong SJ, Zhau HE, von Eschenbach AC, Chung LW (1990) Fibroblast-mediated acceleration of human epithelial tumor growth in vivo. Proc Natl Acad Sci USA 87:75–79

Chang WY, Wilson MJ, Birch L, Prins GS (1999) Neonatal estrogen stimulates proliferation of periductal fibroblasts and alters the extracellular matrix composition in the rat prostate. Endocrinology 140:405–415

Chaudhuri S, Koprowska I, Rowinski J (1975) Different agglutinability of fibroblasts underlying various precursor lesions of human uterine cervical carcinoma. Cancer Res 35:2350–2354

Chiquet-Ehrismann R, Mackie EJ, Pearson CA, Sakakura T (1986) Tenascin: an extracellular matrix protein involved in tissue interactions during fetal development and oncogenesis. Cell 47:131–139

Chung LW, Chang SM, Bell C, Zhau HE, Ro JY, von Eschenbach AC (1989) Co-inoculation of tumorigenic rat prostate mesenchymal cells with non- tumorigenic epithelial cells results in the development of carcinosarcoma in syngeneic and athymic animals. Int J Cancer 43:1179–1187

Cohen RJ, Glezerson G, Taylor LF, Grundle HA, Naude JH (1993) The neuroendocrine cell population of the human prostate gland. J Urol 150:365–368

Cooke P, Buchanan D, Young P, Setiawan T, Brody J, Korach K, Taylor J, Lubahn D, Cunha G (1997) Stromal estrogen receptors (ER) mediate mitogenic effects of estradiol on uterine epithelium. Proc Natl Acad Sci USA 94:6535–6540

Cooke PS, Young P, Cunha GR (1991a) Androgen receptor expression in developing male reproductive organs. Endocrinology 128:2867–2873

Cooke PS, Young P, Hess RA, Cunha GR (1991b) Estrogen receptor expression in developing epididymis, efferent ductules and other male reproductive organs. Endocrinology 128:2874–2879

Cooper M, Pinkus H (1977) Intrauterine transplantation of rat basal cell carcinoma: a model for reconversion of malignant to benign growth. Cancer Res 37:2544–2552

Cornil I, Theodorescu D, Man S, Herlyn M, Jambrosic J, Kerbel RS (1991) Fibroblast cell interactions with human melanoma cells affect tumor cell growth as a function of tumor progression. Proc Natl Acad Sci USA 88:6028—6032

Couse JF, Korach KS (1999) Estrogen receptor null mice: what have we learned and where will they lead us? Endocr Rev 20:358–417

Couse JF, Lindzey J, Grandien K, Gustafsson JA, Korach KS (1997) Tissue distribution and quantitative analysis of estrogen receptor-alpha (ERalpha) and estrogen receptor-beta (ERbeta) messenger ribonucleic acid in the wild-type and ERalpha-knockout mouse. Endocrinology 138:4613–4621

Coussens LM, Hanahan D, Arbeit JM (1996) Genetic predisposition and parameters of malignant progression in K14-HPV16 transgenic mice. Am J Pathol 149:1899–1917

Cunha GR (1972) Epithelio-mesenchymal interactions in primordial gland structures which become responsive to androgenic stimulation. Anat Rec 172:179–196

Cunha GR (1984) Androgenic effects upon prostatic epithelium are mediated via trophic influences from stroma. In: Kimball FA, Buhl AE, Carter DB (eds) New approaches to the study of benign prostatic hyperplasia. Liss, New York, pp 81–102

Cunha GR (1994) Role of mesenchymal-epithelial interactions in normal and abnormal development of the mammary gland and prostate. Cancer 74:1030–1044

Cunha GR, Young P (1991) Inability of Tfm (testicular feminization) epithelial cells to express androgen-dependent seminal vesicle secretory proteins in chimeric tissue recombinants. Endocrinology 128:3293–3298

Cunha GR, Reese BA, Sekkingstad M (1980) Induction of nuclear androgen-binding sites in epithelium of the embryonic urinary bladder by mesenchyme of the urogenital sinus of embryonic mice. Endocrinology 107:1767–1770

Cunha GR, Fujii H, Neubauer BL, Shannon JM, Sawyer LM, Reese BA (1983a) Epithelial-mesenchymal interactions in prostatic development. I. Morphological observations of prostatic induction by urogenital sinus mesenchyme in epithelium of the adult rodent urinary bladder. J Cell Biol 96:1662–1670

Cunha GR, Sekkingstad M, Meloy BA (1983b) Heterospecific induction of prostatic development in tissue recombinants prepared with mouse, rat, rabbit, and human tissues. Differentiation 24:174–180

Cunha GR, Donjacour AA, Cooke PS, Mee S, Bigsby RM, Higgins SJ, Sugimura Y (1987) The endocrinology and developmental biology of the prostate. Endocr Rev 8:338–362

Cunha GR, Alarid ET, Turner T, Donjacour AA, Boutin EL, Foster BA (1992) Normal and abnormal development of the male urogenital tract: role of androgens, mesenchymal-epithelial interactions and growth factors. J Androl 13:465–475

Cunha GR, Hayward SW, Wang Y-Z (2002) Role of stroma in carcinogenesis of the prostate. Differentiation 60:473—485

Cunha GR, Hayward SW, Wang YZ, Ricke WA (2003) Role of the stromal microenvironment in carcinogenesis of the prostate. Int J Cancer 107:1–10

Dawe CJ (1972) Epithelial-mesenchymal interactions in relation to the genesis of polyoma virus-induced tumours of mouse salivary gland. In: Tarin D (ed) Tissue interactions in carcinogenesis. Academic Press, London, pp 305–358

De Cosse J, Gossens CL, Kuzma JF (1973) Breast cancer: induction of differentiation by embryonic tissue. Science 181:1057–1058

De Cosse JJ, Gossens CL, Kuzma JF, Unsworth BR (1975) Embryonic inductive tissues that cause histological differentiation of murine mammary carcinoma in vitro. J Natl Cancer Inst 54:913–921

Donjacour AA, Cunha GR (1988) Seminal vesicle mesenchyme induces prostatic morphology and secretion in urinary bladder epithelium. J Cell Biol 107:609a

Donjacour AA, Cunha GR (1993) Assessment of prostatic protein secretion in tissue recombinants made of urogenital sinus mesenchyme and urothelium from normal or androgen-insensitive mice. Endocrinology 131: 2342–2350

Donjacour AA, Cunha GR (1995) Induction of prostatic morphology and secretion in urothelium by seminal vesicle mesenchyme. Development 121:2199–2207

Donjacour AA, Thomson AA, Cunha GR (2003) FGF-10 plays an essential role in the growth of the fetal prostate. Dev Biol 261:39–54

Egeblad M, Werb Z (2002) New functions for the matrix metalloproteinases in cancer progression. Nat Rev Cancer 2:161–174

Ellis MJ, Singer C, Hornby A, Rasmussen A, Cullen KJ (1994) Insulin-like growth factor mediated stromal-epithelial interactions in human breast cancer. Breast Cancer Res Treat 31:249–261

Foster BA, Gingrich JR, Kwon ED, Madias C, Greenberg NM (1997) Characterization of prostatic epithelial cell lines derived from transgenic adenocarcinoma of the mouse prostate (TRAMP) model. Cancer Res 57:3325–3330

Foster BA, Evangelou A, Gingrich JR, Kaplan PJ, DeMayo F, Greenberg NM (2002) Enforced expression of FGF-7 promotes epithelial hyperplasia whereas a dominant negative FGFR2iiib promotes the emergence of neuroendocrine phenotype in prostate glands of transgenic mice. Differentiation 60:624–632

Frazier KS, Grotendorst GR (1997) Expression of connective tissue growth factor mRNA in the fibrous stroma of mammary tumors. Int J Biochem Cell Biol 29:153–161

Fujii H, Cunha GR, Norman JT (1982) The induction of adenocarcinomatous differentiation in neoplastic bladder epithelium by an embryonic prostatic inductor. J Urol 128:858–861

Fukamachi H, Mizuno T, Kim YS (1986) Morphogenesis of human colon cancer cells with fetal rat mesenchymes in organ culture. Experientia 42:312–315

Fukamachi H, Mizuno T, Kim YS (1987) Gland formation of human colon cancer cells combined with foetal rat mesenchyme in organ culture: an ultrastructural study. J Cell Sci 87:615–621

Gabbert H (1985) Mechanisms of tumor invasion: evidence from in vivo observations. Cancer Metastasis Rev 4:293–309

Gao J, Arnold JT, Isaacs JT (2001) Conversion from a paracrine to an autocrine mechanism of androgen- stimulated growth during malignant transformation of prostatic epithelial cells. Cancer Res 61:5038–5044

Gregoire M, Lieubeau B (1995) The role of fibroblasts in tumor behavior. Cancer Metastasis Rev 14:339–350

Hayashi N, Cunha GR (1991) Mesenchyme-induced changes in neoplastic characteristics of the Dunning prostatic adenocarcinoma. Cancer Res 51:4924–4930

Hayashi N, Cunha GR, Wong YC (1990) Influence of male genital tract mesenchymes on differentiation of Dunning prostatic adenocarcinoma. Cancer Res 50:4747–4754

Hayashi N, Cunha GR, Parker M (1993) Permissive and instructive induction of adult rodent prostatic epithelium by heterotypic urogenital sinus mesenchyme. Epithelial Cell Biol 2:66–78

Hayashi N, Tsuji M, Sugimura Y, Kawamura J, Cunha GR (1996) Change in the morphological and functional cytodifferentiation induced by seminal vesicle mesenchyme in cell suspensions of rat Dunning prostatic adenocarcinoma cells. Int J Cancer 68:788–794

Hayes RB, de Jong FH, Raatgever J, Bogdanovicz J, Schroeder FH, van der Maas P, Oishi K, Yoshida O (1992) Physical characteristics and factors related to sexual development and behaviour and the risk for prostatic cancer. Eur J Cancer Prev 1:239–245

Hayward S, Wang Y, Cao M, Hom Y, Zhang B, Grossfeld G, Sudilovsky D, Cunha G (2001) Malignant transformation in a non-tumorigenic human prostatic epithelial cell line. Cancer Res 61:8135–8142

Hayward SW, Baskin LS, Haughney PC, Cunha AR, Foster BA, Dahiya R, Prins GS, Cunha GR (1996a) Epithelial development in the rat ventral prostate, anterior prostate and seminal vesicle. Acta Anat 155:81–93

Hayward SW, Baskin LS, Haughney PC, Foster BA, Cunha AR, Dahiya R, Prins GS, Cunha GR (1996b) Stromal development in the ventral prostate, anterior prostate and seminal vesicle of the rat. Acta Anat 155:94–103

Hayward SW, Cunha GR, Dahiya R (1996c) Normal development and carcinogenesis of the prostate: a unifying hypothesis. Ann NY Acad Sci 784:50–62

Hayward SW, Dahiya R, Cunha GR, Bartek J, Despande N, Narayan P (1995) Establishment and characterization of an immortalized but non-tumorigenic human prostate epithelial cell Line: BPH-1. In Vitro 31A:14–24

Hayward SW, Rosen MA, Cunha GR (1997) Stromal-epithelial interactions in normal and neoplastic prostate. Br J Urol 79 [Suppl 2]:18–26

Hayward SW, Haughney PC, Rosen MA, Greulich KM, Weier HU, Dahiya R, Cunha GR (1998) Interactions between adult human prostatic epithelium and rat urogenital sinus mesenchyme in a tissue recombination model. Differentiation 63:131–140

Higgins SJ, Young P, Brody JR, Cunha GR (1989) Induction of functional cytodifferentiation in the epithelium of tissue recombinants. I. Homotypic seminal vesicle recombinants. Development 106:219–234

Ingber DE, Jamieson JD (1982) Tumor formation and malignant invasion: role of basal lamina. In: Liotta L, Hart I (eds) Tumor invasion and metastasis. Nijhoff, The Hague, pp 335–357

Isaacs JT (1984) Antagonistic effect of androgen on prostatic cell death. Prostate 5:545–557

Itoh T, Tanioka M, Yoshida H, Yoshioka T, Nishimoto H, Itohara S (1998) Reduced angiogenesis and tumor progression in gelatinase A-deficient mice. Cancer Res 58:1048–1051

Itoh T, Tanioka M, Matsuda H, Nishimoto H, Yoshioka T, Suzuki R, Uehira M (1999) Experimental metastasis is suppressed in MMP-9-deficient mice. Clin Exp Metastasis 17:177–181

Kinbara H, Cunha GR (1995) Ductal heterogeneity in rat dorsal-lateral prostate. Prostate 28:58–64

Krtolica A, Campisi J (2002) Cancer and aging: a model for the cancer promoting effects of the aging stroma. Int J Biochem Cell Biol 34:1401–1414

Kuiper GG, Enmark E, Pelto-Huikko M, Nilsson S, Gustafsson JA (1996) Cloning of a novel receptor expressed in rat prostate and ovary. Proc Natl Acad Sci USA 93:5925–5930

Kurita T, Young P, Brody J, Lydon JP, O'Malley BW, Cunha GR (1998) Stromal progesterone receptors mediate the inhibitory effects of progesterone on estrogen-induced uterine epithelial cell (UtE) proliferation. Endocrinology 139:4708–4713

Kurita T, Wang Y-Z, Donjacour AA, Zhao C, Lydon JP, O'Malley BP, Isaacs JT, Dahiya R, Cunha GR (2001) Paracrine regulation of apoptosis by steroid hormones in the male and female reproductive system. Cell Death Differ 8:192–200

Kurita T, Medina RT, Mills AA, Cunha GR (2004) Role of p63 and basal cells in the prostate. Development 131:4955–4964

Kyprianou N, Bruckheimer EM, Guo Y (2000) Cell proliferation and apoptosis in prostate cancer: significance in disease progression and therapy. Histol Histopathol 15:1211–1223

Lau KM, Leav I, Ho SM (1998) Rat estrogen receptor-alpha and -beta, and progesterone receptor mRNA expression in various prostatic lobes and microdissected normal and dysplastic epithelial tissues of the Noble rats. Endocrinology 139:424–427

Lee C (1981) Physiology of castration-induced regression in rat prostate. In: Karr JP, Sandberg AA, Murphy GP (eds) The prostatic cell: structure and function, part A. Liss, New York, pp 145–159

Lee C, Sensibar JA, Dudek SM, Hiipakka RA, Liao ST (1990) Prostatic ductal system in rats: regional variation in morphological and functional activities. Biol Reprod 43:1079–1086

Lipschutz JH, Fukami H, Yamamoto M, Tatematsu M, Sugimura Y, Kusakabe M, Cunha GR (1999) Clonality of urogenital organs as determined by analysis of chimeric mice. Cells Tissues Organs 165:57–66

Lopes ES, Foster BA, Donjacour AA, Cunha GR (1996) Initiation of secretory activity of rat prostatic epithelium in organ culture. Endocrinology 137:4225–4234

Lynch CC, Matrisian LM (2002) Matrix metalloproteinases in tumor-host cell communication. Differentiation 70:561–573

Marker PC, Stephan JP, Lee J, Bald L, Mather JP, Cunha GR (2001) fucosyltransferase1 and H-type complex carbohydrates modulate epithelial cell proliferation during prostatic branching morphogenesis. Dev Biol 233:95–108

Marker PC, Donjacour AA, Dahiya R, Cunha GR (2003) Hormonal, cellular, and molecular control of prostatic development. Dev Biol 253:165–174

Masson R, Lefebvre O, Noel A, Fahime ME, Chenard MP, Wendling C, Kebers F, LeMeur M, Dierich A, Foidart JM, Basset P, Rio MC (1998) In vivo evidence that the stromelysin-3 metalloproteinase contributes in a paracrine manner to epithelial cell malignancy. J Cell Biol 140:1535–1541

McGarvey TW, Stearns ME (1995) Keratinocyte growth factor and receptor mRNA expression in benign and malignant human prostate. Exp Mol Pathol 63:52–62

McKeehan WL, Kan M, Hou J, Wang F, Adams P, Mansson P-E (1991) Heparin-binding (fibroblast) growth factor/receptor gene expression in the prostate. In: Karr J, Coffey DS, Smith RG, Tindall DJ (eds) Molecular and cell biology of prostate cancer. Plenum Press, New York, pp 115–126

McKeehan WL, Wang F, Kan M (1998) The heparan sulfate-fibroblast growth factor family: diversity of structure and function. Prog Nucleic Acid Res Mol Biol 59:135–176

McNeal JE (1983) The prostate gland: morphology and pathobiology. Monogr Urol 4:3–37

Moeller H, Blank B, Lander K, Mates G (1987) Ontogeny of the androgen receptor in rat ventral prostate during sexual development. Res Exp Med 187:287–294

Momose H, Uchida K, Kakinuma H, Yura Y, Oyasu R (1990) Involvement of urine in epithelial-stromal interactions in urinary bladder carcinogenesis. Cancer Lett 53:91–96

Nakamura T, Matsumoto K, Kiritoshi A, Tano Y, Nakamura T (1997) Induction of hepatocyte growth factor in fibroblasts by tumor-derived factors affects invasive growth of tumor cells: in vitro analysis of tumor-stromal interactions. Cancer Res 57:3305–3313

Norman JT, Cunha GR, Sugimura Y (1986) The induction of new ductal growth in adult prostatic epithelium in response to an embryonic prostatic inductor. Prostate 8:209–220

Oishi K, Romijn JC, Schroeder FH (1981) The surface character of separated prostatic cells and cultured fibroblasts of prostatic tissue as determined by concanavalin-a hemadsorption. Prostate 2:11–21

Olumi AF, Grossfeld GD, Hayward SW, Carroll PR, Tlsty TD, Cunha GR (1999) Carcinoma-associated fibroblasts direct tumor progression of initiated human prostatic epithelium. Cancer Res 59:5002–5011

Phillips J, Hayward S, Wang Y, Vasselli J, Pavlovich C, Padilla-Nash H, Pezullo J, Ghadimi B, Grossfeld G, Rivera A, Linehan W, Cunha G, Ried T (2001) The consequences of chromosomal aneuploidy on gene expression profiles in a cell line model for prostate carcinogenesis. Cancer Res 61:8143–8149

Picard O, Rolland Y, Poupon MF (1986) Fibroblast-dependent tumorigenicity of cells in nude mice: implication for implantation of metastases. Cancer Res 46:3290–3294

Podlasek CA, Seo RM, Clemens JQ, Ma L, Maas RL, Bushman W (1999) Hoxa-10 deficient male mice exhibit abnormal development of the accessory sex organs. Dev Dyn 214:1–12

Ponten F, Ren Z, Nister M, Westermark B, Ponten J (1994) Epithelial-stromal interactions in basal cell cancer: the PDGF system. J Invest Dermatol 102:304–309

Price D (1963) Comparative aspects of development and structure in the prostate. Natl Cancer Inst Monogr 12:1–27

Prins G, Birch L (1995) The developmental pattern of androgen receptor expression in rat prostate lobes is altered after neonatal exposure to estrogen. Endocrinology 136:1303–1314

Prins G, Birch L (1997) Neonatal estrogen exposure up-regulates estrogen receptor expression in the developing and adult rat prostate lobes. Endocrinology 138:1801–1809

Prins G, Birch L, Greene G (1991) Androgen receptor localization in different cell types of the adult rat prostate. Endocrinology 129:3187–3199

Prins GS, Jung MH, Vellanoweth RL, Chatterjee B, Roy AK (1996) Age-dependent expression of the androgen receptor gene in the prostate and its implication in glandular differentiation and hyperplasia. Dev Genet 18:99–106

Prins G, Marmer M, Woodham C, Chang W, Kuiper G, Gustafsson JA, Birch L (1998) Estrogen receptor-beta messenger ribonucleic acid ontogeny in the prostate of normal and neonatally estrogenized rats. Endocrinology 139:874–83

Prins G, Birch L, Couse JF, Choi I, Katzenellenbogen B, Korach KS (2001) Estrogen imprinting of the developing prostate gland is mediated through stromal estrogen receptor alpha: studies with alpha-ERKO and beta-ERKO mice. Cancer Res 61:6089–6097

Pupa SM, Menard S, Forti S, Tagliabue E (2002) New insights into the role of extracellular matrix during tumor onset and progression. J Cell Physiol 192:259–267

Redler P, Lustig ES (1968) Differences in the growth-promoting effect of normal and peritumoral dermis on epidermis in vitro. Dev Biol 17:679–691

Risbridger G, Wang H, Frydenberg M, Cunha GR (2001a) The metaplastic effects of estrogen on prostate epithelium: proliferation of cells with basal cell phenotype. Endocrinology 142:2443–2450

Risbridger G, Wang H, Young P, Kurita T, Wong Y, Lubahn D, Gustafsson J-A, Cunha G (2001b) Evidence that epithelial and mesenchymal estrogen receptor-a mediates effects of estrogen on prostatic epithelium. Dev Biol 229:432–442

Ronnov-Jessen L, Petersen OW, Bissell MJ (1996) Cellular changes involved in conversion of normal to malignant breast: importance of the stromal reaction. Physiol Rev 76:69–125

Ross R, Bernstein L, Judd H, Hanisch R, Pike M, Henderson B (1986) Serum testosterone levels in healthy young black and white men. J Natl Cancer Inst 76:45–48

Schor SL, Schor AM, Rushton G (1988) Fibroblasts from cancer patients display a mixture of both foetal and adult-like phenotypic characteristics. J Cell Sci 90:401–407

Sciavolino PJ, Abrams EW, Yang L, Austenberg LP, Shen MM, Abate-Shen C (1997) Tissue-specific expression of murine Nkx3.1 in the male urogenital system. Dev Dyn 209:127–138

Seljelid R, Jozefowski S, Sveinbjornsson B (1999) Tumor stroma. Anticancer Res 19:4809–4822

Shattuck-Brandt RL, Lamps LW, Heppner-Goss KJ, DuBois RN, Matrisian LM (1999) Differential expression of matrilysin and cyclooxygenase-2 in intestinal and colorectal neoplasms. Mol Carcinog 24:177–187

Signoretti S, Waltregny D, Dilks J, Isaac B, Lin D, Garraway L, Yang A, Montironi R, McKeon F, Loda M (2000) p63 is a prostate basal cell marker and is required for prostate development. Am J Pathol 157:1769–1775

Singer C, Rasmussen A, Smith HS, Lippman ME, Lynch HT, Cullen KJ (1995) Malignant breast epithelium selects for insulin-like growth factor II expression in breast stroma: evidence for paracrine function. Cancer Res 55:2448–2454

Staack A, Donjacour AA, Brody J, Cunha GR, Carroll P (2003) Mouse urogenital development: a practical approach. Differentiation 71:402–413

Sternlicht MD, Lochter A, Sympson CJ, Huey B, Rougier JP, Gray JW, Pinkel D, Bissell MJ, Werb Z (1999) The stromal proteinase MMP3/stromelysin-1 promotes mammary carcinogenesis. Cell 98:137–146

Sugimura Y, Cunha GR, Bigsby RM (1986a) Androgenic induction of deoxyribonucleic acid synthesis in prostatic glands induced in the urothelium of testicular feminized (Tfm/y) mice. Prostate 9:217–225

Sugimura Y, Cunha GR, Donjacour AA (1986b) Morphogenesis of ductal networks in the mouse prostate. Biol Reprod 34:961–971

Takeda H, Chang C (1991) Immunohistochemical and in situ hybridization analysis of androgen receptor expression during the development of the mouse prostate gland. J Endocrinol 129:83–89

Tam NNC, Phil M, Wang YZ, Wong YC (1997) The influence of mesenchyme of neonatal seminal vesicle and embryonic urogenital sinus on the morphologic and functional cytodifferentiation of Dunning prostatic adenocarcinoma: roles of growth factors and proto-oncogenes. Urol Oncol 3:85–93

Tarin D (1972) Tissue Interactions in Carcinogenesis. Academic Press, London, 483 pp

Thomson AA (2001) Role of androgens and fibroblast growth factors in prostatic development. Reproduction 121:187–195

Thyberg J (1996) Differentiated properties and proliferation of arterial smooth muscle cells in culture. Int Rev Cytol 169:183–265

Timms BG, Mohs TJ, DiDio JA (1994) Ductal budding and branching patterns in the developing prostate. J Urol 151:1427–1432

Turner MA, Darragh T, Palefsky JM (1997) Epithelial-stromal interactions modulating penetration of matrigel membranes by HPV 16-immortalized keratinocytes. J Invest Dermatol 109:619–625

Tuxhorn JA, McAlhany SJ, Dang TD, Ayala GE, Rowley DR (2002) Stromal cells promote angiogenesis and growth of human prostate tumors in a differential reactive stroma (DRS) xenograft model. Cancer Res 62:3298–3307

Uchida K, Samma S, Momose H, Kashihara N, Rademaker A, Oyasu R (1990) Stimulation of urinary bladder tumorigenesis by carcinogen-exposed stroma. J Urol 143:618–621

Van Kempen LCL, Rhee J-S, Dehne K, Lee J, Edwards DR, Coussens LM (2002) Epithelial carcinogenesis: dynamic interplay between neoplastic cells and their microenvironment. Differentiation 60:610–623

Walsh PC (1975) Physiologic basis for hormonal therapy in carcinoma of the prostate. Urol Clin North Am 2:125–131

Wang Y, Hayward S, Cao M, Thayer K, Cunha G (2001a) Cell differentiation lineage in the prostate. Differentiation 68:270–279

Wang Y, Hayward SW, Donjacour AA, Young P, Jacks T, Sage J, Dahiya R, Cardiff R, Day ML, Cunha GR (2000) Hormonal carcinogenesis of the Rb-knockout mouse prostate. Cancer Res 60:6008–6017

Wang Y, Sudilovsky D, Zhang B, Haughney PC, Rosen MA, Wu DS, Cunha TJ, Dahiya R, Cunha GR, Hayward SW (2001b) A human prostatic epithelial model of hormonal carcinogenesis. Cancer Res. 61:6064–6072

Wang YZ, Hayward SW, Cao M, Young P, Cardiff R, Cunha G (2001c) Role of estrogen signaling in prostatic hormonal carcinogenesis. J Urol 165:1320

Weihua Z, Makela S, Andersson LC, Salmi S, Saji S, Webster JI, Jensen EV, Nilsson S, Warner M, Gustafsson JA (2001) A role for estrogen receptor beta in the regulation of growth of the ventral prostate. Proc Natl Acad Sci USA 98:6330–6335

Werb Z, Ashkenas J, MacAuley A, Wiesen JF (1996) Extracellular matrix remodeling as a regulator of stromal-epithelial interactions during mammary gland development, involution and carcinogenesis. Braz J Med Biol Res 29:1087–1097

Wilson CL, Heppner KJ, Labosky PA, Hogan BL, Matrisian LM (1997) Intestinal tumorigenesis is suppressed in mice lacking the metalloproteinase matrilysin. Proc Natl Acad Sci USA 94:1402–1407

Wong YC, Tam NNC (2002) Dedifferentiation of stromal smooth muscle as a factor in prostate carcinogenesis. Differentiation 60:633

Wong YC, Cunha GR, Hayashi N (1992) Effects of mesenchyme of embryonic urogenital sinus and neonatal seminal vesicle on the cytodifferentiation of the Dunning tumor: ultrastructural study. Acta Anat 143:139–150

Wright JH, McDonnell S, Portella G, Bowden GT, Balmain A, Matrisian LM (1994) A switch from stromal to tumor cell expression of stromelysin-1 mRNA associated with the conversion of squamous to spindle carcinomas during mouse skin tumor progression. Mol Carcinog 10:207–215

Xue Y, van der Laak J, Smedts F, Schoots C, Verhofstad A, de la Rosette J, Schalken J (2000) Neuroendocrine cells during human prostate development: does neuroendocrine cell density remain constant during fetal as well as postnatal life? Prostate 42:116–123

Yan G, Fukabori Y, Nikolaropoulos S, Wang F, McKeehan WL (1992) Heparin-binding keratinocyte growth factor is a candidate stromal to epithelial cell andromedin. Mol Endocrinol 6:2123–2128

Yan G, Fukabori Y, McBride G, Nikolaropoulos S, McKeehan W (1993) Exon switching and activation of stromal and embryonic FGF/FGF receptor genes in prostate epithelial cells accompanies stromal independence and malignancy. Mol Cell Biol 13:4513–4522

Yee D, Paik S, Lebovic GS, Marcus RR, Favoni RE, Cullen KJ, Lippman ME, Rosen N (1989) Analysis of insulin-like growth factor I gene expression in malignancy: evidence for a paracrine role in human breast cancer. Mol Endocrinol 3:509–517

Yeh S, Tsai MY, Xu Q, Mu XM, Lardy H, Huang KE, Lin H, Yeh SD, Altuwaijri S, Zhou X, Xing L, Boyce BF, Hung MC, Zhang S, Gan L, Chang C (2002) Generation and characterization of androgen receptor knockout (ARKO) mice: an in vivo model for the study of androgen functions in selective tissues. Proc Natl Acad Sci USA 99:13498–13503

Neoplastic Growth Through the Developmental Stages of the Organism

A. Macieira-Coelho

1 Developmentally and Nondevelopmentally Related Events in the Natural History of Cancers

A morbidity study was performed in Sweden at the Malmö General Hospital (Berge and Lundberge 1977) using the systematic screening for the presence of neoplastic growth in all organs as found in autopsies from all causes of death over a 10-year period in an area covered by the hospital. The study included those tumors that remained silent and those from patients who had been cured of their disease. The incidence of tumors was never above 43%, and only prostate cancer increased progressively with age and did not level off. The prevalence of most cancers peaked at different times through the human life span and then declined. On the other hand, the incidence of stomach, small intestine, and rectum cancers remained constant during the second half of life. Occurrence of cancer of the colon in men peaked between ages 30 to 34 and 75 to 79, and then leveled off. In women it was steady at all ages. These studies showed that neoplastic growth occurs throughout life, with each type of cancer having a different age distribution; this also becomes apparent from data collected *intra vitam* of the clinical incidence of each cancer.

The two great divides in histological origin of tumors through the human life span suggest a developmental determinant in the initiation of neoplasia. Neoplastic growth in younger age groups is predominantly mesenchymal while that of maturity and postmaturity is mostly epithelial. Moreover, many childhood cancers have the features of tumors produced experimentally by infection of animals with viruses. On the other hand, the adult-type epithelial tumors are more frequently associated with prolonged exposure to environmental chemical carcinogens (Doll 1978). In animals several types of tumors are characteristic of young rather than old animals (Cotchin 1975), in particular those caused by viruses, such as leukemia in cats (Hardy et al. 1973). Young

A. Macieira-Coelho (e-mail: macieira-coelho@wanadoo.fr)
INSERM, 73 bis, rue du Maréchal Foch, 78000 Versailles, France

Progress in Molecular and Subcellular Biology
Alvaro Macieira-Coelho (Ed.)
Development Biology of Neoplastic Growth

animals indeed seem to be more susceptible to oncogenic viruses than their older counterparts (Gross 1951).

Genetics create another divide: for almost any type of human cancer there is a heritable and a non-heritable form, the former having an earlier onset (Anderson 1975).

When one reviews the clinical incidence of cancers it is apparent that each cancer has a preferential incidence related to one of the developmental periods of the human life.

1.1 Childhood Cancers

Childhood cancers kill more children under 15 years of age in the United States than any cause except accidents. Some cancers of this period that can be considered as development disorders are described in this volume.

Cancers of the reticuloendothelial system, the central nervous system, and the connective tissues are relatively common in children and adolescents (Adami et al. 1993), while widespread cancers of adults of epithelial origin (i.e., lung, breast, gastrointestinal tract) are absent. These latter cancers are rare in patients under the age of 30 years, reflecting differences in etiology and pathogenesis and suggesting developmentally related events in the initiation of the tumor. Child tumors can originate inter alia from remnants of embryonic cells, from genetic defects, or from prenatal, embryonic, or postnatal exposure to infectious, chemical, or physical agents. Indeed, prenatal events have to be taken into account as shown by the incidence of cancers in offspring of parents engaged in agricultural activities. The use of pesticides is associated with cancers at an early age, whereas cancers in offspring of parents engaged in animal husbandry, in particular poultry farming, are more prevalent in adolescence and young adulthood (Kristensen et al. 1996).

The classical example of embryonic exposure is that of children from mothers treated during pregnancy with diethylstilbestrol.

Children are particularly vulnerable to the action of physical agents as revealed by the data collected from Hiroshima and Nagasaki victims (Tokuoka and Inai 1990) and from the Chernobyl accident, where children became the first victims of fallout (Balter 1996). A dramatic increase in thyroid cancer in children was observed after the latter.

Neuroblastoma is found in situ in the adrenal glands of 1 in 200 newborn infants and may undergo spontaneous regression or transformation to a benign ganglioneuroma (Maurer 1978); development into malignant growth occurs with the highest median incidence in children 2 years of age. Wilms' tumor, associated with loss of sequences on chromosome 11, occurs at ca. 2–3 years of age. Loss of sequences on chromosome 11 is also found in bladder cancer patients 50–84 years old (Fearon et al. 1985) and in young and older patients with other cancers. Non-Hodgkin lymphoma occurs in children most commonly between the age of 2 and 12 years (Maurer 1978). Rhabdomyosarcoma is most frequent before the age of 10, and osteosarcoma and Ewing's sarcoma during the second decade of life.

1.2 Cancers in Adolescents

In adolescence, the most common cancers in order of frequency are lymphomas, tumors of the central nervous system, acute leukemias, bone and soft tissue sarcomas, germ cell tumors, malignant melanoma, and thyroid carcinoma (Stiller 1994). Among leukemias, acute lymphoblastic leukemia (ALL) has a peak incidence in children and adolescents that then declines to age 34 (Cohen et al. 1981). However, the most common cell type of childhood ALL is rare in adults (Teppo 1990). The minimum risk for ALL is at age 30, while for acute non-lymphocytic leukemia it is at age 10 (Stevens 1986). Most ALL in children has a reciprocal translocation between chromosome 11q23 and one of several partner chromosomes (Ford et al. 1993). The pandemic of the Asian influenza in 1956 in Finland was followed by an increase in the incidence of leukemia among children whose mothers had been pregnant at the time (Teppo 1990). No such increase was observed after the later influenza epidemics in the 1960s. These findings illustrate the interaction of viral, environmental, and developmental factors in the pathogeny of this disease. Acute myeloid leukemia occurs in adolescents and older people but not usually in young children (Butturini and Gale 1989).

1.3 Cancers of the Reproductive Stage

Risk for melanoma is concentrated among the young; it is apparent by age 15, much earlier than for non-melanoma skin cancer (Ballou and Dix 1989), and then the risk rises to become constant beyond age 35.

Testicular cancer (TC) strikes mostly young adults; the age-specific incidence peaked in all countries at ages 25 to 34 from 1943 through 1989, but the geographical variation was found to be considerable, with a maximal incidence in Denmark and the lowest in Finland and Poland (Adami et al. 1994). There is an early peak in children under 5 years with a sharp decline between ages 5 and 14. The data indicate that environmental influences are strong. There is an increased risk of TC among boys whose mothers were exposed to an excess of estrogens during pregnancy (Teppo 1990). Abnormal weight gain and nausea during pregnancy, often reported by the mothers, suggest excessive estrogen levels from an exogenous source (Henderson et al. 1988). The occurrence of TC is rapidly increasing; the fact that it is increasing fastest among single men suggests still other etiological factors (Newell et al. 1987).

Cervical intraepithelial neoplasia occurs mainly in young women and the incidence is increasing in the younger age group (20–34 years) in regions as different as New Zealand, Australia, and Canada (Green 1978). Carcinoma of the cervix is now considered a venereal disease due to human papilloma virus. Etiological observations show very clearly that the cervix of the younger woman is more susceptible to the development of squamous carcinoma than that of the older woman (Carmichael 1990). In an epidemiological study in the UK, it was most common in women under 45; the trend suggests that women over 50 with a history of negative results on adequate smear testing may be safely discharged from further screening (Wijngaarden and Duncan 1993).

A pattern of an increased cancer risk in younger individuals is becoming apparent for different type of cancers. In Sweden there is evidence of a continuous increase in the risk of developing cancers in younger birth cohorts (Adami et al. 1993). In Iceland a prominent peak in the incidence of Hodgkin's disease (HD) was found in the 21–30 years age group, which fell beyond that age (Agnarsson et al. 1987).

HD has a bimodal incidence in several countries; in Saudi Arabia two distinct peaks were observed at 18 and 48 years of age for both sexes (Mughal et al. 1985). In other countries the second peak occurs around age 75. In the USA the disease has declined among older persons of both sexes. The bimodal distribution suggests that HD is composed of at least two disease entities, one of which occurs in young adults and one (or several) in elderly patients. In some cases the Epstein-Barr virus plays a part in the production of the disease; there is a consistent finding of an increased incidence of HD among persons who have been diagnosed with mononucleosis (Glimelius and Endblad 1990).

The mean age for chondrosarcoma is 37 years (Azzarelli et al. 1986); for intraoral carcinoma in Nigeria the peak incidence was in the fourth and fifth decades of life, considerably younger than in Europe and America (Adekeye et al. 1985).

In women in Western countries, the most common cancers are those of the breast, large bowel, lung, and corpus uteri. In men, those of the lung, prostate, large bowel, and bladder are most common. These represent more than half of all cancers in both sexes and become manifest mainly during the second half of life (Cutler et al. 1975). The incidence for the breast and the uterus starts to increase relatively early in life and reaches substantial levels by age 40. In men the four most common cancers increase at later ages. Since 80% of human cancers are environmentally induced, a conclusion endorsed by the WHO, these cancers must belong to that category (Muir 1990).

The natural history of breast cancers has been extensively studied; it shows that developmental events play a strong role in the initiation and progression of these tumors. Breast tissue is reorganized through the developmental stages of women, i.e., embryonic, prepubertal, pubertal, pregnancy, lactation, and sexual involution. Epithelial ducts and branching morphogenesis are controlled by factors coded by genes expressed in stromal cells. The whole extracellular matrix contributes to this process. The instructive effect of the mesenchyme on branching is specific since embryonic mammary epithelium exposed to salivary gland mesenchyme assumes a salivary-like branching (Kratochwill 1969). Many of the factors essential for mammary gland development are also associated with cancer and many of these factors are of stromal origin (Hennighausen and Robinson 2001). Overexpression of stromal regulators can lead to uncontrolled branching and cancer (Li et al. 2000); this is the case when the secreted factor Wnt-1 is overexpressed in the mouse mammary gland (Li et al. 2000). Deregulation of matrix metalloproteinases, which play a role in the developmental stages of the mammary gland, occurs during the initiation and progression of tumors (Wiseman and Werb 2002).

The incidence of cancer of the breast is rising in women 25–44 years old, which may be related to an increased tendency to delay child bearing (Krieger 1988). There are regional differences in the prevalence and age incidence of cancers of the breast. It is found in 55.2 per 100,000 in Sweden and in 28.5 per 100,000 in Cuba where the age profile is shifted toward younger ages (Skoog et al. 1987). The age-specific incidence in Japan decreases after age 50 in contrast to the incidence in the USA and northern Europe where it increases after age 50 (Mohle-Boetani 1990). Like that of other cancers, the incidence in Thailand is very low.

Different histological types have different age distributions, the peri-menopausal break on the age-incidence curve is most marked for carcinomas of predominantly lobular origin and virtually absent for carcinomas of predominantly ductal origin (Stalsberg and Thomas 1993). In males the age distribution of papillary carcinoma is similar to that of all types combined, whereas that of mucinous carcinoma is skewed towards older ages, as in females. Anaplastic duct breast carcinomas in women are of early onset type (Berg and Robbins 1968). A clear distinction should be made between heritable, early onset, bilateral breast tumors, and the unilateral disease (Anderson 1975).

The determinants of neoplastic growth in the breast are complex and probably the result of the interaction between developmental modifications and the environment. The pre- and post-menopausal bimodal distribution of patients with mammary carcinoma suggests the existence, concerning causes, of at least two types (Waard et al. 1975). Nutritional factors in terms of energy, rather than of a specific effect of fat or other nutrients, are very significant in favoring breast malignancies through endocrine mechanisms where estrogens play a major part (Waard and Trichopoulos 1988). The well-established influence of alcohol consumption on breast tumor incidence is also probably due to the increase in endogenous estrogen levels (Willett 2000). Habits early in life influence breast malignancies, and there is growing evidence that for breast cancers the early stages begin at a young age, at least at puberty and during adolescence (Waard and Trichopoulos 1988). The initiation of mammary neoplasia in women can precede clinical detection by 20 to 30 years (Foulds 1958). There is also evidence that intrauterine exposure to high estrogen concentrations may be important in the etiology of breast malignancies

(Ekbom et al. 1992). A direct correlation seems also to be found between birth weight, excessive weight gain in adolescence, and risk (Ziegler 1998).

Cancers of the mammary gland depend strongly on hormonal history, i.e., exposure to endogenous and exogenous hormones, development and length of sexual maturity, pregnancy, lactation, etc. Breast cancer patients have shown subnormal levels of dehydroepiandrosterone and its sulfate (Brownsey et al. 1972); moreover, the concentrations of both steroids were found to be age invariant in the cancer patients (Zumoff et al. 1981). Reproductive factors play a significant role. An early age at menarche, a late age at first full-term pregnancy, a low parity, and a late menopause, all increase the risk of breast malignancies. These parameters are a reflection of the total number of ovulatory cycles being the most important risk factor (Olsson 1990). Breast cancer patients have a shorter average menstrual cycle length than patients with benign breast disease or controls. MacMahon et al. (1982) have shown that estrogen excretion levels in post-menarche girls depend on age at menarche. Women with early menarche have a longer and higher level exposure to estrogens causing an increased proliferation in the mammary gland and thus a higher susceptibility to DNA damage. The implications of cell proliferation in the susceptibility to cancer of the mammary gland is also observed in animals; mammary DNA synthesis of high mammary tumor strains of mice is very high even at the end of pregnancy (Nagasawa 1989). There are other events associated with an early menarche, such as a more profound decrease in the sensitivity of the hypothalamic–pituitary axis to the negative feedback of circulating steroids, implying modifications in the concentration of follicle-stimulating hormone (Vikho and Apter 1986). Artificial menopause, particularly at an early age, protects against breast cancer (Feinleib 1968). A first pregnancy before the age of 23 halves the risk of breast cancer compared to the nulliparous, and the effect is life-long (Olsson 1990). This has been attributed to a slowdown of precancerous lesions by mitotic arrest and terminal differentiation of mammary gland epithelium generated by an early first pregnancy (Russo and Russo 1980). The longer the total period of low mammary DNA synthesis, the smaller the risk (Nagasawa 1989). A high number of children are also protective (Bergkvist et al. 1987).

Animal studies have shown that the susceptibility of the mammary gland to carcinogenic effects varies with the developmental stage. Daily injections of estradiol or diethylstilbestrol for the first 5 days of perinatal life of female BALB/c mice inhibited gland growth on day 6, but the treatment stimulated growth after 4 weeks (Tomooka and Bern 1982). Estradiol treatment between days 1 and 5 induced the most mammary abnormalities at 12 months, however, these declined markedly when treatment began after day 5 (Bern et al. 1983). In other words, long-term effects of perinatal exposure to hormones on advanced-age tumor growth depend on perinatal age (Nagasawa 1989).

1.4 Cancers of the Postreproductive Stage

The postmenopausal type mammary cancer was found to be associated with obesity, hypertension, and decreased glucose tolerance. Body mass rather than overweight was found to be the risk factor (Waard 1975). Colloid cancer is one of the slower-growing breast carcinomas found in postmenopausal women. Though these cancers apparently befall older women, they may originate at the same age as those of earlier onset, faster growing, anaplastic duct carcinomas, and merely take longer to become clinically apparent (Berg and Robbins 1968).

Cancer of the corpus uteri is common in obese, tall, nulliparous women and frequently parallels the incidence of breast cancer (Muir 1990). Incidence rates fall when the use of estrogen by postmenopausal women ceases. Use of oral contraceptives prior to menopause reduces the risk of endometrial cancer significantly (Key and Pike 1988). After menopause when the ovarian production of estrogen ceases, a direct association between circulating androgen levels and risk of endometrial cancer was found, mediated mainly through their conversion to estrogens (Lukanova et al. 2004).

Other neoplasias where hormonal history plays a role are those of the ovary, as would be expected. The mean and median ages at diagnosis are 55.3 and 56 years, respectively. As with mammary cancers, the age incidence was related to the histology. Tumors from younger women (<40 years) were more likely to have a mucinous histology while those from older patients were more likely to be serous or endometrioid.

Findings were also consistent with the incessant-ovulation hypothesis: nulliparity increases the risk, whereas factors preventing ovulation such as pregnancy and use of oral contraceptives are protective. The repeated cycles of cell divisions that epithelial ovarian cells undergo with ovulation may increase the probability of mutational events forming the basis for malignant transformation (Pieretti and Turker 1997). The rate of increase in incidence of ovarian cancers diminishes after menopause.

After age 60 there is again an increase in incidence of ALL (Cohen et al. 1981). A sharp rise in incidence at and following middle age with a peak in the 70–80 years age group is seen with non-Hodgkin's lymphoma; it was found to be associated with the use of herbicides (Hoar et al. 1986).

Cancers of the large bowel have increased in several regions, the phenomenon being attributed to an increased consumption of animal fat (Muir 1990). Other factors, though, play a role in the risk for these cancers. Increasing parity is associated with a decline in risk of colon cancer in women older than 50 years at diagnosis. On the other hand, there is an increased risk of colorectal cancer with increasing age at first birth, a finding that has also been observed with other cancers, as mentioned above (Broeders et al. 1996).

Gastric cancers are more frequent during the second half of life; their decline in Japan and Western countries shows that they are strongly determined by environmental factors. In Japan the curve of the age parameter in females showed a shoulder around age 40 and a subsequent upward slope to age 80, compatible with two different entities. The histologically diffuse type is prominent in the younger age groups, while the intestinal type is dominant in the older group; in males it is unimodal (Hamajima and Lee 1987).

The frequency of gastric cancers is decreasing in countries where the survival of the population is improving. Immigration studies showed that the decrease in incidence was more pronounced in second generation than in first generation Japanese, suggesting that exposures in early life may determine the risk of cancers of the stomach. Second generation Japanese in America are taller than their parents, suggesting a better diet, which in turn is considered to explain the decreased risk of stomach cancer (Hamajima and Lee 1987). Since *Helicobacter piloris* is responsible for most gastric cancers, it is not surprising that improved hygiene and food consumption have caused a decline in the frequency of these cancers.

Carcinoma of the prostate is another cancer characteristic of later life. This is probably one of the most prevalent of all cancers; in the Malmö studies mentioned above, the incidence was found to increase progressively with age; it nearly doubles every decade over age 50, from 10% in men in the 50 s to 70% in men in the 80 s (Gittes 1991). However, there is only a 6 to 8% chance that a man will have a clinically detected prostatic cancer in his lifetime. These data indicate that at least nine of ten such cancers remain undetected. Although the disease manifests itself mainly during the second half of human life, its history starts way back during adolescence. It is more frequent in individuals with high testosterone levels during the reproductive period, especially the early reproductive ages. Muscle mass, more than the amount of tissue fat, correlates with the risk of the disease, again implicating the role of androgens (Olsson 1990). However, a high level of circulating testosterone at the time of diagnosis is not associated with an increased prostate cancer risk (Stattin et al. 2004). Moreover, this cancer does not occur in eunuchs castrated early in life; these data point to early events influencing this carcinoma that manifests late in the life.

Men with an inherited predisposition have a higher probability of clinically manifest disease and are likely to develop the cancer at an early age (Olsson 1990). A tumor suppressor gene was found mutated in 71% of these tumors (Narla et al. 2001) further work is needed to ascertain if this mutation is correlated with the cases of early onset.

1.5 Cancers During Senescence

Carcinoma of the prostate is the only cancer whose incidence increases with time without leveling off or declining, both in data collected *intra vitam* (Cutler et al. 1974) and postmortem (Berge and Lundberge 1977). With all the other cancers characteristic of the second half of life and whose incidence increases with age, the frequency levels off or declines after age 75–80. This is patent in studies from different origins: mammary cancers increase to reach 25.4% of the American population at ages 65–74, then decline to 17.1% at ages 75–84 and to 5.6% beyond 84 (Hankey 1994); in the USA, at age 65, 30% of all deaths are due to neoplastic growth, at age 80 only 12% (Brody 1983); the age-specific incidence/100,000 population for all cancers combined peaks at age 75–80 and then declines (Piantanelli 1988); in centenarians the frequency of neoplastic disease is negligible (Allard and Robin 2000).

In general, the only explanation proposed is that of a cohort effect, the long-surviving individuals being those resistant to neoplastic disease, in spite of the fact that resistance to neoplasia seems to carry a low weight in the determination of human longevity (National Center of Health Statistics 1973). However, the decline in the clinical incidence of cancers during human senescence can also be explained in terms of the developmental changes occurring during the last window of the human life span, as described below.

2 Putative Mechanisms of Developmentally Related Events in the Natural History of Cancers

The age range for the manifestation of each type of cancer suggests a relationship with the developmental stages of the organism; in some cancers the age incidence seems more dependent on the time of exposure to the carcinogen, but the susceptibility to carcinogens varies with the developmental stage. Since, according to the World Health Organization, 80% of all cancers are environmentally associated (Muir 1990), in most instances tumor growth should be viewed as the result of deviations from normal development caused by environmental aggressions on a favorable genetic background. The word environment encompasses the air we breathe, the water we drink, the food we eat, exposure at work, the use of tobacco and alcoholic beverages, nature and quantity of food consumed, and culturally influenced habits such as age of first coitus and full-term pregnancy, and the use of oral contraceptives (Muir 1990).

The deviations from normal development can have other causes, e.g., genetic, where the environment has an additive effect.

Examples were given above in regard to the susceptibility of developmental stages to viruses and chemical carcinogens – and ionizing radiation should be added. Children are particularly vulnerable to the action of physical agents; after the Chernobyl incident children became the first victims of fallout; a dramatic increase in thyroid cancer in children was observed after the accident (Balter 1996). The female breast is particularly sensitive to cancer induction by ionizing radiation at puberty and adolescence (Land et al. 1980). An excess risk of radiation-related cancer of the breast was ascertained in females who were under 10 years of age when the bombs fell on Hiroshima and Nagasaki (Tokunaga et al. 1987).

2.1 Cell-Related Mechanisms

2.1.1 Cellular Susceptibility to Carcinogens

The response to carcinogens varies through the developmental stages of a cell population life cycle. Experiments in vitro showed that at the cellular level the susceptibility to malignant transformation is correlated with the developmental stage of a cell population. Some mitotic cell compartments go through a programmed development of new cell types evolving to a terminal differentiation necessary for homeostasis. With aging there is a shift in cell types, with the predominance of cells in a more advanced stage of the differentiation cycle. Human fibroblasts have been particularly well studied in this respect. Serially proliferating human fibroblasts go through several stages leading to a terminal postmitotic cell that seems to correspond to a terminal differentiation cycle (Martin et al. 1974); evidence has accumulated favoring this view. The finding that the postmitotic cell accumulates in different pathological conditions supports the view that the normal unraveling of the terminal differentiation cycle is necessary for homeostasis. It led to the conclusion that the backlog signals a disturbance of the homeostatic regulation of the differentiation cycle and is part of the pathological picture (Macieira-Coelho 1995, 2001).

The terminal postmitotic cell has been wrongly considered significant for aging of the organism and hence was called senescent cell (Macieira-Coelho 2003). In fact, different markers showed that the number of terminal cells does not increase with aging in vivo. Modifications in chromatin fibers (Macieira-Coelho 1991), telomere length (Allsop et al. 1992), expression of genes involved in cell cycling (Grassilli et al. 1996), and gene profile determined by cDNA microarrays (Park et al. 2001) all showed that the post-mitotic cell is unrelated to aging of the organism.

At least three types of fibroblasts have been identified as this development proceeds. During the postnatal period, the number of early population doubling cells (type I fibroblasts) decreases with a shift to type II and type III fibroblasts, which are more advanced along the differentiation cycle (Franks and Cooper 1972; Russell and Witt 1976; Hennis et al. 1981; Steinhardt 1985).

Four main phases of the fibroblast population were identified by the kinetics of cell proliferation and biochemical parameters (Macieira-Coelho and Taboury 1982). Early (phases I and II) population doubling level (PDL) fibroblasts are susceptible to ionizing radiation (Diatloff and Macieira-Coelho 1982) and to chemical carcinogens (Milo and Casto 1986), while higher PDL cells are refractory to the insult. Fibroblasts of phases I, II, and III are susceptible to the action of a tumor promoter, unlike terminal phase IV cells (Elliot and Katze 1986). On the other hand, SV40 virus more rapidly transforms cells entering phase IV (Jensen et al. 1963) than cells of earlier phases. In regard to the transforming action of ultraviolet radiation, the cells are equally susceptible through the different phases of their life span (McCormick et al. 1986).

These results show that at the cellular level susceptibility to transformation is not a simple linear cumulative phenomenon, but varies with the changing properties of the cell population through its developmental stages. Since during the life span of the organism there is a shift in the cell population, this could constitute one of the mechanisms to explain the variation in the age incidence of cancers.

It has been claimed that phase IV cells are capable of promoting tumor growth (Krtolica et al. 2001). This observation is currently being considered as supporting the correlation between aging and neoplastic growth; since (as mentioned above) there is convincing evidence that the terminal cell is irrelevant for human aging, the finding is also irrelevant for the correlation. On the other hand since post-mitotic cells are increased in some pathologies the results could be relevant for the influence of the latter on tumor initiation.

2.1.2
Stem Cells

Research on cancers has recently focused on the presence of a small fraction of cells with renewal potential that would be responsible for the perpetuation of the neoplastic growth. These cells are called stem cells because they are thought to share similarities with normal stem cells in the mechanisms that regulate self-renewal, and because of the possibility that tumors might arise from normal stem cells (Reya et al. 2001). There are, however, significant differences between normal stem cells and the putative cancer stem cells.

The concept of the former is evolving thanks to new findings that are changing this field continuously. They were thought to be a separate pool of cells present in tissues ready to regenerate the respective cells; although some results are still not universally accepted, this view is becoming increasingly diluted. The question presently asked summarizes the problem: is the stem cell an entity, a cell compartment, or is it just a function depending on the cell environment (Blau et al. 2001)?

It has become apparent that the capacity to be unipotent (to give rise to only one type of differentiated cell), to be multipotent (to give rise to several different differentiated cells), to be pluripotent (to give rise to every cell of the organism), or to be totipotent (to give rise to the whole organism when implanted in the uterus), is a question of context. The restoration of the regenerative potential has been demonstrated in circumstances previously believed to be forbidden. In general, it was thought that once stem cells are triggered to differentiate along a given line, the transit-amplified cells could not turn back. Kondo and Raff (2000) claim to have managed to coax rat oligodendrocyte precursor cells committed to becoming oligodendrocytes or astrocytes, into becoming neurons. It is not only possible to produce pluripotent embryonic stem cells from adult somatic nuclei (Munsie et al. 2000), but also to reconstitute the whole organism with the activated nucleus of an adult somatic cell in an appropriate environment, the oocyte cytoplasm (Gurdon 1962; Wilmut et al. 1997). These findings have opened a completely new perspective of the subject. The restoration of the regenerative potential of a cell at any stage of its life cycle seems to be a possibility in the future. Indeed calves could be cloned reconstructing bovine oocytes by transfer of nuclei obtained from bovine fibroblast populations that were close to the postmitotic terminal stage (95% life span completed) after serial replication in vitro (Lanza et al. 2000). The question remains whether the resultant organism is reset the same way as that resulting from outcrossing and recombination, and goes through normal developmental stages. Several data suggest it does not. Cloned mice for instance, that develop to birth and beyond display major disturbances of gene expression (Humpherys et al. 2002).

Another question that has been raised is whether the infinite proliferation potential of the normal stem cell is universal; some results are still conflicting but it seems that the maintenance or decline of the mammalian stem cell pool during senescence depends on the organ and on the species (reviewed in Macieira-Coelho 2003). Hence a fundamental difference between normal and tumor stem cells would be that in the former the renewal potential is either maintained or it declines with aging, while in tumors there is a drift towards heightened malignancy and the tendency to regenerate an increasingly anaplastic growth.

The problem of the immortality of tumor stem cells raises other questions, since cancer cell populations are not necessarily immortal; indeed, it seems that immortalization can be a late event during the progression of tumors. The time of occurrence of the immortalization step during tumor progression in vivo has been studied with breast cancer cells, and it was found to take place at a late stage during the evolution of the breast cancer studied (Smith et al. 1987). Cells from breast cancer metastases were obtained from one patient over a 2-year period, but despite repeated attempts, only the last specimen obtained at the end of that period reproducibly yielded cell populations with an infinite life span. Cells can also become malignant, for instance after treatment with chemical carcinogens, but still retain the mortal phenotype (Milo and DiPaolo 1978). Cell immortalization is not synonymous with malignancy; in most cases additional events are necessary for the cells to become malignant (Hei et al. 1997).

Attempts to develop permanent cell lines were also made with specimens obtained from the primary site and the peripheral lymph nodes from patients with small-cell lung cancer (Masuda et al. 1991). Twenty-one percent permanent lines were obtained with specimens from the primary sites and 80% from the peripheral lymph nodes. The survival times of the patients whose cells grew continually in culture were significantly shorter than those of the patients whose cells had a finite life span. These results suggest that the tendency of a tumor cell population to yield permanent cell lines is related to a more biologically aggressive form of the disease and to have prognostic value (Masuda et al. 1991). Identical findings were obtained with neuroblastoma (Reynolds et al. 1980) and non-Hodgkin's lymphoma (Tweeddale et al. 1987).

The work of Masuda et al. (1991) also suggests that specimens obtained from metastases yield permanent cell lines more frequently and thus represent a more advanced stage of malignancy. Hence, stem cells are not present from the beginning of the neoplastic growth and seem to be the expression of the progression to an increased malignancy.

In acute myeloid leukemia only a small fraction of the neoplastic cell population is capable of initiating the leukemia in mice, and this fraction has a phenotype CD34+CD38–, similar to normal hematopoietic stem cells (Bonnet and Dick 1997). A similar fraction of cells was identified in solid tumors (Al-Hajj et al. 2003; Singh et al. 2003).

The idea that only certain subpopulations within a tumor are »important« for the maintenance of the neoplastic growth has been challenged. The behavior of a cancer is not necessarily determined by its »most malignant« fraction (Heppner et al. 1983). The identification of stem cells can be fallacious in the sense that by isolating subpopulations through cloning or transplantation, the cells express properties that are not expressed in the tumor mass (Heppner and Miller 1983). The idea of a fraction of cells that would be autonomous, capable by themselves to maintain the uncontrolled growth does not fit the physiopathology of cancers. Propagation of the tumor depends on the interaction of its cellular components and, as described below, on developmental modifications of the stroma and of the whole organism. It is possible, however, that as the tumor progresses a fraction of cells becomes increasingly autonomous and less dependable in micro-environmental interactions.

The suggestion that expansion of somatic stem cells may be the first step in the initiation of some tumors is an interesting possibility (Taipale and Beachy 2001). Many adult tumors originate from tissues (skin, intestine, blood) that proliferate throughout life and where somatic stem cells persist. Two genes, Wnt and Hedgehog (Hh), that regulate growth and patterning during embryonic development and in proliferating adult tissues, are deregulated in some tumors arising from skin and intestine. Hence it was suggested that tumor initiation involves mutations in such genes leading to expansion of aberrant transit-amplified cells. Subsequent mutations would lead to invasive properties and perpetuation of the neoplastic process (Taipale and Beachy 2001). The disorganization of tissue structure through mutations in genes regulating patterning during tissue renewal could be one of the first events in the initiation of some cancers. Indeed, there is strong evidence that conformational modifications can lead to malignant transformation (Macieira-Coelho 2002).

Experiments with transgenic mice buttressed the hypothesis that components of the Wnt signaling pathway transform mammary cell progenitors (Li et al. 2003). Two genes that appear to be preferentially expressed in mammary stem cells were present in preneoplasias and tumors of these mice. Wnt signaling genes are also overexpressed in chronic lymphocytic leukemia (Lu et al. 2004). They may contribute to the defect in apoptosis that characterizes this malignancy.

The retinoblastoma gene constitutes another example of a gene that acts in developmental events, whose deregulation can cause neoplastic growth. Its product functions as a switch determining white versus brown adipocyte differentiation during development (Hansen et al. 2004).

An interesting example of genes implicated in embryonic development that are reactivated during carcinogenesis concerns the gene coding for gastrin-releasing peptide growth factor receptor (GRPR) located on the X chromosome. It is activated in early fetal development in human lungs; when the lungs have matured the receptor normally remains inactive in adult lung tissue of men and 45% of women. GRPR becomes reactivated in smokers, earlier in women than in men, which could be a factor in the higher susceptibility of women to tobacco-induced lung cancer (Shriver et al. 2000).

An attempt was made to identify the mechanism by which oncogenes disrupt the normal developmental program of human hematopoietic cells (Pereira et al. 1998). The retroviral transduction of a myeloid leukemia-associated fusion gene to human cord blood cells could arrest erythroid differentiation and increase the self-renewal capacity of myeloid progenitors.

Adult human mesenchymal stem cells transduced with the telomerase h TERT gene after serial proliferation in vitro were able to originate tumors in mice (Serakinci et al. 2004). This experiment shows that transplantable tumors can be obtained by this method from stem cells, as it has been shown before for other cells – yet does not prove that spontaneous tumors originate from stem cells. I was reported that mouse gastric epithelial cancers originated by chronic infection are derived from bone marrow cells, a possible source of stem cells (Houghton et al 2004).

2.1.3 *The Cellular Environment*

In many forms of neoplasia there is evidence of a wide field of prepared tissue within which neoplasia develops (Willis 1945); more recent data have shown that in many instances the whole organism is modified. At least in some cases, neoplasia from the beginning is a generalized disease of the internal milieu where mesenchymal cells create a field effect that favors cell transformation. Data showing that skin fibroblasts distal from the site where the cancer develops present in vitro deviations from normalcy suggest a participation of the whole organism in the neoplastic process. This was first shown with the skin fibroblasts from osteosarcoma patients (Smith et al. 1976) that displayed a greater ability to grow in immunosuppressed mice than did fibroblasts from non-cancerous donors. The higher susceptibility to transformation by Rous sarcoma virus of dermal fibroblasts from osteosarcoma patients buttressed the findings of Smith and coworkers (Gogusev et al. 1988). Moreover, skin fibroblasts from some patients with carcinomas and from patients at high risk of cancers were susceptible to the stimulatory action of low-dose-rate ionizing radiation on cell division (Diatloff and Macieira-Coelho 1979), and the life span of these cells could be prolonged through variations in the culture conditions (Diatloff and Macieira-Coelho 1982), two features that were not observed in normal postnatal controls.

Connective tissue has a seminal role in the regulation of ontogeny and of homeostasis in the postnatal organism. This is so because the function of this tissue is to integrate cells in tissues, tissues in organs, and organs in the whole organism (Robert and Labat-Robert 1995). In the embryo, defects in connective tissue can lead to a variety of malformations; mouse collagen I, for instance, has an important role in early hemopoiesis and in establishing the mechanical stability of the circulatory system (Lohler et al. 1984). Development of the thymus depends on an interaction between mesenchyme and epithelium (Auerbach 1960), and mouse embryo mammary rudiments fail to develop in the absence of mesenchyme (Kratochwill 1969). Concerning the thymus, early during embryonic development the site where the organ will be located is invaded by an intensive proliferation of mesenchymal cells that will constitute a feeder layer for thymocyte precursors. Involution of the thymus is due in part to the aging of the mesenchymal cell feeder layer subsequent to the exhaustion of its division potential.

The homeostatic inductive role of connective tissue on neighboring cells is essential in the postnatal organism and some events of the aging syndrome are due to dysfunctions of this interaction. In the developing prostate, the mesenchyme is the actual target and mediator of androgenic effects upon the epithelium and stromal–epithelial interactions continue in adulthood (Cunha et al. 1987). Rat adult bladder epithelium placed in contact with urogenital sinus mesenchyme is converted to prostatic epithelium (Cunha et al. 1983), showing the importance of morphogenetic fields in the adult. A reorganization of the prostatic stroma progresses through the postnatal life span with a differentiation of fibroblasts into smooth muscle cells after puberty; during senescence smooth muscle cells become the predominant component (Rumpold et al. 2002). This shift in cell population leads to a decreased production of inhibitors in favor of epithelial growth stimulants. Prostate hypertrophy, a typical entity of the aging syndrome in which the prostate resumes growth late in life, is thought to be due to a change in the inductive activity of the stroma, which has re-expressed embryonic properties (McNeal 1983).

Deviations from normal were observed with skin fibroblasts from breast cancer patients (Azzarone et al. 1984). They consisted in the absence of a correlation between the doubling potential and the donor's age, anchorage independence, colony formation on monolayers of normal epithelial cells, and increased saturation densities. A population of cells capable of growing at higher densities could be separated from the main population; these cells had the ability to invade normal tissue in organ culture.

The deviation from normal of the fibroblasts seems to start long before the neoplastic growth is expressed, since it was observed in a patient with a benign lesion of the breast whose mother had developed a breast cancer. Periodic controls of the patient resulted in the early detection of a carcinoma 3 years after the operation for the benign lesion (Azzarone et al. 1984).

Spontaneous transformation was observed of skin fibroblasts from patients with melanoma; the cells exhibited aneuploid karyotype, anchorage-independent growth, and in nude mice produced progressively growing tumors with morphologic characteristics of sarcoma (Mukherji et al. 1984). This is an extremely rare event with normal human fibroblasts. The instability of the proliferation life span of fibroblasts from melanoma patients was also revealed by the observation that cells from a member of a melanoma-prone family were resistant to Ras-mediated mitotic arrest (Brookes et al. 2002). Increased sensitivity to mutagenic effects (Howell et al. 1984) and to UV radiation (Smith et al. 1982) has also been reported in those cells.

An increased ability to migrate was detected in fetal fibroblasts as compared to normal postnatal cells. The fetal fibroblasts undergo a programmed transition towards the end of their proliferation life span when they manifest the adult-type migration. However, skin fibroblasts from patients with different types of cancers had a migratory behavior falling within the fetal range but did not undergo the transition to the normal adult-type phenotype towards the end of the proliferation life span (Schor et al. 1985). Skin fibroblasts from 17 out of a total of 34 breast cancer patients with no previous family history of this type of cancer also displayed fetal-type migratory behavior; on the other hand, the fibroblasts of 15 out of a total of 16 patients with a positive family history of breast cancer displayed the fetal-type migration (Schor et al. 1986). The abnormal behavior was also observed in unaffected first-degree relatives of the latter patients. However, a significant fraction of cells from donors with benign breast disease or with no disease had fetal behavior, which constitutes a caveat for these findings. The increased migration was found to be due to a stimulating factor with an autocrine mode of action (Schor et al. 1988), which consists of a truncated onco-fetal fibronectin isoform (Schor et al. 2003). Skin fibroblasts from patients at high genetic risk of cancer also behave in part as embryonic cells (Pfeffer et al. 1976). The reactivation of fetal characteristics in tumor stromal fibroblasts is supported by the presence of fibroblast-activation protein α, which is absent in normal adult tissues but transiently expressed in some fetal mesenchymal tissues and abundantly expressed in the stroma of 90% of breast, colorectal, and lung carcinomas (Scanlan et al. 1994).

Cytogenetic investigations also revealed that the mesenchyme of patients with several types of cancer deviates from normal. Chromosomal aberrations were detected, for instance, in skin fibroblast cultures from patients with thyroid cancer (Kritzman et al. 1987). The deviation from normalcy extends beyond the mesenchyme, since mutagen susceptibility in different cancer patients (Hsu et al. 1985) and chromosomal defects in patients with precancerous and cancerous lesions of the cervix uteri (Murty et al. 1985) were observed in peripheral blood cultures. This buttresses the finding that the whole organism is modified.

The increase in the risk of several non-cutaneous cancers in patients with basal cell carcinomas (BCC) is also suggestive of a broad deviation of homeostatic control favoring tumor growth (Milan et al. 2000). Patients who were under 40 years of age at the time of BCC diagnosis had a significantly higher relative risk for a subsequent cancer than the older patients; time since diagnosis did not influence the overall risk.

A defect was found in the cell division cycle of skin fibroblasts of cancer patients (Azzarone and Macieira-Coelho 1987). It consisted in abnormal DNA synthetic activity when the cells became confluent because of a prolongation of the G2 period that created a bottleneck before mitosis; a faulty checkpoint may have been causative. This abnormality could be germane to the previous observations dealing with the higher sensitivity to ionizing radiation in the G2 period of fibroblasts from high-risk cancer patients (Parshad et al. 1985).

The findings that the abnormal growth of the skin fibroblasts was observed 3 years before the manifestation of the cancer (Azzarone et al. 1984), and that the fetal migratory behavior was detected in first-degree relatives of cancer patients (Schor et al. 1986) show that the changes have a relationship of causality *with* rather than a consequence *of* the cancer. The presence of the abnormal fibroblasts in the organism could be due to a developmental deregulation of their terminal differentiation cycle, with accumulation of transit-amplified cells that change the homeostatic control of the microenvironment. Such deregulation would create a field effect favorable for the transformation of epithelial cells.

It is becoming apparent that fibroblasts play a role during postnatal development in maintaining order and patterning in tissues (Chang et al. 2002); hence tissue structural disorganization through deviations from their normal development program could favor the initiation of neoplastic growth. The influence of the spatial organization of tissues in the initiation of some cancers has a bearing with the reactivation of a mesenchymal architectural transcription factor (HMGA2) in oral carcinomas (Miyazawa et al. 2004). HMGA2 is expressed in the developing mesenchyme but is absent in the adult. It was found to be expressed in squamous cell carcinomas, localized to the invasive front where the cells exhibit epithelial–mesenchyme transition.

Schor et al. (1994) suggested that the persistent expression of fetal behavior influences the development of the epithelial tumor because of dysfunction of normal epithelial–mesenchymal interactions. Indeed, it was shown that the implantation of fetal fibroblasts into an adult rat mammary gland induces the hyperplastic growth of the normal epithelium, rendering it more sensitive to neoplastic transformation by carcinogenic agents (Sakakura 1983). Crosstalk between epithelium and stroma plays a role both in the developmental stages of the mammary gland and in malignant transformation. Human breast fibroblasts secrete a factor that stimulates 17β-oestradiol dehydrogenase activity, promoting the reduction of estrone to estradiol. Fibroblasts derived from malignant tumors had a more potent effect on reductive enzyme activity than fibroblasts derived from normal breast tissue (Adams et al. 1988). Tumor stromal fibroblasts also intervene as immune modulators through the secretion of factors such as tenascin-C, thrombospondin-1, matrix metalloproteinases, cytokines, and fibroblast-activation protein α (Silzle et al. 2004). TGF-β is another significant stroma-secreted factor with an effect on the progression of epithelial tumors (Derynck et al. 2001). Conditional inactivation of the TGF-β type II receptor gene in mouse fibroblasts resulted in intraepithelial neoplasia in the prostate and invasive squamous cell carcinoma of the fore-stomach, associated with an increased abundance of stromal cells (Bhowmick et al. 2004). Moreover, stroma genetically modified to overexpress either hepatocyte growth factor or type TGF-β1 promoted the outgrowth of premalignant and malignant epithelial cells from a cell preparation of morphologically normal mammary epithelium. In contrast, normal human fibroblasts allowed only the growth of normal mammary epithelium from the same donor material (Kuperwasser et al. 2004). In spontaneous tumors, however, events are not so straightforward because of the complex interactions of TGF-β. Most tumors do not have inactivated TGF-β receptors, and, surprisingly, suppression of TGF-β signaling protects against tumor progression (Derynck et al. 2001).

Once the transformation process is initiated through the action of a propitious environment, the malignant cells start influencing the surrounding tissue and later the whole organism. The amount of stroma varies between tumors. In carcinomas of the breast, stomach, and pancreas stroma may account for over 90% of the total tumor mass (Dvorak 1986). On the other hand, melanomas possess minimal stroma. It was postulated that the primary tumor induces the malignant transformation of adjacent normal tissue (Slaughter et al. 1953). There is indeed evidence that the stroma of tumors display deviations from normal. Although this has been treated in a separate chapter in this volume, we will refer to some of the data showing these deviations.

Cell overlap, with piling up and forming colonies, was reported in fibroblasts from breast tumors (Delinassios et al. 1983). The morphology, the nonsarcomatous nature, and the normal diploid karyotype determined the stromal origin of the cells. Prostate-carcinoma-associated fibroblasts have an increased capability to form colonies in soft agar as compared to normal human prostate fibroblasts. This could be due to the higher expression of TGF-β1 in the abnormal cells (San Francisco et al. 2004).

Animal studies showed that the stromal fibroblasts are modified through inductive signals received from the epithelial cells. Freshly excised human tumors grafted into nude mice produced tumors containing both human cancer cells and the host's stromal cells (Goldenberg and Pavia 1981). After short-term propagation of these tumors in vitro the murine mesenchymal cells transformed and were tumorigenic in nude mice. Moreover, stroma fibroblasts of experimentally induced epithelial tumors in the rat thyroid have an extended potential for division (Wynford-Thomas et al. 1986). Serum thyroid stimulating hormone led first to hyperplasia, then to multiple benign, and finally to malignant tumors. Fibroblasts from hyperplastic glands had a longer life span than normal fibroblasts in culture; the fibroblasts from tumor-bearing glands did not show signs of growth decline. Moreover, two types of fibroblasts could be identified in breast tissue with benign lesions, each type had either fetal- or adult-type migratory behavior (Schor et al. 1994). In cancer patients both types of fibroblasts displayed the fetal behavior.

Another host response to neoplasia is the occurrence of myofibroblasts in the stroma of scirrhous mammary carcinomas. They contribute to the desmoplasia and retraction that characterize many of these neoplasms (Seemayer et al. 1979). Both epithelial-mesenchyme and mesenchyme-epithelial transitions were observed in different types of neoplastic growth. Bone marrow aspirates were obtained from a patient with chronic myelogenous leukemia with the Ph-1 chromosome marker and from a patient with acute leukemia demonstrating C-group trisomy (Hentel and Hirschhorn 1971). A proportion of fibroblasts, which grew from these aspirates presented the Ph-1 chromosome and C-group trisomy showing that at least some leukemia bone marrow fibroblasts are derived from hemopoietic cells. An immortal fibroblastic cell line was obtained from a breast carcinoma biopsy, which shared a non-random X-chromosome inactivation pattern with the epithelial tumor (Petersen et al. 2003). The cells had the ability to undergo myofibroblast conversion, had residual keratin expression, and formed epithelial microfoci in a reconstituted membrane.

Deregulation of genes promoting epithelial-mesenchymal transition (EMT) during development can cause oncogenic transformation. Notch is one of those genes, which promote EMT both during normal development and oncogenic transformation (Timmerman et al. 2004). Expression of a constitutively active form of Notch-4 in mouse mammary epithelial cells inhibits epithelial differentiation and leads to tumor formation. Notch-1 and Notch-4 are targets for insertion and rearrangement by the mouse mammary tumor virus promoting epithelial mammary malignancy (Politi et al. 2004). Moreover, Notch-1 encodes a trans-membrane receptor that regulates normal T cell development; mutations in two domains of Notch-1 are found in the majority of human T cell acute lymphoblastic leukemias (Weng et al. 2004).

EMT that occurs during embryonic development has also been observed in cancer progression; it can be induced through a molecular switch in the form of an enzyme known as GSK-3β, which alters the function of proteins (Zhou et al. 2004). Normal epithelial cells are anchored to each other and to membranes through the protein E-cadherin, which is not expressed in mesenchymal cells, allowing them to move and migrate. A transcription factor known as Snail was found to control the gene coding for E-cadherin; it turns off the expression of the gene, freeing epithelial cells from their anchorage. GSK-3β regulates the function of Snail through phosphorylation of the protein. Inhibition of GSK-3β results in the upregulation of Snail and downregulation, of E-cadherin. Tumor cells in which GSK-3β activity is repressed become unanchored, thus acquiring the ability to migrate, invade neighboring tissues, and metastasize (Zhou et al. 2004).

The decline of neoplastic growth late during the human life span could be due to functional modifications of the connective tissue accompanying the changes of the growth potential of its cells during aging, with the loss of the capacity to help induce the transformation of epithelial cells or to facilitate the growth of the tumor cells. Indeed, the synthesis of molecules that can act as growth facilitators is modified during cellular aging.

Other possible mechanisms involved in the decline of tumor incidence during senescence concern the structural modifications occurring in connective tissue (Robert 2000). There are structural modifications of the large molecules of the extracellular matrix (elastin, proteoglycans, and collagen) and changes in the relative proportions of the cellular and molecular components; this restructuring may render more difficult cell invasion and the anchorage of cells in sites distant from the primary tumor.

2.2 Molecularly Related Mechanisms

The variation in the potential for chromosome recombination can constitute an explanation for the variation in cancer incidence through the human life span.

There is an interesting correlation between the potential for recombination events in the genome and the potential for cell proliferation. Cells with a higher potential for chromosomal recombination also have a higher growth potential. Human embryonic fibroblasts when cultivated in vitro go through continuous chromosomal rearrangements without any definite pattern becoming apparent; multiple clones arise continuously and compete with each other without anyone overgrowing the others (Chen and Ruddle 1974). These cells have a longer division potential than postnatal fibroblasts, which go through more stable type chromosome rearrangements during serial divisions (Bourgeois et al. 1981). There is indeed a decreased genome plasticity during postnatal development; a significant decline in mutagen-induced sister chromatid exchanges in old individuals could be ascertained in mouse bone marrow cells and human skin fibroblasts (Schneider and Gilman 1979). These findings show that during development and aging the genome loses its plasticity, as expressed in terms of a decline in the potential for chromosome rearrangements, which could constitute a mechanism to protect from deviations of cell proliferation.

Deregulation of the chromosome recombination potential can lead to neoplastic growth. Homozygous mutations in the human ATM gene are responsible for the cancer-prone syndrome ataxia-telangiectasia; the ATM gene mutation is responsible for hyper-recombination and telomere shortening (Fritz et al. 2000). Cells from Bloom's syndrome patients, an inherited disease with a high risk of developing cancer, exhibit elevated levels of somatic crossing-over and sister chromatid exchanges (German 1993). A gene was identified (BLM) that encodes a protein homologous to the one coding for a helicase subfamily, first identified in bacteria and yeast (Rothstein and Gangloff 1995); mutations in BLM confer the common phenotype of hyper-recombination to cells. It was also found that formation of colonies of transformed cells and the prolongation of the doubling potential of skin fibroblasts from retinoblastoma patients by low dose rate ionizing radiation was correlated with the potential for sister chromatid exchanges of the cells of the respective patients (Macieira-Coelho 1994). The presence of the deletion was not enough to confer susceptibility to the carcinogen: other genetic factors that confer a high recombination potential to the genome determine the proneness of these cells to transformation by ionizing radiation. Several genetically determined developmental malformations with a high incidence of cancers have disturbances in the frequency of DNA recombinations as a common denominator. Fanconi's anemia and Werner and Rothmund-Thompson syndromes are examples of such diseases. In the two latter syndromes mutations in helicases are also implicated.

Other data also show that cells with a higher rate of chromosome recombination are more prone to malignant transformation. The skin fibroblasts of melanoma patients, which are susceptible to immortalization, have an increased UV-induced sister chromatid exchange (Knees-Matzen et al. 1991). The probability of transformation of human fibroblasts by ionizing radiation is directly correlated with the chromosome recombination potential of a cell population and is accompanied by an increase in break points involved in chromosomal exchanges; the large majority of the break points involved in exchanges concern the centromeric and telomeric regions (Bourgeois et al. 1981), which are known to be prone to recombination.

The potential for chromosome rearrangements is not identical across species and seems to confer characteristics that have a bearing with patterns of aging and susceptibility to malignant transformation. The mouse genome is an example of a species that is particularly prone to chromosomal recombination events characterized by somatic crossing-over (radial figures) and bridges between chromosomes (Macieira-Coelho and Azzarone 1988), which can explain why the proliferation of mouse fibroblasts can get easily out of control and immortalize. Mouse fibroblasts have a proliferation life span shorter but more unstable than that of human fibroblasts, which have less potential for chromosome recombination (Macieira-Coelho 1994). This could be linked to the structure of mouse telomeres; their chromosomal terminal repeats are many times longer than those of human chromosomes (Kipling and Cook 1990) and their size is also largely unchanged through somatic cell division. The mouse terminal-repeat fragment is highly polymorphic, suggesting an unusually high mutation rate.

The potential for chromosome recombination events expressed by mouse fibroblasts in vitro also has a counterpart in vivo. Although 40 acrocentric chromosomes is the usual diploid number of the mouse species, localized races with 38 to 22 chromosomes resulting from Robertsonian fusions have been found in the wild (Capanna 1973). This property of the mouse genome could be responsible for the high probability of mouse cells to spontaneously immortalize and acquire malignant characteristics, for the high susceptibility to viral, chemical, and physical carcinogens and to oncogenes, and for the facility with which one can induce tumors in mice. The instability of the mouse genome, due to yet unknown properties, is favorable for the species to adapt and survive, but reduces the life span of the individual because of the high rate of mutational events. These data show that developmental characteristics were created through phylogeny that are relevant for the species differences in the incidence of neoplastic growth.

Several observations suggest an association between development, chromosome recombination, and reorganization of chromosome ends. Telomeres seem to act as regulators of recombination events and deregulation of this control can lead to deviations from normal development and to tumor growth. A decline in telomere length of peripheral blood lymphocytes early during the human life span, a plateau during young adulthood, and a gradual attrition later in life, was reported (Frenck et al. 1998). This developmental-related rather than senescence-related modification in telomere structure is compatible with another work showing that human telomeres are modified as early as the embryonic stage in different organs (Ulaner et al. 2001).

The telomere lengths in immature rat testes containing type A spermatogonia were compared with those in adult testes having more differentiated germ cells (Achi et al. 2000). Mean telomere length in the immature testis was significantly shorter in comparison to adult testis. Pachytene spermatocytes exhibited longer telomeres compared to type A spermatogonia. The results indicated that telomere length increases during the development of male germ cells from spermatogonia to spermatozoa and is inversely correlated with the expression of telomerase activity. In adult normal mice, changes in telomere length are tissue specific and seem to be developmentally regulated (Prowse and Greider 1995); long telomere lengths and telomerase activity were detected in the testis about 5–6 weeks after birth, at a time the testis increases dramatically in size due to the production of the first spermatocytes.

Other works support this view that the changes at chromosome ends are critical for normal development. Mice without telomerase, obtained with gene knockout techniques, reproduce normally for several generations (Blasco et al. 1997). These telomerase null mice do not present pathophysiological symptoms of aging (Rudolph et al. 1999); sixth generation null mice, however, are infertile and have an increased embryonic lethality due to neural tube closure defects. Further, long-term renewal of hematopoietic stem cells is compromised, and there is delayed epithelial regeneration of skin wounds.

Although it is not yet known what the real implications of telomere length and structure are for normal development, it is obvious that they vary through ontogeny and phylogeny, that they seem to have a regulatory role in recombination events of the genome, and that dysfunctions in this role can lead to malformations and neoplastic growth. Telomere dysfunction in telomerase-deficient mice promotes the development of epithelial cancers by a process of fusion-bridge breakage that leads to the formation of complex non-reciprocal translocations and epithelial cancers (Artandi et al. 2000). Non-reciprocal translocations at the break points of oncogenes can activate their transformation potential; truncation of proto-oncogenes by recombination with retroviral or cellular genes are necessary to convert proto-oncogenes to transforming genes (Duesberg 1987). This concept fits a proposal made almost 100 years ago that tumors are initiated by chromosomal rearrangements (Bovery 1914).

The telomeres of human fibroblast chromosomes were analyzed with the canonical probe that detects all telomeres, and with a TelBam 11 probe that is specific for a subset of human telomeres (Ben 1997). A reduction in terminal repeat-fragment size was found with the former probe; however, there was no evidence of a loss of TelBamm 11 homologous sequences even in cells at the end of their life span. The heterogeneity in fragment size was much greater when the canonical probe was used, relative to that seen using the TelBam 11 probe. This could be attributable to a greater variability in the location of the restriction enzyme sites rather than to differences in the length of terminal repeats. Ben concluded that variation in terminal repeat length may be related to how much telomeres participate in chromosome rearrangement. Telomere-promoted recombination can lead to degeneration of the telomeric sequence and subsequent loss of the hybridization ability (Ashley and Ward 1993). Indeed, several examples of apparent terminal deletions are actually subtelomeric translocations (Meltzer et al. 1993).

There is a connection between telomerases and retrotransposons; the critical step of telomere addition is strikingly similar to the retrotransposition mechanism used by the non-long-terminal repeat retrotranposons and the group II introns (Eickbush 1997). Insertion of an interstitial telomere repeat in a mammalian chromosome led to chromosome instability, increasing gene rearrangements 30-fold (Kilburn et al. 2001). Moreover, telomeric satellite

DNA functions in regulating frequency and location of chiasmata; the DNA sequences of the telomeric heterochromatic regions of chromosomes interact with other DNA segments of the chromosome in which crossing-over occurs (Lima-de-Faria 1983), and the existence of a gradient of increase in crossing-over in the distal telomere region of a chromosome could be demonstrated (McClintock 1943). The role of the telomere's repeated sequence could be that of regulator of recombination events upon which depends the cell proliferation and differentiation potential.

Moreover, the regulation of telomerase activity is subject to multiple levels of control and to different factors in different cellular contexts and lineages. A variety of oncogenes positively regulate telomerase activity, whereas tumor suppressor genes accomplish the converse (Elenitoba-Johnson 2001).

The decline in the recombination potential of the genome during senescence of the organism could be related to the decreased number of long telomeres available for recombination (Martens et al. 2000) and could contribute to the decline in the incidence of cancers during human senescence.

In summary, modifications in the structure and length of telomeres occur during development and are associated with the rate of chromosome recombination and the long-term division potential of cells. The rate of chromosome recombination must be finely tuned for development to proceed normally in line with the regulatory role of chromosomal recombination in development, as proposed by McClintock (1943), telomeres being probably critical in this regulation. Deviations from this control mechanism can lead to deviations from normal development such as malformations and neoplastic growth. On the other hand, the decrease during aging of the potential for chromosome recombination (Schneider and Gilman 1979; Bourgeois et al. 1981) could be one of the mechanisms responsible for the decline in the incidence of cancers toward the end of the human life span.

References

Achi M, Ravindranath N, Dym M (2000) Telomere length in male germ cells is inversely correlated with telomerase activity. Biol Reprod 63:591–598

Adami HO, Bergstrom R, Sparen P, SchwartzAD, Baron J (1993) Increasing cancer risk in younger birth cohorts in Sweden. Lancet 314:773–777

Adami HO, Bergstrom R, Mohner M, Zatonski W, Storm H et al. (1994) Testicular cancer in nine northern European countries. Int J Cancer 59:33–38

Adams EF, Newton CJ, Tait GH, Braunsberg H, Reed MJ, James VHT (1988) Paracrine influence of human breast stromal fibroblasts on breast epithelial cells: secretion of a polypeptide which stimulates reductive 17β-oestradiol dehydrogenase activity. Int J Cancer 42:119–122

Adekeye EO, Asamoa E, Cohen B (1985) Intraoral carcinoma in Nigeria: a review of 137 cases. Ann R Coll Surg Engl 67:180–182

Agnarsson BA, Olafsdottir K, Benediktsson H (1987) Tumours in Iceland. Acta Pathol Microb Scand Sect A 95:23–28

Al-Hajj M, Wicha MS, Benito-Hernandez A, Morrison SJ, Clarke MF (2003) Prospective identification of tumorigenic breast cancer cells. Proc Natl Acad Sci USA 100:3983–3988

Allard M, Robine JM (2000) Les centenaries français. Serdi, Paris

Allsop RC, Vaziri H, Patterson C, Goldstein S, Younglali EV, Futcher AB et al (1992) Telomere length predicts replicative capacity of human fibroblasts. Proc Natl Acad Sci USA 89:10114–10118

Altman AJ (1983) Malignant diseases of infancy, childhood and adolescence. Saunders, Philadelphia, pp 1–21

Anderson DE (1974) Genetic study of breast cancer: identification of a high risk group. Cancer 34:1090–1097

Anderson DE (1975) Familial susceptibility. In: Fraumeni JF Jr (ed) Persons at high risk of cancer, an approach to cancer etiology and control. Academic Press, New York, pp 43–54

Artandi SE, Chang S, Lee SL, Alson S, Gottlieb GJ, Chin L, DePinho RA (2000) Telomere dysfunction promotes non-reciprocal translocations and epithelial cancers in mice. Nature 406:641–645

Ashley T, Ward DC (1993) A »hot spot« of recombination coincides with an interstitial telomeric sequence in the Armenian hamster. Cytogenet Cell Genet 62:169–176

Auerbach R (1960) Morphogenetic interactions in the development of the mouse thymus gland. Dev Biol 2:271–282

Azzarelli A, Gennari L, Quagliulo V, Bonfanti G, Cerasoli S, Bufalino R (1986) Chondrosarcoma – 55 unreported cases: epidemiology, surgical treatment and prognostic factors. Eur J Surg Oncol 12:165–168

Azzarone B, Macieira-Coelho A (1987) Further characterization of the defects of skin fibroblasts from cancer patients. J Cell Sci 87:155–162

Azzarone B, Scemama P, Macieira-Coelho A (1984) Abnormal properties of skin fibroblasts from patients with breast cancer. Int J Cancer 33:759–770

Ballou JA, Dix D (1989) On the role of aging in cancer incidence: analyses of the skin cancer data. Anticancer Res 9:1237–1240

Balter M (1996) Thyroid cancer. Children become the first victims of fallout. Science 272:357–360

Barz H, Barz D (1984) Altersabhängigkeit der Metastasierung. Eine Untersuchung an über 5000 Karzinomtodesfallen. (English abstract). Arch Geschwultstforsch 54:77–83

Ben P (1997) Aging chromosome telomeres: parallel studies with terminal repeat and telomere associated DNA probes. Mech Ageing Dev 99:153–166

Berg JW, Robbins GF (1968) The histologic epidemiology of breast cancer. Breast Cancer: early and late. Year Book Medical Publishers, Chicago, pp 19–26

Berge T, Lundberge S (1977) Cancer in Malmö 1958–1969. An autopsy study. Acta Pathol Microb Scand [Suppl] 260:1–279

Bergkvist L, Tabar L, Bergström R, Adami HO (1987) Epidemiologic determinants of the mammographic pattern. Am J Epidemiol 126:1075–1081

Bern HA, Mills KT, Jones LA (1983) Critical period for neonatal estrogen exposure in occurrence of mammary gland abnormalities in adult mice. Proc Soc Exp Biol Med 172:239–243

Bhowmick NA, Chytil A, Plieth D, Gorska AE, Dumont N, Shappell S, Washington MK, Neilson EG, Moses HL (2004) TGF-β signaling in fibroblasts modulates the oncogenic potential of adjacent epithelia. Science 303:848–851

Blasco MA, Lee HW, Hande MP, Smaper E, Lansdorp PM, DePinho RA, Greider CW (1997) Telomere shortening and tumor formation by mouse cells lacking telomerase. Cell 91:25–34

Blau HM, Brazelton TR, Weimann JM (2001) The evolving concept of a stem cell: entity or function? Cell 105:829–841

Bonnet D, Dick JE (1997) Human acute myeloid leukaemia is organized as a hierarchy that originates from a primitive hematopoietic cell. Nat Med 3:730–737

Bourgeois CA, Raynaud, N, Diatloff-Zito C, Macieira-Coelho A (1981) Effect of low dose rate ionizing radiation on the division potential of cells in vitro. VIII. Cytogenetic analysis of human fibroblasts. Mech Ageing Dev 17:225–235

Bovery T (1914) Zur Frage der Entstehung maligner Tumoren. Fisher, Jena, Germany

Brody S (1983) Limited importance of cancer and competing risk theories in aging. J Clin Exp Gerontol 5:141–154

Broeders MJM, Lambe M, Baron IA, Leon DA (1996) History of childbearing and colorectal cancer risk in women aged less than 60: an analysis of Swedish routine registry data 1960–1984. Int J Cancer 66:170–175

Brookes S, Rowe J, Ruas M, Llanos S, Clark PA, Lomax M, James CM, Vatcheva R, Bates S, Vousden KH, Parry D, Gruis N, Smit N, Bergman W, Peters G (2002) INK4a-deficient human diploid fibroblasts are resistant to Ras-induced senescence. EMBO J 21:2936–2945

Brownsey B, Cameron EH, Griffiths K, Gleave EN, Forest APM, Campbell H (1972) Plasma dehydroepiandrosterone sulfate levels in patients with benign and malignant breast disease. Eur J Cancer 8:131–136

Butturini A, Gale RP (1989) Age of onset and type of leukemia. Lancet II:789–791

Capanna E (1973) Cytotaxonomy and vertebrate evolution. Academic Press, London

Carmichael JA (1990) Relationship of age to cancer of the cervix. In: Macieira-Coelho A and Nordenskjold B (eds) Cancer and aging, CRC Press, Boca Raton, pp 239–244

Chang HY, Chi JT, Dudoit S, Bondre C, van de Rijn M, Botstein D, Brown PO (2002) Diversity, topographic differentiation, and positional memory in human fibroblasts. Proc Natl Acad Sci USA 99:12877–12882

Chen TR, Ruddle FH (1974) Chromosome changes revealed by the Q-banding staining method during cell senescence of WI-38. Proc Soc Exp Biol Med 147:533–537

Cohen P, Connetta B, Dix D (1981) The incidence of hematologic tumours: a cellular model for the age dependence. J Theor Bio1 90:427–436

Cotchin E (1975) Spontaneous tumors in young animals. Proc Soc Exp Biol Med 68:653–657

Cunha GR, Fuji H, Neubauer BL, Shannon JM, Sawyer L, Reese BA (1983) Epithelial-mesenchymal interactions in prostatic development. I. Morphologic observations of prostatic induction by urogenital sinus mesenchyme in epithelium of the adult rodent urinary bladder. J Cell Biol 90:1662–1670

Cunha GR, Donjacour AA, Cooke PS, Mee S, Bigsby RM, Higgins SS, Sugimura Y (1987) The endocrinology and developmental biology of the prostate. Endocr Rev 8:338–362

Cutler GR (1975) Transcription of unique and reiterated DNA sequences in mouse liver and brain tissues as a function of age. Exp Gerontol 10:37–60

Delinassios JG, Kottardis SD, Garas J (1983) Uncontrolled growth of tumour stromal fibroblasts in vitro. Exp Cell Biol 51:201–209

Derynck R, Akhurst RJ, Balmain A (2001) TGF-beta signaling in tumor suppression and cancer progression. Nat Genet 29:351–360

Diatloff C, Macieira-Coelho A (1979) Effect of low-dose-rate irradiation on the division potential of cells in vitro. V. Human skin fibroblasts from donors with a high risk of cancer. J Natl Cancer Inst 63:55–59

Diatloff C, Macieira-Coelho A (1982) Effect of growth arrest on the doubling potential of human fibroblasts in vitro: a possible influence of the donor. In Vitro 18:606–610

Doll R (1978) An epidemiological perspective of the biology of cancer. Cancer Res 38:3573–3583

Duesberg P (1987) Cancer genes: rare recombination instead of activated oncogenes (a review). Proc Natl Acad Sci USA 84:2117–2124

Dvorak HF (1986) Tumors: wounds that do not heal. N Engl J Med 315:1651–1659

Eickbush TE (1997) Telomerase and retrotransposons: which came first. Science 277:911–912

Ekbom A, Trichopoulos D, Adami HO, Hsieh CC Lan SJ (1992) Evidence of prenatal influences on breast cancer risk. Lancet 340:1015–1018

Elenitoba-Johnson KS (2001) Complex regulation of telomerase activity. Implications for cancer therapy. Am J Pathol 159:405–410

Elliot MS, Katze JR (1986) Inhibition of queuine uptake in diploid human fibroblasts by phorbol-12, 13-didecanoate. J Biol Chem 261:13019–13025

Fearon ER, Feinberg AP, Hamilton SH and Vogelstein B (1985) Loss of genes on the short arm of chromosome 11 in bladder cancer. Nature 318:377–378

Feinleib M (1968) Breast cancer and artificial menopause: a cohort study. J Natl Cancer Inst 41:315–329

Ford AM, Ridge SA, Cabrera ME, Mahmoud H, Steel CM, Chan LC and Greaves, M (1993) In utero rearrangements in the trithorax-related oncogene in infant leukaemias. Nature 363:358–359

Foulds L (1958) The natural history of cancer. J Chron Dis 8:2–37

Franks LM, Cooper TW (1972) The origin of human embryo lung cells in culture. A comment on cell differentiation. Int J Cancer 9:19–29

Frenck RW Jr, Blackburn, EH, Shannon KM (1998) The rate of telomere sequence loss in human leukocytes varies with age. Proc Natl Acad Sci USA 95:5607–5610

Fritz E, Friedl AA, Zwacka RM, Eckardt-Schupp F, Meyn MS (2000) The yeast TEL1 gene partially substitutes for human ATM in suppression hyper-recombination, radiation-induced apoptosis and telomere shortening in A-T cells. Mol Biol Cell 11:2605–2616

Furugori E, Hirayama R, Nakamura KI, Kammori M, Esaki Y, Takubo K (2000) Telomere shortening in gastric carcinoma with aging despite telomerase activation. J Cancer Res Clin Oncol 126:481–485

German J (1993) Bloom syndrome: a Mendelian prototype of somatic mutational disease. Medicine 72:393–404

Gittes RF (1991) Carcinoma of the prostate. New Engl J Med 324:236–245

Glimelius B, Enblad G (1990) Hodgkin's disease. In: Macieira-Coelho A, Nordenskjold B (eds) Cancer and aging, CRC Press, Boca Raton, pp 228–237

Gogusev J, Teutsch B, Morin MT, Mongiat F, Haguenau F, Suskind G, Rabotti GF (1988) Inhibition of HLA class I antigen and mRNA expression induced by Rous sarcoma virus in transformed human fibroblasts. Proc Natl Acad Sci USA 85:203–207

Goldenberg DM, Pavia RA (1981) Malignant potential of murine stromal cells after transplantation of human tumors into nude mice. Science 212:65–67

Grassilli E, Bellesia D, Salomoni P, Croce MA, Sikora E, Radiszewska E et al (1996) C-fos/C-jun expression and AP-1 activation in skin fibroblasts from centenarians. Biochem Biophys Res Commun 226:517—523

Green GH (1978) Cervical cancer and cytology screening in New Zealand. Br J Obstetr Gynaecol 85:881–886

Gross L (1951) Spontaneous leukemia developing in C3H mice following inoculation in infancy with AK-leukemic extracts or AK-embryos. Proc Soc Exp Biol Med 76:27–31

Gurdon JB (1962) The developmental capacity of nuclei taken from intestinal epithelium cells of feeding tadpoles. J Embryol Exp Morphol 10:622–640

Hamajima N, Lee JAR (1987) Relationships of age, period, and birth cohort for stomach cancer mortality in Japan. Jpn J Cancer Res (Gann) 78:547–558

Hankey BF (1994) Age distribution of breast cancer cases. J Natl Cancer Inst 86:1441

Hansen JB, Jorgensen C, Petersen RK, Hallenborg P, de Matteis R, Boye HA, Petrovic N, Enerbäck s, Nedergaard J, Cinti S, de Riele H, Kristiansen K (2004) Retinoblastoma protein function as a molecular switch determining white versus brown adipocyte differentiation Proc Natl Acad Sci USA 101:4112–4117

Hardy WD, Old LJ, Hess PW, Essex M, Cotter S (1973) Horizontal transmission of feline leukemia virus. Nature 244:266–268

Hei TK, Wu LJ, Piao CQ (1997) Malignant transformation of immortalized human bronchial cells by asbestos. Environ Health Perspect 105:1058–1088

Henderson BE, Ross RK, Pike MC (1988) Epidemiology of testicular cancers. In: Skinner DG, Lieskovsky G (eds) Diagnosis and management of genitourinary cancers. Saunders, Philadelphia, pp 46–57

Hennighausen L, Robinson GW (2001) Signaling pathways in mammary gland development. Dev Cell 1:467–475

Hennis HL, Braid HL, Vincent RA Jr (1981) Unscheduled DNA synthesis in cells of different shape in fibroblast cultures from donors of various ages. Mech Ageing Dev 16:355–361

Hentel J, Hirschhorn K (1971) The origin of some bone marrow fibroblasts. Blood 38:81–86

Heppner GH, Miller BE (1983) Tumor heterogeneity: biological implications and therapeutic consequences. Cancer Metastasis Rev 2:5–23

Heppner GH, Miller BE, Miller FR (1983) Tumor subpopulation interaction in neoplasms. Biochem Biophys Acta 695:215–226

Hoar SK, Blair A, Holmes FF, Boysen CD, Robel RJ, Hoover R, Fraumeni JF Jr (1986) Agricultural herbicide use and risk of lymphoma and soft-tissue sarcoma. J Am Med Assoc 256:1141–1147

Houghton JM, Stoicov C, Nomura S, Rogers AB, Carlson J, Li H, Cai X, Fox JG, Goldenring JR, Wang TC (2004) Gastric cancer orginating from bone marrow-derived cells. Science 306:1568–1572

Howell JN, Greene MH, Corner RC, Maher VM, McCormick JJ (1984) Fibroblasts from patients with hereditary cutaneous malignant melanoma are abnormally sensitive to the mutagenic effect of simulated sunlight and 4-nitroquinoline 1-oxide. Proc Natl Acad Sci USA 81:1179–1183

Hsu TC, Cherry LM, Samaan NA (1985) Differential mutagen susceptibility in cultured lymphocytes of normal individuals and cancer patients. Cancer Genet Cytogenet 17:307–313

Humpherys D, Eggan K, Akutsu H, Friedman A, Hochedlinger K, Yanagimachi R, Lander ES, Golub TR, Jaenisch R (2002) Abnormal gene expression in cloned mice derived from embryonic stem cell and cumulus cell nuclei. Proc Natl Acad Sci USA 99:12889–12894

Jensen F, Koprowsky H, Pontén J (1963) Rapid transformation of human fibroblast cultures by simian virus 40. Proc Natl Acad Sci USA 50:343–348

Kaesberg P, Ershler WB (1990) The effect of aging on the development and progression of cancer: from the laboratory animal to the clinic. In: Macieira-Coelho A, Nordenskjöld B (eds) Cancer and aging. CRC Press, Boca Raton, pp 263–282

Key TJA, Pike MC (1988) The dose-effect relationship between »unopposed« oestrogens and endometrial mitotic rate: its central role in explaining and predicting endometrial cancer risk. Br J Cancer 57:205–210

Kilburn AE, Shea MJ, Sargent RG, Wilson JH (2001) Insertion of a telomere-repeat sequence into a mammalian gene causes chromosome instability. Mol Cell Biol 21:126–135

Kipling D, Cook HJ (1990) Hypervariable ultra-long telomeres in mice. Nature 347:400–402

Kondo T, Raff M (2000) Oligodendrocyte precursor cells reprogrammed to become multipotential CNS stem cells. Science 289:1754–1757

Knees-Matzen S, Roser M, Reimers U, Ehlert U, Weichenthal M, Breitbar EW, Rüdiger HW (1991) Increased UV-induced sister-chromatid exchange in cultured fibroblasts of first degree relatives of melanoma patients. Cancer Genet Cytogenet 53:265–170

Kratochwil K (1969) Organ specificity in mesenchymal induction demonstrated in the embryonic development of the mammary gland of the mouse. Dev Biol 20:46–71

Krieger N (1988) Rising incidence of breast cancer. J Natl Cancer Inst 80:2

Kristensen P, Andersen A, Irgens LM, Bye AS, Sundheim L (1996) Cancer in offspring of parents engaged in agricultural activities in Norway: incidence and risk factors in the farm environment. Int J Cancer 65:39–50

Kritzman DB, Pathak S, Samaan NA, Hickey RC (1987) Genetic instability in fibroblasts of patients with thyroid cancers. Int J Cancer 39:179–181

Krtolica A, Parrinello S, Lockett S, Desprez PY, Campisi J (2001) Senescent fibroblasts promote epithelial cell growth and tumorigenesis: a link between cancer and aging. Proc Natl Acad Sci USA 98:12072–12077

Kuperwasser C, Chavarria T, Wu M, Magrane G, Gray JW, Carey L, Richardson A, Weinberg RA (2004) Reconstruction of functionally normal and malignant human breast tissues in mice. Proc Natl Acad Sci USA 101:4966–4971

Land CE, Boice JD, Shore RE, Norman JE, Tokunaga M (1980) Breast cancer risk from low-dose exposures to ionizing radiation: results of parallel analysis of three exposed populations of women. J Natl Cancer Inst 65:353–376

Lanza RP, Cibelli JB, Blackwell C, Cristofalo VJ, Francis MK, Baerlocher GM, Mak J, Schertzer M, Chavez EA, Sawyer N, Lansdorp PM, West MD (2000) Extension of cell life-span and telomere length in animals cloned from senescent somatic cells. Science 288:665–669

Li Y, Hively WP, Varmus HE (2000) Use of MMTV-Wnt-1 transgenic mice for studying the genetic basis of breast cancer. Oncogene 19:1002–1006

Li Y, Welm B, Podsypanina K, Huang S, Chamorro M, Zhang X, Rowlands T, Egelblad M, Cowin P, Werb Z, Tan LK, Rosen JM, Varmus HE (2003) Evidence that transgenes encoding components of the Wnt signaling pathway preferentially induce mammary cancers from progenitor cells. Proc Natl Acad Sci USA 26:15853–15858

Lima-de-Faria A (1983) Molecular evolution and organization of the chromosome. Elsevier, Amsterdam

Lohler J, Timpl R, Jaenisch R (1984) Embryonic lethal mutation in mouse collagen I gene causes rupture of blood vessels and is associated with erythropoietic and mesenchymal cell death. Cell 38:597–607

Lu D, Zhao Y, Tawatao R, Cottam HB, Sen M, Leoni LM, Kipps TJ, Corr M, Carson DA (2004) Activation of the Wnt signaling pathway in chronic lymphocytic leukemia. Proc Natl Acad Sci USA 101:3118–3123

Lukanova A, Lundin E, Micheli A, Arslan A, Ferrari P, Rinaldi S, Hrogh V, Lenner P, Shore RE, Bessy C, Muti P, Riboli E, Koenig KL, Levitz M, Stattin P, Berrino F, Hallmans G, Kaaks R, Toniolo P, Zeleniuch-Jacquotte A (2004) Circulating levels of sex steroid hormones and risk of endometrial cancer in postmenopausal women. Int J Cancer 108:424–432

Macieira-Coelho A (1991) Chromatin reorganization during senescence of proliferating cells. Mutat Res 256:81–104

Macieira-Coelho A (1994) Genome reorganization through cell division. Implications for aging of the organism and cancer development. Ann NY Acad Sci 719:108–128

Macieira-Coelho A (1995) The last mitoses of the human fibroblast proliferative life span. Physiological implications. Mech Ageing Dev 82:91–104

Macieira-Coelho A (2001) Relevance of in vitro studies for aging of the organism. Z Gerontol Geriatr 34:429–436

Macieira-Coelho (2002) The biology of conformation in the regulation of the senescent and transformed cell phenotypes. Semin Cancer Biol 12:165–171

Macieira-Coelho A (2003) Biology of aging. Progress in molecular and subcellular biology, vol 30. Springer, Berlin Heidelberg New York

Macieira-Coelho A, Azzarone B (1988) The transition from primary culture to spontaneous immortalization in mouse fibroblast populations. Anticancer Res 8:669–676

Macieira-Coelho A, Taboury F (1982) A re-evaluation of the changes in proliferation in human fibroblasts during aging in vitro. Cell Tissue Kinet 15:213–224

MacMahon B, Trichopoulos D, Brown J, Andersen AP, Cole P, de Waard F, Kauraniemi T, Polychronopoulou A, Ravnihar B, Stormby N, Westlund K (1982) Age at menarche, urine estrogens and breast cancer risk. Int J Cancer 30:427–431

Martens UM, Chavez EA, Poon SS, Schmoor C, Landsdorp PM (2000) Accumulation of short telomeres in human fibroblasts prior to replicative senescence. Exp Cell Res 256:291–299

Martin GM, Sprague CA, Norwood TH, Pendergrass WR (1974) Clonal selection, attenuation, and differentiation in an in vitro model of hyperplasia. Am J Pathol 74:137–154

Masuda N, Fukuoka M, Matsui K, Kudoh S, Negoro S, Takifuji M, Fujisue N, Morino H, Nakagawa K, Nishioka M, Takada M (1991) Establishment of tumor cell lines as an independent prognostic factor for survival time in patients with small-cell lung cancer. J Natl Cancer Inst 83:1743–1747

Maurer HM (1978) Current concepts in cancer. Solid tumors in children. N Engl J Med 299:1345–1348

McClintock B (1943) Maize genetics. Yearbook Carnegie Inst 42:148–152

McCormick JJ, Kately-Kohler M, Watanabe M, Mahler VM (1986) Abnormal sensitivity of human fibroblasts from Xeroderma pigmentosum variants to transformation to anchorage independence by ultraviolet radiation. Cancer Res 46:489–493

McNeal JE (1983) The prostate gland morphology and pathobiology. Monogr Urol 4:3–14

Meltzer PS, Guan XY, Trent JM (1993) Telomere capture stabilizes chromosome breakage. Nat Genet 4:252–255

Milo GE, Casto BC (1986) Conditions for transformation of human fibroblast cells: an overview. Cancer Lett 31:1–15

Milo GE, DiPaolo JA (1978) Neoplastic transformation of human diploid cells in vitro after chemical carcinogen treatment. Nature 275:130–132

Milan T, Pukkala E, Verkasalo PK, Kaprio J, Jansén CT, Koskenvuo M, Teppo L (2000) Subsequent primary cancers after basal-cell carcinoma: a nationwide study in Finland from 1953 to 1995. Int J Cancer 87:283–288

Miyazawa J, Mitoro A, Kawashiri S, Chada KK, Imai K (2004) Expression of mesenchyme-specific gene HMGA2 in squamous cell carcinomas of the oral cavity. Cancer Res 64:2024–2029

Mohle-Boetani J (1990) Age at breast cancer diagnosis as a predictor of subsequent survival. In: Macieira-Coelho A, Nordenskjold B (eds) Cancer and aging, CRC Press, Boca Raton, pp 245–254

Mughal TI, Robinson WA, Padmos MA (1985) Adult Hodgkin's disease in Saudi Arabia. Eur J Surg Oncol 11:41–45

Muir CS (1990) Geographical pattern of cancer: role of environment. In: Macieira-Coelho A, Nordenskjöld B (eds) Cancer and aging. CRC Press, Boca Raton, pp 187–206

Mukherji B, MacAlister TJ, Guha A, Gillies CG, Jeffers DC, Slocum SK (1984) Spontaneous in vitro transformation of human fibroblasts. J Natl Cancer Inst 73:583–593

Munsie MJ, Michalska AE, O'Brien CM, Trounson AO, Pera MF, Mountford PS (2000) Isolation of pluripotent embryonic stem cells from reprogrammed adult mouse somatic cell nuclei. Curr Biol 10:989–992

Murty VS, Mitra AB, Luthra UK (1985) Spontaneous chromosomal aberrations in patients with precancerous and cancerous lesions of the cervix uteri. Cancer Genet Cytogenet 17:347–354

Narla G, Heath KE, Reeves HL, Li D, Giono LE, Kimmelman AC, Glicksman MJ, Narla J, Eng FJ, Chan AM, Ferrari AC, Martignetti JA, Friedman SL (2001) KLF6, a candidate tumor suppressor gene mutated in prostate cancer. Science 294:2563–2566

National Center of Health Statistics (1973) Some demographic aspects of aging in the United States. US Public Health Service and US Bureau of the Census, Washington, DC

Newell GR, Spitz MR, Sider JG, Pollack ES (1987) Incidence of testicular cancer in the United States related to marital status, histology and ethnicity .J Natl Cancer Inst 78:881–885

Olsson H (1990) Reproductive and hormonal factors in relation to cancer occurrence in the breast, prostate, testis, uterine corpus, ovary, uterine cervix, and thyroid gland. In: Macieira-Coelho A, Nordenskjold B (eds) Cancer and aging, CRC Press, Boca Raton, pp 109–126

Pieretti M, Turker MS (1997) Mutation, ageing, and ovarian cancer. Lancet 349:700–701

Rothstein R, Gangloff S (1995) Hyper-recombination and Bloom's syndrome: microbes again provide clues about cancer. Genome Res 5:421–426

Park WY, Hwang CI, Kang MJ, Seo JY, Chung JH, Kim YS et al (2001) Gene profile of replicative senescence is different from progeria or elderly donor. Exp Cell Res 282:934–939

Parshad R, Sanford KK, Jones GM (1985) Chromosomal radiosensitivity during the G2 cell-cycle period of skin fibroblasts from individuals with familiar cancer. Proc Natl Acad Sci USA 82:5400–5403

Pereira DS, Dorrell C, Ito CY, Gan OI, Murdoch B, Rao V, Zou JP, Reddy ES, Dick JE (1998) Retroviral transduction of TLS-ERG initiates a leukemogenic program in normal human hematopoietic cells. Proc Natl Acad Sci USA 95:8239–8244

Petersen OW, Nielsen HL, Gudjonsson T, Villadsen R, Rank F, Niebhur E, Bissell MJ, Ronnov-Jessen L (2003) Epithelial to mesenchyme transition in human breast cancer can provide a nonmalignant stroma. Am J Pathol 162:391–402

Pfeffer L, Lipkin M, Stutman D, Kopelovitch L (1976) Growth abnormalities of cultured skin fibroblasts derived from individuals with hereditary adenomatosis of the colon and rectum. J Cell Physiol 89:29–38

Piantanelli L (1988) Cancer and aging: from the kinetics of biological parameters to the kinetics of cancer incidence and mortality. Ann NY Acad Sci 521:99–109

Politi K, Feirt N, Kitajewski J (2004) Notch in mammary gland development and breast cancer. Semin Cancer Biol 14:341–347

Prowse KR, Greider CW (1995) Developmental and tissue-specific regulation of mouse telomerase and telomere length. Proc Natl Acad Sci USA 92:4818–4820

Reya T, Morrison SJ, Clarke MF, Weissman IL (2001) Stem cells, cancer, and cancer stem cells. Nature 414:105–111

Reynolds CP, Frenkel EP, Smith RG (1980) Growth characteristics of neuroblastoma in vitro correlate with patient survival. Trans Assoc Am Physicians 93:203–211

Robert L (2000) Cellular and molecular mechanisms of aging and related diseases. Pathol Oncol Res 6:3–9

Robert L, Labat-Robert J (1995) Extracellular matrix. In: Macieira-Coelho A (ed) Molecular basis of aging. CRC Press, Boca Raton, pp 460–492

Rothstein R, Gangloff S (1995) Hyper-recombination and Bloom's syndrome: microbes again provide clues about cancer. Genome Res 5:421–426

Rudolph KL, Chang S, Lee HW, Blasco M, Gottlieb GJ, Greider C, DePinho RA (1999) Longevity, stress response, and cancer in aging telomerase-deficient mice. Cell 96:701–712

Rumpold H, Untegasser G, Maderbascher S, Berger P (2002) The development of benign prostatic hyperplasia by transdifferentiation of prostatic stromal cells. Exp Gerontol 37:1001–1004

Russell J, Witt W (1976) Cell size and growth characteristics of cultured fibroblasts isolated from normal and keloid tissue. Plast Reconstr Surg 57:207–212

Russo J, Russo IH (1980) Influence of differentiation and cell kinetics on the susceptibility of the rat mammary gland to carcinogenesis. Cancer Res 40:2677–2687

Sakakura T (1983) Epithelial-mesenchymal interactions in mammary gland development and its perturbation in relation to tumorigenesis. In: Rich MA, Hagar JC, Furmanski P (eds) Understanding breast cancer. Decker, New York, pp 261–284

San Francisco IF, DeWolf WC, Peehl DM, Olumi AF (2004) Expression of transformation growth factor-beta 1 and growth in soft agar differentiate prostate carcinoma-associated fibroblasts from normal prostate fibroblasts. Int J Cancer 112:213–218

Scanlan MJ, Raj BKM, Calvo B, Garin-Chesa P, Sanz-Moncasi MP, Healey JH, Old LJ, Rettig WJ (1994) Molecular cloning of fibroblast activation protein α, a member of the serine protease family selectively expressed in stromal fibroblasts of epithelial cancers. Proc Natl Acad Sci USA 91:5657–5661

Schneider EL, Gilman B (1979) Sister chromatid exchanges and aging. III. The effect of donor age on mutagen-induced sister chromatid exchanges in human diploid fibroblasts. Hum Genet 46:57–62

Schor SL, Schor AM, Durning P, Rushton G (1985) Skin fibroblasts obtained from cancer patients display foetal-like migratory behaviour on collagen gels. J Cell Sci 73:235–244

Schor SL, Haggie JA, Durning, Howell A, Smith L, Sellwood RAS, Crowther D (1986) Occurrence of a fetal fibroblast phenotype in familial breast cancer. Int J Cancer 37:831–836

Schor SL, Schor AM, Grey AM, Rushton G (1988) Foetal and cancer patient fibroblasts produce an autocrine migration-stimulating factor not made by normal adult cells. J Cell Sci 90:391–399

Schor AM, Rushton G, Ferguson JE, Howell A, Redford J, Schor SL (1994) Phenotypic heterogeneity in breast fibroblasts: functional anomaly in fibroblasts from histologically normal tissue adjacent to carcinoma. Int J Cancer 59:25–32

Schor SL, Ellis IR, Jones SJ, Baillie R, Seneviratne K, Clausen J, Motegi K, Vojtesek B, Kankova K, Furrie E, Sales MJ, Schor AM, Kay RA (2003) Migration-stimulating factor: a genetically truncated onco-fetal fibronectin isoform expressed by carcinoma and tumor-associated stromal cells. Cancer Res 15:8827–8826

Seemayer TA, Lagace R, Schurch W, Tremblay G (1979) Myofibroblasts in the stroma of invasive and metastatic carcinoma: a possible host response to neoplasia. Am J Surg Pathol 3:525–533

Serakinci N, Guldberg P, Burns JS, Abdallah B, Schrödder H, Jensen T, Kassem M (2004) Adult human mesenchymal stem cell as a target for neoplastic transformation. Oncogene 23:5095–5098

Shriver SP, Bourdeau HA, Gubish CT, Tirpak DL, Davis AL, Luketich JD, Siegfried JM (2000) Sex-specific expression of gastrin-releasing peptide receptor: relationship to smoking history and risk of lung cancer. J Natl Cancer Inst 92:24–33

Silzle T, Randolph GJ, Kreutz M, Kunz-Schughart LA (2004) The fibroblast: sentinel cell and local immune modulator in tumor tissue. Int J Cancer 108:173–180

Singh KS, Clarke ID, Terasaki M, Bonn VE, Hawkins C, Squire J, Dirks PB (2003) Identification of a cancer stem cell in human brain tumors. Cancer Res 63:5821–5828

Skoog L, Wallgren A, Pascual MR, Macias A, Perez R, Lage A (1987) Factors associated with prognosis in human breast cancer. Neoplasma 34:587–594

Slaughter DP, Southwick HW, Smejkal W (1953) Field cancerization in oral stratified squamous epithelium. Cancer 6:963–968

Smith HS, Owens RB, Hiller AJ, Nelson-Rees WA, Johnston JO (1976) The biology of human cells in tissue culture. I. Characterization of human cells derived from osteogenic sarcomas. Int J Cancer 17:219–234

Smith HS, Wolman SR, Dairkee SH, Hancock MC (1987) Immortalization in culture: occurrence at a late stage in the progression of breast cancer. J Natl Cancer Inst 78:611–615

Smith PJ, Greene MH, Devlin DA, MCKeen EA, Paterson MC (1982) Abnormal sensitivity to UV-radiation in cultured skin fibroblasts from patients with hereditary cutaneous malignant melanoma and dysplastic nevus syndrome. Int J Cancer 30:39–45

Stalsberg H, Thomas DB (1993) Age distribution of histologic types of breast carcinoma. Int J Cancer 54:1–7

Stattin P, Lumme S, Tenkanen L, Alfthan H, Jellum E, Hallmans G, Thoresen S, Hakulinen T, Luostarinen T, Lehtinen M, Dillner J, Stenman U-H, Hakama M (2004) High levels of circulating testosterone are not associated with increases prostate cancer risk: a pooled prospective study. Int J Cancer 108:418–424

Steinhardt M (1985) Effect of donor age on clonal differentiation of human skin fibroblasts in vitro. Gerontology 31:27–38

Stevens RG (1986) Age and risk of acute leukaemia. J Natl Cancer Inst 76:845–848

Stiller CA (1994) Cancer in adolescence. Br Med J 308:1382–1383

Taipale J, Beachy PA (2001) The hedgehog and Wnt signaling pathways in cancer. Nature 411:349–354

Teppo L (1990) Childhood cancer- Incidence and etiological considerations. In: Macieira-Coelho A, Nordenskjold B (eds) Cancer and aging. CRC Press, Boca Raton, Florida, pp 207–216

Timmerman LA, Grego-Bessa J, Raya A, Bertran E, Pérez-Pomares, Diez J, Aranda S, Palomo S, McCormick F, Izpisua-Belmomnte JC, de la Pompa JL (2004) Notch promotes epithelial-mesenchymal transition during cardiac development and oncogenic transformation. Genes Dev 18:99–115

Tokunaga M, Land CE, Yamamoto t, Asano M, Tokuoka S, Ezaki H, Nishimori I (1987) Incidence of female breast cancer among atomic bomb survivors, Hiroshima-Nagasaki, 1950–1980. Radiat Res 112:243–250

Tokuoka S, Inai K (1990) Late radiation effects in atomic bomb survivors: a review. In: Macieira-Coelho A, Nordenskjold B (eds) Cancer and aging. CRC Press, Boca Raton, pp 163–186

Tomooka Y, Bern HA (1982) Growth of mouse mammary glands after neonatal sex hormone treatments. J Natl Cancer Inst 69:1347–1352

Tweeddale ME, Lim B, Jamal N (1987) The presence of clonogenic cells in high-grade malignant lymphoma: a prognostic factor. Blood 69:1307–1314

Ulaner GA, Hu JF, Vu TH, Giudice LC, Hoffman AR (2001) Tissue-specific alternate splicing of human telomerase reverse transcriptase (HTERT) influences telomere lengths during human development. Int J Cancer 91:644–649

Vihko R, Apter DL (1986) The epidemiology and endocrinology of the menarche in relation to breast cancer. Cancer Surv 5:561–571

Waard F de (1975) Breast cancer incidence and nutritional status with particular reference to body weight and height. Cancer Res 35:3351–3356

Waard F de, Trichopoulos D (1988) A unifying concept of the aetiology of breast cancer. Int J Cancer 41:666–669

Weng AP, Ferrando AA, Lee W, Morris JP IV, Silverman LB, Sanchez-Irizarry C, Blacklow SC, Look AT, Aster JC (2004) Activating mutations of Notch-1 in human T cell acute lymphoblastic leukemia. Science 306:269–270

Wijngaarden WJ van, Duncan ID (1993) Rationale for stopping cervical screening in women over 50. Br Med J 306:967–971

Willis RA (1945) Further studies on the mode of origin of carcinomas of the skin. Cancer Res 5:469–479

Willett WC (2000) Diet and cancer: past, present and future. Proceedings of the 91 st Meeting Am Ass Cancer Res, pp 6–7

Wilmut I, Schnieke AE, McWhir J, Kind AJ, Campbell HS (1997) Viable offspring derived from fetal and adult mammalian cells. Nature 385:810–813

Wiseman BS, Werb Z (2002) Stromal effects on mammary gland development and breast cancer. Science 296:1046–1049

Wynford-Thomas D, Smith P, Williams ED (1986) Prolongation of fibroblast life span associated with epithelial rat tumor development. Cancer Res 46:3125–3127

Ziegler J (1998) Exposure and habits early in life may influence breast cancer risk. J Natl Cancer Inst 90:187–188

Zhou BP, Deng J, Xia W, Xu J, Li YM, Gundaz M, Hung M-C (2004) Dual regulation of Snail by GSK-3 mediated phosphorylation in control of epithelial-mesenchymal transition. Nat Cell Biol 6:931–940

Zumoff B, Levin J, Rosenfeld RS, Markham M, Strain GW, Fukushima DK (1981) Abnormal 24-hr mean plasma concentrations of dehydroisoandrosterone and dehydroisoandrosterone sulfate in women with primary operable cancer. Cancer Res 41:3360–3368

Subject Index

Printing: Krips bv, Meppel
Binding: Stürtz, Würzburg